Molecular Biology Intelligence Unit

Nonsense-Mediated mRNA Decay

Lynne E. Maquat, Ph.D.
Department of Biochemistry and Biophysics
School of Medicine and Dentistry
University of Rochester
Rochester, New York, U.S.A.

CRC Press is an imprint of the
Taylor & Francis Group, an **informa** business

Nonsense-Mediated mRNA Decay

Molecular Biology Intelligence Unit

CRC Press
Taylor & Francis Group
6000 Broken Sound Parkway NW, Suite 300
Boca Raton, FL 33487-2742

First issued in paperback 2019

ISBN 13: 978-1-58706-296-4 (pbk)

Visit the Taylor & Francis Web site at
http://www.taylorandfrancis.com

and the CRC Press Web site at
http://www.crcpress.com

Library of Congress Cataloging-in-Publication Data

A C.I.P. Catalogue record for this book is available from the Library o f Congress.

About the Editor...

LYNNE E. MAQUAT, Ph.D., is Professor of Biochemistry and Biophysics at the University of Rochester School of Medicine and Dentistry in Rochester, New York. She has studied nonsense-mediated mRNA decay since 1980, beginning with characterizations of patients having the hemolytic anemias β°-thalassemia and triosephosphate isomerase deficiency. She is currently President of the RNA Society and member of the Public Information Committee of the American Society for Cell Biology. Dr. Maquat is on the editorial board of *Molecular & Cellular Biology*, *RNA*, *RNA Biology*, and *Faculty of 1000*, and was formerly chair of the National Institutes of Health CDF-1 study section. At the University of Rochester, she is Director of Graduate Women in Science. She has been selected by the New York State Commissioner of Health as an exemplary scientist and received the Davey Award for outstanding cancer research. Her extracurricular activities include trekking in Ladakh, Nepal, Tibet and other remote places, and fund-raising for the arts community.

Dedication

To Mark and Lily, previous and current members of my lab, co-workers in NMD and related fields, and readers who just wish to know.

CONTENTS

EDITOR

Lynne E. Maquat
Department of Biochemistry and Biophysics
School of Medicine and Dentistry
University of Rochester
Rochester, New York, U.S.A.
Email: lynne_maquat@urmc.rochester.edu
Chapters 4, 17

CONTRIBUTORS

Robert T. Abraham
Signal Transduction Program
The Burnham Institute
La Jolla, California, U.S.A
and
Oncology Research
Wyeth Pharmaceutical Research
Pearl River, New York, U.S.A.
Email: abrahar@wyeth.com
Chapter 16

Nadia Amrani
Department of Molecular Genetics and Microbiology
University of Massachusetts Medical School
Worcester, Massachusetts, U.S.A.
Chapter 2

Philip Anderson
Department of Genetics
University of Wisconsin
Madison, Wisconsin, U.S.A.
Email: andersn@wisc.edu
Chapter 11

Claus M. Azzalin
Swiss Institute for Experimental Cancer Research
École Polytechnique Fédérale de Lausanne
and
National Center of Competence in Research "Frontiers in Genetics"
Lausanne, Switzerland
Chapter 19

Kristian E. Baker
Center for RNA Molecular Biology
Case Western Reserve University School of Medicine
Cleveland, Ohio, U.S.A.
Email: kristian.baker@case.edu
Chapter 1

David M. Bedwell
Department of Microbiology
Gregory Fleming James Cystic Fibrosis Research Center
University of Alabama at Birmingham
Birmingham, Alabama, U.S.A.
Email: dbedwell@uab.edu
Chapter 10

Isabelle Behm-Ansmant
European Molecular Biology Laboratory
Heidelberg, Germany
Chapter 12

Steven E. Brenner
Department of Molecular and Cell Biology
and
Department of Plant and Microbial Biology
University of California
Berkeley, California, U.S.A.
Email: brenner@compbio.berkeley.edu
Chapter 14

Harry C. Dietz
McKusick-Nathans Institute
of Genetic Medicine
Howard Hughes Medical Institute
Johns Hopkins University School
of Medicine
Baltimore, Maryland, U.S.A.
Email: hdietz@jhmi.edu
Chapter 8

Ming Du
Department of Microbiology
Gregory Fleming James Cystic Fibrosis
Research Center
University of Alabama at Birmingham
Birmingham, Alabama, U.S.A.
Chapter 10

Pamela J. Green
Delaware Biotechnology Institute
University of Delaware
Newark, Delaware, U.S.A.
Email: green@dbi.udel.edu
Chapter 13

Jayanthi P. Gudikote
Department of Immunology
The University of Texas M.D. Anderson
Cancer Center
Houston, Texas, U.S.A.
Chapter 6

Feng He
Department of Molecular Genetics
and Microbiology
University of Massachusetts
Medical School
Worcester, Massachusetts, U.S.A.
Chapter 3

Matthias W. Hentze
Molecular Medicine Partnership Unit
and
European Molecular Biology Laboratory
Heidelberg, Germany
Chapter 9

Jill A. Holbrook
Department of Pediatric Oncology,
Hematology and Immunology
University of Heidelberg
and
Molecular Medicine Partnership Unit
Heidelberg, Germany
Email:
HolbrookJill@med.uni-heidelberg.de
Chapter 9

Xin Hong
Department of Biology
Indiana University
Bloomington, Indiana, U.S.A.
Chapter 15

Elisa Izaurralde
European Molecular Biology Laboratory
Heidelberg, Germany
Email: izaurralde@embl-heidelberg.de
Chapter 12

Allan Jacobson
Department of Molecular Genetics
and Microbiology
University of Massachusetts
Medical School
Worcester, Massachusetts, U.S.A.
Email: jacobson@umassmed.edu
Chapters 2, 3

Isao Kashima
Department of Molecular Biology
Yokohama City University School
of Medicine
Yokohama, Japan
Chapter 7

Handan Kaygun
Department of Biology
Program in Molecular Biology
and Biotechnology
University of North Carolina
Chapel Hill, North Carolina, U.S.A.
Chapter 18

Kim M. Keeling
Department of Microbiology
Gregory Fleming James Cystic Fibrosis Research Center
University of Alabama at Birmingham
Birmingham, Alabama, U.S.A.
Chapter 10

Yoon Ki Kim
Department of Biochemistry and Biophysics
School of Medicine and Dentistry
University of Rochester
Rochester, New York, U.S.A.
Chapter 17

Adrian R. Krainer
Cold Spring Harbor Laboratory
Cold Spring Harbor, New York, U.S.A.
Email: krainer@cshl.edu
Chapter 20

Andreas E. Kulozik
Department of Pediatric Oncology, Hematology and Immunology
University of Heidelberg
and
Molecular Medicine Partnership Unit
Heidelberg, Germany
Email: andreas.kulozik@med.uni-heidelberg.de
Chapter 9

Liana F. Lareau
Department of Molecular and Cell Biology
University of California
Berkeley, California, U.S.A.
Chapter 14

Joachim Lingner
Swiss Institute for Experimental Cancer Research
École Polytechnique Fédérale de Lausanne
and
National Center of Competence in Research "Frontiers in Genetics"
Lausanne, Switzerland
Email: joachim.lingner@isrec.ch
Chapter 19

Jens Lykke-Andersen
Molecular, Cellular and Developmental Biology
University of Colorado at Boulder
Boulder, Colorado, U.S.A.
Email: jens.lykke-andersen@colorado.edu
Chapter 5

Michael Lynch
Department of Biology
Indiana University
Bloomington, Indiana, U.S.A.
Email: milynch@indiana.edu
Chapter 15

William F. Marzluff
Department of Biology
Program in Molecular Biology and Biotechnology
Department of Biochemistry and Biophysics
University of North Carolina
Chapel Hill, North Carolina, U.S.A.
Email: marzluff@med.unc.edu
Chapter 18

Gabriele Neu-Yilik
Department of Pediatric Oncology, Hematology and Immunology
University of Heidelberg
and
Molecular Medicine Partnership Unit
Heidelberg, Germany
Chapter 9

Shigeo Ohno
Department of Molecular Biology
Yokohama City University School of Medicine
Yokohama, Japan
Email: ohnos@med.yokohama-cu.ac.jp
Chapter 7

Vasco Oliveira
Signal Transduction Program
The Burnham Institute
La Jolla, California, U.S.A.
Chapter 16

Roy Parker
Howard Hughes Medical Institute
and
Department of Molecular and Cellular Biology
University of Arizona
Tucson, Arizona, U.S.A.
Chapter 1

Sophie Redon
Swiss Institute for Experimental Cancer Research
École Polytechnique Fédérale de Lausanne
and
National Center of Competence in Research "Frontiers in Genetics"
Lausanne, Switzerland
Chapter 19

Douglas G. Scofield
Department of Biology
Indiana University
Bloomington, Indiana, U.S.A.
Chapter 15

Neda A. Sharifi
McKusick-Nathans Institute of Genetic Medicine
Baltimore, Maryland, U.S.A.
Chapter 8

Guramrit Singh
Molecular, Cellular and Developmental Biology
University of Colorado at Boulder
Boulder Colorado, U.S.A.
Chapter 5

David A.W. Soergel
Biophysics Graduate Group
University of California,
Berkeley, California, U.S.A.
Chapter 14

Ambro van Hoof
Department of Microbiology and Molecular Genetics
University of Texas Health Science Center - Houston
Houston, Texas, U.S.A.
Email: ambro.van.hoof@uth.tmc.edu
Chapter 13

Miles F. Wilkinson
Department of Immunology
The University of Texas M.D. Anderson Cancer Center
Houston, Texas, U.S.A.
Email: mwilkins@mdanderson.org
Chapter 6

Akio Yamashita
Department of Molecular Biology
Yokohama City University School of Medicine
Yokohama, Japan
Chapter 7

Zuo Zhang
Cold Spring Harbor Laboratory
Cold Spring Harbor, New York, U.S.A.
Chapter 20

PREFACE

Of all the fascinating topics in molecular biology, why would anyone be interested in nonsense? Nonsense literally makes no sense, as those of us who have spent any time in the field fully appreciate from the many light-hearted comments we have heard (again, and again). It turns out, though, precisely because nonsense generally encodes nothing constructive, it is potentially dangerous and something cells have evolved to eliminate.

The nonsense to which I refer takes the form of translation termination codons—most notably, premature translation termination codons, or PTCs. The means by which cells eliminate PTC-containing mRNA is called nonsense-mediated mRNA decay (NMD).

NMD is a type of quality control that typifies all organisms that have been analyzed, with the exception of bacteria. The initial stimulus to study NMD derived from correlations first made in the late 1970s and early 1980s using *Saccharomyces cerevisiae* and humans: mRNAs that prematurely terminate translation are abnormally short-lived.

This is the first book devoted to NMD. The rationale for such a book is the enormous information that studies of NMD have provided on the intricacies of posttrancriptional gene expression. The first five sections of the book are divided according to organism and begin with chapters on *S. cerevisiae* and mammals, from which most NMD data derive. Chapters within these sections discuss the two basic ways cells differentiate between a termination codon that elicits NMD and one that does not. NMD in *S. cerevisiae, D. melanogaster* and, probably, *C. elegans* and plants appears to be triggered by an abnormally long distance between a termination codon and a feature of the downstream untranslated region. In contrast, NMD in mammals is generally triggered by a sufficiently large distance between a termination codon and a downstream, post-splicing exon junction complex of proteins.

Several unifying themes arise repeatedly throughout these first five sections. One is that core Upf factors are conserved among all species active in NMD. Another is the importance of particular mRNA-associated proteins to NMD. There have been many surprises along the way. For example, 5' PTCs generally elicit mRNA decapping and, thus, NMD more efficiently than 3' PTCs in yeast but not in mammals. As another example, while NMD targets newly synthesized as well as steady-state mRNA in yeast, it is essentially restricted to newly synthesized mRNA in mammals, hence the distinction between a pioneer round of translation for the purpose of mRNA surveillance and steady-state translation for the purpose of protein synthesis. Additionally, yeast lack the SMG factors that function in NMD in worms, flies and mammals; the relationship between plant proteins and metazoan SMG proteins remains uncertain.

The book makes clear that studies of NMD provide a unique opportunity to examine the coupling of mRNA translation and decay, and relationships between mRNP structure and function. The importance of NMD is evident from descriptions of its influence on the expression of genes that are essential for many cellular processes, including cell differentiation, cell maintenance and cell death. If justification for efforts that deepen our understanding of NMD requires a contribution toward improved health for mankind, there is the report that 12% of all mutations in the Human Gene Database are nonsense mutations. This realization suggests that understanding the various steps of the NMD pathway will be useful in the design and implementation of disease therapies.

A sixth section of the book is devoted to evolutionary aspects of NMD. That the position of introns within pre-mRNA could influence the half-life of product mRNA implies that NMD likely influenced the evolution of mammalian gene structure. Furthermore, considering that NMD is not 100% efficient so that a fraction of NMD targets escape decay, the prediction that an estimated one-third of alternatively spliced human mRNAs are NMD targets raises the important issue about the primary purpose of NMD. Does NMD degrade mainly mRNAs that encode either nonfunctional or deleterious proteins as a means to eliminate errors in gene expression or, alternatively, mRNAs that encode useful protein isoforms as a means of maximizing the genetic potential of DNA?

The final section of the book exemplifies how cells can utilize an NMD factor in one or more other metabolic pathways, some of which are not obviously related to NMD.

This book is meant to be a one-stop source of information that fuels the fires of future experimentation. Thus, it necessarily presents a snap-shot of what is currently known or proposed to be. Many, many questions about NMD remain unanswered. Assuredly, the snap-shot provided here will evolve into a more expansive and intricate picture as additional studies are undertaken.

Lynne E. Maquat, Ph.D.
University of Rochester Medical Center
Rochester, New York, U.S.A.

Acknowledgements

This book would not have been possible without each of my colleagues who readily contributed a chapter. I am grateful to Ron Landes for suggesting the topic and providing a venue for publication. I would also like to thank Cynthia Conomos and, especially, Celeste Carlton at Landes Bioscience for expert editorial assistance, as well as my assistant at the University of Rochester, Katie Simmons, for organizational help.

Section I

Saccharomyces cerevisiae

Chapter 1

Features of Nonsense-Mediated mRNA Decay in *Saccharomyces cerevisiae*

Kristian E. Baker* and Roy Parker

Abstract

Nonsense-mediated mRNA decay (NMD) is an evolutionarily conserved cellular mechanism that reduces errors in eukaryotic gene expression by eliminating mRNAs that undergo aberrant translation termination. Two processes must be implemented for NMD to be achieved: recognition of an mRNA as 'aberrant', and subsequent targeting of the substrate for accelerated degradation. Studies of NMD in *Saccharomyces cerevisiae* have led to a working model for RNA discrimination wherein premature and normal translation termination events are distinct due to the spatial relationship between the termination codon and downstream cis-acting sequences and proteins that bind to these sequences. Improper translation termination then leads to both translational repression and an increased susceptibility of the mRNA to multiple pathways of decay. Findings from yeast have provided an important framework for understanding the more complex events that occur during NMD in organisms such as *Drosophila* and mammals.

Nonsense-Mediated Decay Targets Aberrant RNA Transcripts

Eukaryotic cells exhibit quality control mechanisms that recognize and degrade mRNAs that either have not completed nuclear pre-mRNA processing or fail to encode a proper polypeptide. Such aberrant mRNAs are degraded rapidly, presumably to preclude the accumulation of nonfunctional RNA and/or the encoded deviant protein products that could have adverse effects on the cell. One such quality control system that serves to increase the fidelity of gene expression is referred to as NMD (recently reviewed in refs. 1-3).

The unifying theme for most NMD substrates is an alteration in the normal spatial relationship between the translation termination codon and additional features of the RNA. The first substrates identified for NMD encoded premature translation termination codons (PTCs) within the coding region (for example see refs. 4,5). However, it is now appreciated that the NMD pathway degrades a wide variety of aberrant mRNAs, some of which represent routine transcription events while others arise due to mutations or defects in pre-mRNA processing (see chapters by He and Jacobson, Sharifi and Dietz, and Brenner). Such NMD substrates include transcripts with an upstream open translational reading frame (ORF),[6] an extended 3' UTR (which alters the relationship between the termination codon and the 3' poly(A) tail),[7-10] bicistronic mRNAs,[11] and mRNAs with additional, and utilized, out-of-frame AUG translation initiation codons that lead to the premature termination of translation (as a consequence of translational recoding or leaky ribosome scanning).[12] In addition, in mammalian cells, NMD also targets certain groups of 'normal' mRNAs, and possibly one-third of

*Corresponding Author: Kristian E. Baker—Center for RNA Molecular Biology, Case Western Reserve University - School of Medicine, 10900 Euclid Avenue, Cleveland, Ohio 44106-4973, U.S.A. Email: kristian.baker@case.edu

Nonsense-Mediated mRNA Decay, edited by Lynne E. Maquat. ©2006 Eurekah.com.

all alternatively spliced transcripts (see chapter by Soergel et al).[13-15] In combination, these results suggest NMD serves not only as a surveillance mechanism but also may play a broad role in the regulation of gene expression.[11,16-19]

Translational Termination: Discrimination between Normal versus Aberrant mRNA

A key issue for understanding the specificity of NMD in *Saccharomyces cerevisiae* is to understand the manner in which NMD substrates are identified. Several observations now argue that the distinction between transcripts that are NMD substrates and transcripts that bypass recognition is a biochemical difference in translation termination itself, which then elicits downstream consequences that influence the fate of the transcript. First, depending on the context, a ribosome positioned at a termination codon can undergo several fates, including termination, readthrough, or resumption of scanning after termination.[20,21] This demonstrates that translation termination events can be different. Second, the Upf1, 2, and 3 proteins, which are required for NMD, have been shown to coimmunoprecipitate with the eukaryotic translation termination factors, eRF1 and/or eRF3, providing a direct link between the termination complex and the NMD machinery.[22,23] Moreover, this interaction appears functionally important since *upf1*Δ, *upf2*Δ, or *upf3*Δ strains may have increased frequency of translational readthrough (i.e., nonsense suppression) at termination codons, although this point is still under debate.[24,25] Finally, a critical observation is that a ribosome toeprint generated at a normal translation termination codon is distinct from the toeprint of a ribosome terminating at a premature termination codon, iû a manner dependent upon Upf1p (see chapter by Amrani and Jacobson).[26] Together, these results indicate that translation termination can be modulated, and when termination is aberrant, distinct events occur in an Upf1p-dependent manner to trigger NMD. Moreover, a clear implication of these results for yeast is that features that influence NMD in other organisms, including mammals, will also affect the nature of translation termination.

Translation Termination and mRNP Organization

An unresolved issue is how the position of the translation termination codon relative to other features of the mRNA leads to the distinct translation termination events that manifest as NMD. Two general types of models have been put forth to address this issue. One model is that a proper translation termination event might be associated with interactions that involve one or more proteins bound proximally to the stop codon, and that NMD is triggered when termination occurs in the absence of these stabilizing interactions (Fig. 1A). An alternative model is that the coding region contains a protein mark that, if not removed by an elongating ribosome, remains on the mRNA and subsequently acts negatively to trigger NMD (Fig. 1B). As discussed below, a reasonable working model is that both of these types of interactions contribute to the distinction between normal and aberrant translation termination in *S. cerevisiae* and metazoans.

Several observations from yeast studies suggest that the poly(A) tail and the associated binding protein, Pab1p, may be a positive factor that affects the nature of termination, and its absence may influence NMD. First, the poly(A) tail is generally close to the termination codon and could provide contextual information. Second, Pab1p interacts with eRF3, can influence the nature of translation termination, and may even promote ribosome recycling.[27] Third, it was recently shown that tethering Pab1p proximal to a PTC recruits eRF3 and serves to stabilize nonsense-containing mRNAs (see chapter by Amrani and Jacobson for details).[26] However, Pab1p may not be the sole mark that distinguishes normal and aberrant translation termination since *pab1*Δ strains also show accelerated degradation of mRNAs with premature stop codons compared to normal mRNAs.[28]

Some experiments in yeast have suggested that specific sequences within the gene coding region, referred to as downstream sequence elements (DSEs), are required to trigger NMD when found 3' of a translation termination codon.[29] Such elements have been proposed to bind Hrp1p, which in turn interacts with Upf1p and may recruit additional NMD factors to

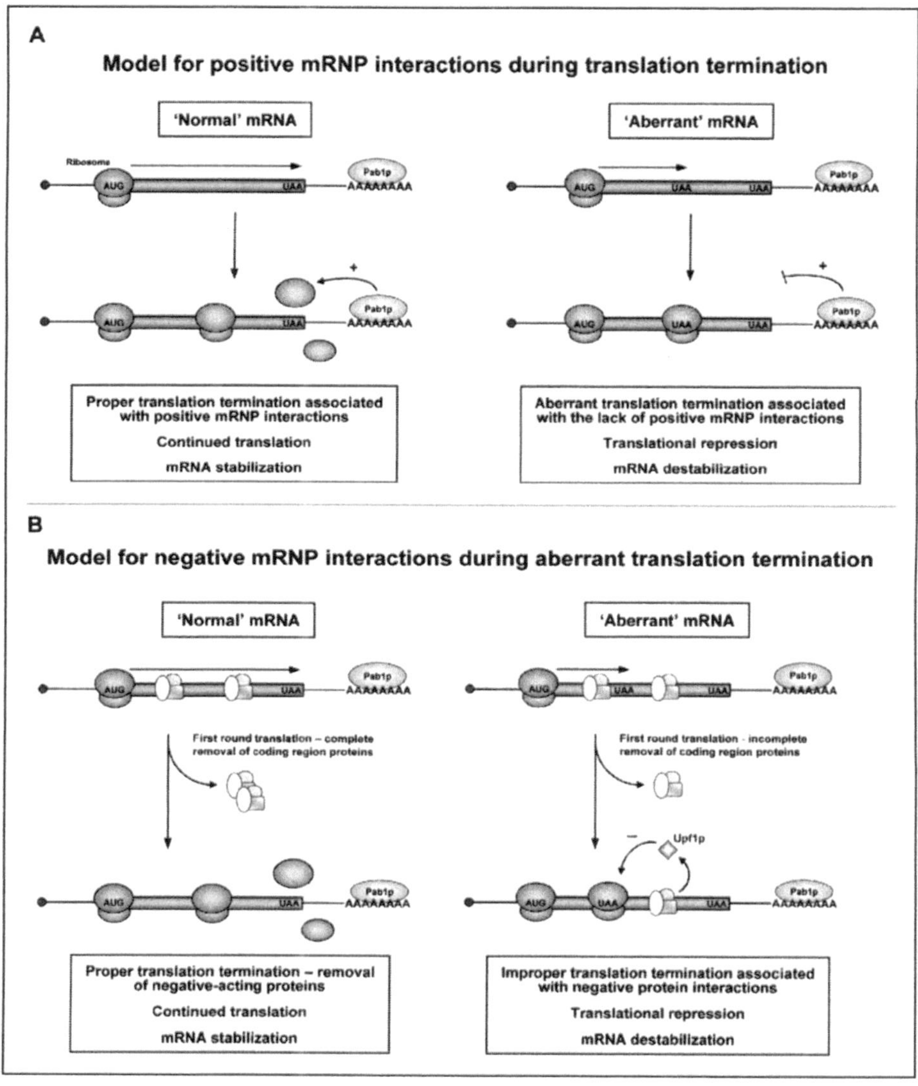

Figure 1. Distinct translation termination events for 'normal' versus 'aberrant' mRNA. A) 'Normal' translation termination occurs when mRNA stabilizing interactions take place between the terminating ribosome and factors bound to the mRNA 3' UTR and/or poly(A) tail, such as Pab1p in *S. cerevisiae* (left panel). Premature translation termination is perceived as aberrant due to the absence of such stabilizing interactions (right panel). B) Protein(s) bound to the mRNA coding region during translation (e.g., downstream element-bound Hrp1p in yeast or an EJC in mammalian cells) act negatively to trigger NMD through an interaction with Upf1p (right panel). Complete translation of the coding region leads to removal of bound proteins and stabilization of the mRNA (left panel).

the mRNA and thereby trigger decay.[30] However, there are undoubtedly additional features of yeast mRNAs that distinguish normal and aberrant transcripts since a nonsense codon downstream of a well characterized DSE can still trigger NMD.[31]

In mammalian cells, the primary determinant for triggering NMD is a protein complex deposited at exon-exon junctions during pre-mRNA splicing whose presence 50-55 nucleotides downstream of a stop codon negatively affects the stability of the mRNA and, perhaps, also influences the nature of translation termination (see chapter by Maquat).[32-36] Importantly, an interaction between the exon junction complex (EJC) protein, Y14, and the NMD factor, Upf3, is required to trigger mRNA decay, thus providing a physical and functional link between the EJC and NMD.[37] Interestingly, while EJCs play a role in NMD, EJCs can also be positive effectors of translation when positioned upstream of the stop codon. Specifically, the presence of either an EJC or a tethered Upf protein within the coding region serves to enhance mRNA 3' end formation and increases the translational yield of the mRNA.[38-40] This raises the possibility that the EJC may serve to dictate normal termination. An attractive possibility that is similar to the positive role of Pab1p in yeast, is that in mammals, the poly(A) tail might interact with the last EJC because of exon-definition and thereby also provide an important mark that enhances the function of the mRNA.[21] Thus, a reasonable model for NMD in yeast and metazoans is that translation termination can be influenced by both positive and negative acting features of the mRNP. Moreover, because in the absence of Upf1p, all termination events appear to be 'normal', the mRNP features affecting NMD presumably work by influencing whether Upf1p can interact with the termination complex, and thereby trigger recognition of the mRNA as aberrant.

Aberrant mRNA Definition Leads to Translational Repression and Accelerated RNA Degradation

Recognition of an mRNA as a substrate by the NMD machinery leads to a series of changes in the behavior of the mRNA. Primarily, recognition as nonsense containing increases the susceptibility of the mRNA to multiple routes of mRNA degradation that are also utilized for normal mRNA turnover. Additionally, translation repression of the mRNA also occurs in an Upf1p-dependent manner. These observations imply that recognition of an mRNA as aberrant leads to dramatic alterations to the NMD substrate.

Two general degradation pathways for normal mRNAs have been identified in yeast and are conserved in eukaryotic cells (Fig. 2A; recently reviewed in refs. 41,42). In both, decay of the mRNA begins with shortening of the 3' poly(A) tail through the exonucleolytic activity of the major cytoplasmic deadenylase, the Ccr4/Pop2/Not protein complex. Deadenylation can be followed by 3' to 5' degradation of the mRNA by a multiprotein assembly of exonucleases found in the cytoplasm, the exosome. Alternatively, deadenylation is more commonly followed by removal of the methyl guanosine cap from the 5' terminus of the mRNA by the yeast decapping enzyme complex, Dcp1/Dcp2p, leaving the mRNA exposed to 5' to 3' exonucleolytic degradation catalyzed by Xrn1p.

Rapid decay of NMD substrates occurs predominantly through the accelerated rate of mRNA decapping that occurs irrespective of prior removal of the poly(A) tail.[31,43] These events result in the mRNA being degraded primarily by rapid deadenylation-independent decapping (Fig. 2B). When decapping or 5' to 3' decay is blocked, NMD can promote faster decay of mRNAs through the 3' to 5' decay pathway (Fig. 2B).[31,43-45] The accelerated deadenylation and 3'-5' decay of NMD substrates still requires Upf1 and utilizes both the cytoplasmic exosome, as well as the Ski protein complex believed to regulate exosome activity. Kinetic analysis argues that NMD accelerates 3' to 5' decay primarily by the accelerated deadenylation rate, although there may be a small effect on the rate of 3' to 5' decay of the mRNA body.[31,45]

The susceptibility of nonsense-containing mRNA to nucleolytic events at both ends after PTC recognition is not interdependent. Deadenylation of the mRNA is not required for accelerated decapping to occur, as decay rates of NMD substrates remain accelerated in yeast strains lacking the deadenylase protein, Ccr4p (D. Cao and R. Parker, unpublished observations). It remains to be determined what role accelerated deadenylation plays in the degradation of NMD

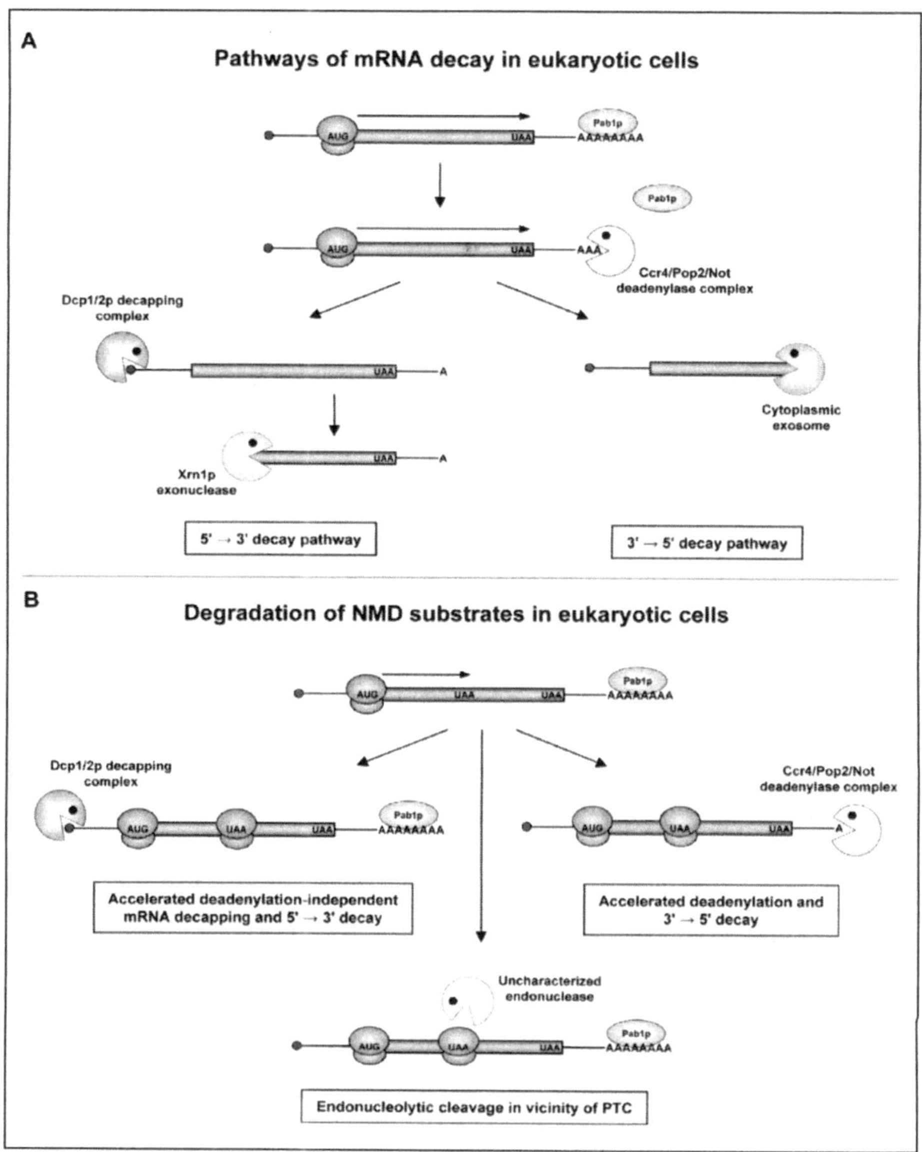

Figure 2. Pathways for the degradation of 'normal' and nonsense-containing mRNAs. A) Degradation of 'normal' mRNA in eukaryotic cells is initiated with removal of the 3' poly(A) tail by the deadenylase complex. Deadenylation is most commonly followed by removal of the mRNA methyl guanosine cap (small ball) and 5' to 3' exonucleolytic digestion by Xrn1p (left panel). Alternatively, the mRNA can be degraded 3' to 5' by the activities of members of the cytoplasmic exosome complex (right panel). B) NMD targets mRNA substrates for accelerated deadenylation-independent decapping and subsequent 5' to 3' degradation (left panel), accelerated deadenylation and 3' to 5' decay (right panel), or, as demonstrated using *Drosophila*, endonucleolytic digestion in the vicinity of the PTC (center).

substrates considering it can be bypassed. One hypothesis suggests that accelerated deadenylation is a passive consequence of the removal of an inhibitory factor which normally precludes the cytoplasmic deadenylase from accessing the poly(A) tail, such as Pab1p. It is therefore of interest to determine if rapid deadenylation is linked to the translational repression that is observed for NMD substrates.

In mammals, both mRNA decapping and 3' to 5' decay are enhanced for NMD targets. Specifically, NMD increases the rate of both deadenylation and decapping,[46-48] and depletion of decapping activity, components putatively required for 3' to 5' degradation of mRNAs, or the cytoplasmic deadenylase poly(A) ribonuclease, leads to reduced kinetics of decay of PTC-containing mRNA.[47] Additionally, NMD can also target an mRNA for enhanced rates of endonucleolytic cleavage. For example, in *Drosophila*, the decay of substrates for NMD is initiated by an endonucleolytic cleavage event in the vicinity of the PTC (Fig. 2B).[49,50] The activity catalyzing the endonucleolytic cleavage remains undefined. However, it is anticipated that its discovery may uncover a similar mechanism in yeast and/or other metazoans. Furthermore, in erythrocytes, NMD can trigger faster susceptibility of β-globin mRNA to cleavage by the PMR1-like endonuclease.[51,52] Thus, even though there exists significant conservation of the proteins required for NMD, there are several different mechanisms by which NMD accelerates RNA degradation (Fig. 2B).

The multiple mechanisms of mRNA degradation triggered by NMD imply that PTC recognition leads to a dramatic change in the function of the mRNA. An appealing possibility is that NMD leads to targeting of the mRNA to a nontranslating pool of transcripts. Since translation factors and translation itself appear to protect mRNAs from destruction, this would generally increase the instability of the transcript. Consistent with this view, measuring the translation efficiency of nonsense-containing mRNAs demonstrated that Upf1p plays a role in repressing the translation of the mRNA, independent of mRNA degradation.[53] Importantly, the translation repression of NMD substrates not only increases their decay rates, but also reduces the production of aberrant proteins.

Decay of NMD Substrates May Occur in Cytoplasmic Foci Termed P Bodies

Recently several proteins implicated in general mRNA turnover have been discovered to be localized to discrete cytoplasmic foci referred to as P bodies (also referred to as Dcp- or GW-bodies in metazoans).[54-57] Data indicate that decapping and 5' to 3' degradation of cellular RNAs can occur in P bodies.[54,57] While factors involved in the decay of NMD substrates are concentrated in subcellular compartments, Upf proteins have been found to cofractionate with polysomes, consistent with the requirement for mRNA translation in NMD substrate recognition.[58,59] The disparate localization of proteins involved in NMD can be reconciled if the first step of mRNA recognition occurs on polysomes, while decay of NMD substrates takes place in P bodies. An implication of this hypothesis is that recognition of an mRNA as aberrant leads to rapid movement of substrate mRNA from the polysome pool to cytoplasmic foci harboring a concentration of decay activities, consistent with the observed translational repression of mRNAs that are substrates for NMD.

Fluorescently tagged Upf1p, Upf2p, and Upf3p are visualized as uniformly distributed throughout the cytoplasm in wild-type yeast; however, these proteins colocalize with P body components when decay is blocked by inhibiting mRNA decapping (U. Sheth and R. Parker unpublished observations). These findings are consistent with NMD involving a polysome-associated recognition step followed by an RNA targeting step that leads to mRNA decay in P bodies. It is unclear how PTC recognition may lead to targeting of an NMD substrate to yeast P bodies, however, recent structural studies in mammalian cells of the NMD factor SMG7 has offered a mechanistic link between NMD substrate recognition and targeting. The protein structure of SMG7 indicates that it harbors a phosphoserine binding pocket which is essential for binding of Upf1 and required for NMD, as wells as a C-terminal domain that targets bound mRNAs for decay.[60,61] Interestingly, overexpression of SMG7 causes Upfl to accumulate in mammalian P bodies.[61] These results further suggest that P bodies may play a role in the decay of NMD substrates.

Position-Dependent NMD

The relative position of a PTC within the mRNA ORF has profound differences on the stability of an NMD substrate in yeast. For example, a 5' proximal PTC triggers faster decay of an NMD substrate compared to a PTC located more distal to the 5' end of the mRNA.[4,29,62] It was initially believed that the position-dependent rate of decay represented differences in the efficiency of PTC recognition, a model commonly referred to as 'Leaky Surveillance'. Recent results indicate, however, that PTC recognition is an absolute process, and that differences in the decay rate of NMD substrates represent position-dependent differences in the rate at which decapping is activated.[31]

The molecular mechanism dictating the position-dependent acceleration in mRNA decapping rate is unclear. One explanation suggests that the polarity actually describes the distance to the 3' end of the mRNA and the opportunity of the terminating ribosome to interact with one or more proteins bound to the 3' end of the mRNA.[31] Consistent with this, yeast 3' UTR sequence have a relatively uniform length of approximately 100 nucleotides. Interestingly, while the rate of decapping is dependent upon the position of the PTC, the acceleration of deadenylation is unchanged regardless of the position.[31] These findings indicate that PTC recognition is rapid and efficient, and that the downstream event of mRNA decapping may be rate limiting in the decay of NMD substrates. How the mRNA decapping rate is modulated, and whether the distance between the PTC and the 5' end of the mRNA is an important determinant of polarity, remain undetermined.

It is unclear if position-dependent NMD is broadly conserved. NMD substrates in *Drosophila* display polarity and, similar to yeast, the underlying mechanism may involve the spatial relationship between the termination codon and the 3' UTR, despite the unique mechanism of endonucleolytic decay of PTC-containing mRNA (see chapter by Behm-Asmant and Izaurralde).[50,63] In contrast, neither *Caenorhabditis elegans* or mammals generally elicit position-dependent decay rates for NMD substrates. However, some mRNAs in mammals do show polarity,[64,65] raising the possibility that the fundamental mechanism of polarity in NMD may be conserved.

Physical Interactions between NMD, Translation and Decay Machineries

An emerging concept is that NMD leads to both active translational repression and the direct recruitment of ribonucleases. The reduction in translation could be due to aberrant translation termination per se, or may be a more direct result of the function of the Upf proteins. For example, proper termination may enhance translational efficiency by promoting ribosome recycling to the initiation complex, while aberrant termination would fail to complete this function and lead to reduced translation. In contrast, the human homologue to Upf2p has been found to physically interact with eIF4A and eIF1 and may act to inhibit their function.[66] Moreover, because mutations in the yeast eIF1 protein inhibit NMD,[67] this interaction may be part of an active mechanism by which aberrant mRNAs are translationally repressed.

Several observations also provide evidence for a more direct interaction between the Upf proteins and the mRNA degradation machinery. For example, yeast Upf1p shows a strong two-hybrid interaction with Dcp2p, the catalytic subunit of the decapping enzyme.[68] Similarly, the mammalian decapping enzyme, Dcp2, and the 5'-3' exonucleases Rat1 and Xrn1 coimmunoprecipitate with Upf proteins.[47,69] Biochemical interactions have also been detected between Upfl and factors involved in 3' to 5' degradation of mRNAs. In yeast, a physical interaction between yeast Upf1p and Ski7p, a protein associated with the cytoplasmic exosome, has been detected.[45] Similarly, in mammalian cells, Upfl coimmunoprecipitates with components, of the exosome, and poly(A) ribonuclease, a putative deadenylase.[47] These physical interactions suggest that Upf proteins may play an additional, and more direct role, in targeting a nonsense-containing RNA for decay by recruiting nucleases to the mRNA.

Spatial and Temporal Recognition of NMD Substrates

When during mRNA biogenesis substrates are sensitive to NMD varies among species. In mammals, NMD appears restricted to newly synthesized mRNA, and substrate recognition is postulated to occur during the first, or pioneer, round of translation (see chapter by Maquat).[70-72] Supporting this, NMD substrates in mammals are bound by the cap-binding complex of proteins CBP80 and CBP20, while eIF4E-bound mRNAs are substrates for steady state translation and are refractory to NMD.[70,71] In yeast, NMD appears to target both newly synthesized and steady state PTC-containing mRNAs. In contrast to mammals, both Cbc1p/2p (which is orthologous to metazoan CBP80/20) and eIF4E can be found associated with early mRNPs (i.e., pre-mRNPs), and the association of either can mediate decay of an NMD substrate.[73,74] In addition, eIF4E-bound mRNAs are targeted for NMD even in cells lacking Cbc1p.[73] Moreover, yeast NMD substrates do not escape degradation to become immune, and are available to NMD recognition at each round of translation.[75] Thus, unlike NMD in mammals, yeast mRNAs undergoing steady state translation are subject to NMD, and the status of the cap-binding protein(s) is not an important determinant of substrate recognition.[73,74]

In yeast, recent findings utilizing conditional mutants defective for mRNA transport revealed that PTC-containing RNAs in the nucleus are insensitive to NMD.[74] These results suggest that NMD in yeast occurs exclusively in the cytoplasm. In mammalian cells, NMD targets are primarily nucleus-associated with some also targeted in the cytoplasm.[76,77] While still unclear, nucleus-associated NMD likely occurs during transport of the mRNA across the nuclear membrane and requires the involvement of cytoplasmic ribosomes (see chapter by Maquat).

Conclusions

Experiments in yeast now suggest a simple model for NMD, wherein differences in the mRNP context of translation termination influence whether Upf1p interacts with the termination complex to elicit downstream consequences to the transcript. If Upf1p is active during a translation termination event, the mRNA is translationally repressed and subject to increased rates of deadenylation and deadenylation-independent decapping. Interestingly, the translationally repressed mRNA might be rapidly relocated from its polysome pool to P bodies for degradation. Critical to a deeper understanding of the process of NMD will be a description of the molecular events that occur during aberrant termination, Upf1p recruitment, translational repression and destruction of the mRNA.

Acknowledgements

These authors are supported by funding to Roy Parker by the Howard Hughes Medical Institute.

References

1. Baker KE, Parker R. Nonsense-mediated mRNA decay: Terminating erroneous gene expression. Curr Opin Cell Biol 2004; 16:293-299.
2. Conti E, Izaurralde E. Nonsense-mediated mRNA decay: Molecular insights and mechanistic variations across species. Curr Opin Cell Biol 2005; 17:316-325.
3. Maquat LE. Nonsense-mediated mRNA decay: Splicing, translation and mRNP dynamics. Nat Rev Mol Cell Biol 2004; 5:89-99.
4. Losson R, Lacroute F. Interference of nonsense mutations with eukaryotic messenger mRNA stability. Proc Natl Acad Sci USA 1979; 76:5134-5137.

5. Maquat LE, Kinniburgh AJ, Rachmilewitz EA et al. Unstable β-globin mRNA in mRNA-deficient β°-thalassemia. Cell 1981; 27:543-553.
6. Oliveira CC, McCarthy JE. The relationship between eukaryotic translation and mRNA stability. A short upstream open reading frame strongly inhibits translational initiation and greatly accelerates mRNA degradation in the yeast Saccharomyces cerevisiae. J Biol Chem 1995; 270:8936-8943.
7. Zaret KS, Sherman F. Mutationally altered 3' ends of yeast CYC1 mRNA affect transcript stability and translational efficiency. J Mol Biol 1984; 177:107-135.
8. Pulak RA, Anderson P. Structures of spontaneous deletions in Caenorhabditis elegans. Mol Cell Biol 1988; 8:3748-3754.
9. Pulak R, Anderson P. mRNA surveillance by the Caenorhabditis elegans smg genes. Genes Dev 1993; 7:1885-1897.
10. Muhlrad D, Parker R. Aberrant mRNAs with extended 3' UTRs are substrates for rapid degradation by mRNA surveillance. RNA 1999; 5:1299-1307.
11. He F, Li X, Spatrick P et al. Genome-wide analysis of mRNAs regulated by the nonsense-mediated and 5' to 3' mRNA decay pathways in yeast. Mol Cell 2003; 12:1439-1452.
12. Welch EM, Jacobson A. An internal open reading frame triggers nonsense-mediated decay of the yeast SPT mRNA. EMBO J 1999; 18:6134-6145.
13. Mendell JT, Sharifi NA, Meyers JL et al. Nonsense surveillance regulates expression of diverse classes of mammalian transcripts and mutes genomic noise. Nat Genet 2004; 36:1073-1078.
14. Lewis BP, Green RE, Brenner SE. Evidence for the widespread coupling of alternative splicing and nonsense-mediated mRNA decay in humans. Proc Natl Acad Sci USA 2003; 100:189-192.
15. Green RE, Lewis BP, Hillman RT et al. Widespread predicted nonsense-mediated mRNA decay of alternatively-spliced transcripts of human normal and disease genes. Bioinformatics 2003; 19:I118-I121.
16. Lelivelt MJ, Culbertson MR. Yeast Upf proteins required for RNA surveillance affect global expression of the yeast transcriptome. Mol Cell Biol 1999; 19:6710-6719.
17. Culbertson MR. RNA surveillance. Unforeseen consequences for gene expression, inherited genetic disorders and cancer. Trends Genet 1999; 15:74-80.
18. Hentze MW, Kulozik AE. A perfect message: RNA surveillance and nonsense-mediated decay. Cell 1999; 96:307-10.
19. Frischmeyer PA, Dietz HC. Nonsense-mediated mRNA decay in health and disease. Hum Mol Genet 1999; 8:1893-1900.
20. Janzen DM, Geballe AP. Modulation of translation termination mechanisms by cis- and trans-acting factors. Cold Spring Harb Symp Quant Biol 2001; 66:459-467.
21. Hilleren P, Parker R. Mechanisms of mRNA surveillance in eukaryotes. Annu Rev Genet 1999; 33:229-260.
22. Czaplinski K, Ruiz-Echevarria MJ, Paushkin SV et al. The surveillance complex interacts with the translation release factors to enhance termination and degrade aberrant mRNAs. Genes Dev 1998; 12:1665-1677.
23. Wang W, Czaplinski K, Rao Y et al. The role of Upf proteins in modulating the translation read-through of nonsense-containing transcripts. EMBO J 2001; 15:880-890.
24. Bidou L, Stahl G, Hatin I et al. Nonsense-mediated decay mutants do not affect programmed -1 frameshifting. RNA 2000; 6:952-961.
25. Keeling KM, Lanier J, Du M et al. Leaky termination at premature stop codons antagonizes nonsense-mediated mRNA decay in S. cerevisiae. RNA 2004; 10:691-703.
26. Amrani N, Ganesan R, Kervestin S et al. A faux 3'-UTR promotes aberrant termination and triggers nonsense-mediated mRNA decay. Nature 2004; 432:112-118.
27. Mangus DA, Evans MC, Jacobson A. Poly(A)-binding proteins: Multifunctional scaffolds for the post-transcriptional control of gene expression. Genome Biol 2003; 4:223.1-233.14.
28. Caponigro G, Parker R. Multiple functions for the poly(A)-binding protein in mRNA decapping and deadenylation in yeast. Genes Dev 1995; 9:2421-2432.
29. Peltz SW, Brown AH, Jacobson A. mRNA destabilization triggered by premature translational termination depends on at least three cis-acting sequence elements and one trans-acting factor. Genes Dev 1993; 7:1737-1754.
30. Gonzalez CI, Ruiz-Echevarria MJ, Vasudevan S et al. The yeast hnRNP-like protein Hrp1/Nab4 marks a transcript for nonsense-mediated mRNA decay. Mol Cell 2000; 5:489-499.
31. Cao D, Parker R. Computational modeling and experimental analysis of nonsense-mediated decay in yeast. Cell 2003; 113:533-545.
32. Lejeune F, Maquat LE. Mechanistic links between nonsense-mediated mRNA decay and pre-mRNA splicing in mammalian cells. Curr Opin Cell Biol 2005; 17:309-315.

33. Cheng J, Belgrader P, Zhou X et al. Introns are cis effectors of the nonsense-codon-mediated reduction in nuclear mRNA abundance. Mol Cell Biol 1994; 14:6317-6325.
34. Zhang J, Sun X, Qian Y et al. Intron function in the nonsense-mediated decay of β-globin mRNA: Indications that pre-mRNA splicing in the nucleus can influence mRNA translation in the cytoplasm. RNA 1998; 4:801-815.
35. Zhang J, Sun X, Qian Y et al. At least one intron is required for the nonsense-mediated decay of triosephosphate isomerase mRNA: A possible link between nuclear splicing and cytoplasmic translation. Mol Cell Biol 1998; 18:5272-5283.
36. Thermann R, Neu-Yilik G, Deters A et al. Binary specification of nonsense codons by splicing and cytoplasmic translation. EMBO J 1998; 17:3484-3494.
37. Gehring NH, Neu-Yilik G, Schell T et al. Y14 and hUpf3b form an NMD-activating complex. Mol Cell 2003; 11:939-949.
38. Matsumoto K, Wassarman KM, Wolfe AP. Nuclear history of a pre-mRNA determines the translational activity of cytoplasmic mRNA. EMBO J 1998; 17:2107-2121.
39. Nott A, Meislin SH, Moore MJ. A quantitative analysis of intron effects on mammalian gene expression. RNA 2003; 9:607-617.
40. Nott A, Le Hir H, Moore MJ. Splicing enhances translation in mammalian cells: An additional function of the exon junction complex. Genes Dev 2004; 18:210-222.
41. Parker R, Song H. The enzymes and control of eukaryotic mRNA turnover. Nat Struct Mol Biol 2004; 11:121-127.
42. Coller J, Parker R. Eukaryotic mRNA decapping. Annu Rev Biochem 2004; 73:861-890.
43. Muhlrad D, Parker R. Premature translational termination triggers mRNA decapping. Nature 1994; 370:578-581.
44. Mitchell P, Tollervey D. An NMD pathway in yeast involving accelerated deadenylation and exosome-mediated 3' - 5' degradation. Mol Cell 2003; 11:1405-1413.
45. Takahashi S, Araki Y, Sakuno T et al. Interaction between Ski7p and Upf1p is required for nonsense-mediated 3'-to-5' mRNA decay in yeast. EMBO J 2003; 22:3951-3959.
46. Chen CY, Shyu AB. Rapid deadenylation triggered by a nonsense codon precedes decay of the RNA body in a mammalian cytoplasmic nonsense-mediated decay pathway. Mol Cell Biol 2003; 23:4805-4813.
47. Lejeune F, Li X, Maquat LE. Nonsense-mediatd mRNA decay in mammalian cells involves decapping, deadenylating, and exonucleolytic activities. Mol Cell 2003; 12:675-687.
48. Couttet P, Grange T. Premature termination codons enhance mRNA decapping in human cells. Nucl Acids Res 2004; 32:488-494.
49. Gatfield D, Unterholzner L, Ciccarelli FD et al. Nonsense-mediated mRNA decay in Drosophila: At the intersection of the yeast and mammalian pathways. EMBO J 2003; 22:3960-3970.
50. Gatfield D, Izaurralde E. Nonsense-mediated mRNA decay is initiated by endonucleolytic cleavage in Drosophila. Nature 2004; 429:575-578.
51. Stevens A, Wang Y, Bremer K et al. β-globin mRNA decay in erythroid cells: UG site-preferred endonucleolytic cleavage that is augmented by a premature termination codon. Proc Natl Acad Sci USA 2002; 99:12741-12746.
52. Bremer KA, Stevens A, Schoenberg DR. An endonuclease activity similar to Zenopus PMR1 catalyzes the degradation of normal and nonsense-containing human β-globin mRNA in erythroid cells. RNA 2003; 9:1157-1167.
53. Muhlrad D, Parker R. Recognition of yeast mRNAs as "nonsense containing" leads to both inhibition of mRNA translation and mRNA degradation: Implication for the control of mRNA decapping. Mol Biol Cell 1999; 10:3971-3978.
54. Sheth U, Parker R. Decapping and decay of messenger RNA occur in cytoplasmic processing bodies. Science 2003; 300:805-808.
55. van Dijk E, Cougot N, Meyer S et al. Human Dcp2: A catalytically active mRNA decapping enzyme located in specific cytoplasmic structures. EMBO J 2002; 21:6915-6924.
56. Ingelfinger D, Arndt-Jovin DJ, Luhrmann R et al. The human LSm1-7 proteins colocalize with the mRNA-degrading enzymes Dcp1/2 and Xrn1 in distinct cytoplasmic foci. RNA 2002; 8:1489-1501.
57. Cougot N, Babajko S, Seraphin B. Cytoplasmic foci are sites of mRNA decay in human cells. J Cell Biol 2004; 165:31-40.
58. Atkin AL, Altumura N, Leeds P et al. The majority of yeast UPF1 colocalizes with polyribosomes in the cytoplasm. Mol Biol Cell 1995; 6:611-625.

59. Atkin AL, Schenkman LR, Eastham M et al. Relationship between yeast polyribosomes and Upf proteins required for nonsense mRNA decay. J Biol Chem 1997; 272:22163-22172.
60. Fukuhara N, Ebert J, Unterholzner L et al. SMG7 is a 14-3-3-like adaptor in the nonsense-mediated mRNA decay pathway. Mol Cell 2005; 17:537-547.
61. Unterholzner L, Izaurralde E. SMG7 acts as a molecular link between mRNA surveillance and mRNA decay. Mol Cell 2004; 16:587-596.
62. Hagan KW, Ruiz-Echevarria MJ, Quan Y et al. Characterization of cis-acting sequences and decay intermediates involved in nonsense-mediated mRNA turnover. Mol Cell Biol 1995; 15:809-823.
63. Brogna S. Nonsense mutations in the alcohol dehydrogenase gene of Drosophila melanogaster correlate with an abnormal 3' end processing of the corresponding pre-mRNA. RNA 1999; 5:562-573.
64. Baumann B, Potash MJ, Kohler G. Consequences of frameshift mutations at the immunoglobin heavy chain locus of the mouse. EMBO J 1985; 4:351-359.
65. Wang J, Gudikote JP, Wilkinson MF. Boundary-independent polar nonsense-mediated decay. EMBO Rep 2002; 3:274-279.
66. Mendell JT, Medghalchi SM, Lake RG et al. Novel Upf2p orthologues suggest a functional link between translation initiation and nonsense surveillance complexes. Mol Cell Biol 2000; 20:8944-8957.
67. Cui Y, Gonzalez CI, Kinzy TG et al. Mutations in the MOF2/SUI1 gene affect both translation and nonsense-mediated mRNA decay. RNA 1999; 5:794-804.
68. He F, Jacobson A. Identification of a novel component of the nonsense-mediated mRNA decay pathway by use of an interacting protein screen. Genes Dev 1995; 9:437-454.
69. Lykke-Andersen J. Identification of a human decapping complex associated with hUPF proteins in nonsense-mediated decay. Mol Cell Biol 2002; 22:8114-8121.
70. Ishigaki Y, Li X, Serin G et al. Evidence for a pioneer round of mRNA translation: mRNAs subject to nonsense-mediated decay in mammalian cells are bound by CBP80 and CBP20. Cell 2001; 106:607-617.
71. Lejeune F, Ishigaki Y, Li X et al. The exon junction complex is detected on CBP80-bound but not eIF4E-bound mRNA in mammalian cells: Dynamics of mRNP remodeling. EMBO J 2002; 21:3536-3545.
72. Lejeune F, Ranganathan AC, Maquat LE. eIF4G is required for the pioneer round of translation in mammalian cells. Nat Struct Mol Biol 2004; 11:992-1000.
73. Gao Q, Dax B, Sherman F et al. Cap-binding protein 1-mediated and eukaryotic translation initiation factor 4E-mediated pioneer rounds of translation in yeast. Proc Natl Acad Sci USA 2005; 102:4258-4263.
74. Kuperwasser N, Brogna S, Dower K et al. Nonsense-mediated decay does not occur within the yeast nucleus. RNA 2004; 10:1907-1915.
75. Maderazo AB, Belk JP, He F et al. Nonsense-containing mRNAs that accumulate in the absence of a functional nonsense-mediated mRNA dcay pathway are destabilized rapidly upon its restitution. Mol Cell Biol 2003; 23:842-851.
76. Baserga SJ, Benz EJ. Beta-globin nonsense mutation: Deficient accumulation of mRNA occurs despite normal cytoplasmic stability. Proc Natl Acad Sci USA 1992; 89:2935-2939.
77. Cheng J, Maquat LE. Nonsense codons can reduce the abundance of nuclear mRNA without affecting the abundance of pre-mRNA or the half-life of cytoplasmic mRNA. Mol Cell Biol 1993; 13:1892-1902.

Chapter 2

All Termination Events Are Not Equal:

Premature Termination in Yeast Is Aberrant and Triggers NMD

Nadia Amrani and Allan Jacobson*

Abstract

Nonsense-mediated mRNA decay (NMD) is triggered by premature translation termination, but the features distinguishing that event from normal termination are unknown. One model for NMD in yeast suggests that decay-inducing factors bound to mRNA during early processing events are routinely removed by elongating ribosomes, but maintain an mRNA association when termination is premature, thereby triggering rapid turnover. Recent experiments challenge this notion and point to a model which posits that mRNA decay is activated by the intrinsically aberrant nature of premature termination. Toeprinting assays that delineate ribosome positioning demonstrate that premature translation termination in yeast extracts is indeed aberrant. Aberrant termination depends on prior nonsense codon recognition and is eliminated in extracts derived from cells lacking the principal NMD factors, Upf1p or Nmd2p (Upf2p), or by flanking the nonsense codon with a normal 3'-untranslated region (UTR). Tethered poly(A)-binding protein (Pab1p), used as a mimic of a normal 3'-UTR, recruits the termination factor Sup35p (eRF3) and stabilizes nonsense-containing mRNAs. These findings indicate that efficient termination and mRNA stability are dependent on a properly configured 3'-UTR and lead to a model specifying roles for Upf1p, Nmd2p, and Upf3p that are exclusive to premature termination.

Quality Control of Translation Termination

Several mechanisms are responsible for maintaining fidelity in the flow of genetic information, including those that monitor the integrity of RNA synthesis and processing, tRNA aminoacylation, AUG site selection, and peptide elongation.[1-7] At translation termination, quality control is implemented by release factors that discriminate sense codons from nonsense codons before triggering polypeptide hydrolysis.[8] In addition, a specialized mRNA decay mechanism, NMD or mRNA surveillance, serves to rid the cell of transcripts that lack complete open reading frames.[9-11] While such transcripts can arise de novo from genes in which a mutation or an error in transcription or processing has given rise to a premature nonsense codon, there are also a large number of endogenous substrates of the NMD pathway. In the yeast *Saccharomyces cerevisiae*, these include: inefficiently spliced pre-mRNAs that enter the cytoplasm with their introns intact, mRNAs in which the ribosome has bypassed the initiator AUG and commenced translation further downstream, some mRNAs containing upstream open reading frames (uORFs), mRNAs subject to frameshifting, bicistronic mRNAs, transcripts of pseudogenes

*Corresponding Author: Allan Jacobson—Department of Molecular Genetics and Microbiology, University of Massachusetts Medical School, Worcester, Massachusetts 01655-0122, U.S.A. Email: allan.jacobson@umassmed.edu

Nonsense-Mediated mRNA Decay, edited by Lynne E. Maquat. ©2006 Eurekah.com.

and transposable elements, and mRNAs with abnormal extensions of their 3'-untranslated regions (UTRs) (see chapter by He and Jacobson). As will be discussed below, targeting of these transcripts for NMD not only minimizes the generation of potentially deleterious polypeptide fragments, but also appears to liberate and recycle ribosomes and other components of the translation apparatus from termination complexes that would otherwise be inefficiently dissociated.

Factors Regulating Translation Termination and NMD

Termination of protein synthesis occurs when the elongating ribosome encounters an in-frame UAA, UAG, or UGA codon. This step involves two classes of release factors (RFs). Class I RFs recognize stop codons within the ribosomal A site and trigger the hydrolysis of the ester bond connecting the polypeptide chain and the tRNA in the P site. Class II RFs are GTPases that stimulate class I RF activity and confer GTP dependency upon the termination process.[12-14] In eukaryotes, translation termination is mediated by two interacting proteins, eRF1 and eRF3, which act as class I and class II factors and are respectively named Sup45 protein (p) and Sup35p in yeast.[15-17] GTP hydrolysis is required to couple eRF1 recognition of the termination signal in mRNA to efficient polypeptide chain release,[18] thus indicating that discrimination of sense from nonsense codons by eRF1 mimics the recognition of a cognate codon by a tRNA.

In addition to its interaction with eRF1, eRF3 also interacts with the poly(A)-binding protein (PABP; Pab1p in yeast). The Pab1p-interacting site on Sup35p is located in the N-terminal part of the protein and, reciprocally, the C-terminal domain of Pab1p interacts with Sup35p. Interactions between cytoplasmic PABP and eRF3 have been observed in yeast, frog, and mammalian cells[19-22] and these interactions appear to influence translation termination and mRNA decay. In yeast, Pab1p overexpression has been shown to enhance termination efficiency in cells harboring a sup35 (eRF3) mutation.[20] Moreover, PABP:eRF3 interaction has been shown to promote multiple rounds of ribosome recycling in cis in vitro and to play a role in the translation-dependent shortening of the poly(A) tail.[22,23]

Yeast factors that regulate NMD, identified in screens for translational suppressors or two-hybrid interactors with known factors, are principally those encoded by the *UPF1*, *NMD2*(*UPF2*), and *UPF3* genes.[24-29] They have been characterized extensively and are conserved in all eukaryotes examined. The *UPF1* gene encodes a 109-kD protein with seven conserved motifs common to members of helicase superfamily I.[26,30,31] Not surprisingly, purified Upfl protein (p) has RNA-binding, as well as RNA-dependent ATPase and RNA helicase activities.[32-35] An N-terminal Zn^{++}-finger-like domain of Upf1p interacts with a C-terminal domain of Nmd2p, an acidic 127-kD protein with multiple MIF4G (middle portion of eukaryotic translation initiation factor (eIF)4G) domains.[24,28,29,36-38] In turn, a MIF4G domain in the C-terminal half of Nmd2p interacts with Upf3p, a basic 45-kD protein.[24,39,40]

A Role for the Upf/Nmd Factors in Translation Termination

Mutations in the yeast *UPF/NMD* genes not only lead to the stabilization of nonsense-containing mRNAs, they also promote nonsense suppression.[18,26,34,35,41,42] Importantly, these effects can be separated from changes in mRNA levels by mutation, by independent alterations in the abundance of nonsense-containing mRNAs, and by normalization of suppression data to RNA content.[18,34,35,41] Mutations in the release factors, Sup35p and Sup45p, also promote nonsense suppression[18,43] and these effects are additive with *UPF/NMD* gene mutations,[18] indicating distinct functions in termination. The apparent role of the Upf/Nmd factors in the promotion of termination fidelity is underscored by in vitro binding experiments suggesting that Sup35p bridges their interactions with the termination complex. Upf1p, Nmd2p, and Upf3p all interact with the release factor Sup35p, and Upf1p can also interact with Sup45p.[44,45] Upf1p binds to a specific region of Sup35p, but Nmd2p, Upf3p, and Sup45p all compete for binding to Sup35p and may thus share a common interaction domain.[45] Interaction with either release factor inhibits the ATPase activity of Upf1p, and interaction with Sup35p prevents formation of a Upf1p:RNA complex.[44,45] Since Sup35p and RNA may compete for binding to Upf1p, factors such as ATP, which is capable of decreasing Upf1p affinity for RNA, might promote release factor binding[34,35,46] or regulate Upf1p interaction with the release factors in the presence of competing RNAs.[44] The identification of a mutant form of Upf1p that can bind but not hydrolyze ATP, and which is active in translation termination but inactive in mRNA decay, suggests that ATP may be a cofactor that switches Upf1p between its translation termination and NMD activities.[34,35,46]

Pathways for Decay of Yeast Nonsense-Containing mRNAs: A Requirement for Ongoing Translation

Wild-type yeast mRNAs are degraded through the decapping-dependent 5' to 3' pathway and the exosome-mediated 3' to 5' pathway.[47-51] The initial decay event in both pathways is the shortening of the poly(A) tail to a length of 10 to 12 nucleotides (nt).[47,52,53] After poly(A) shortening, transcripts can be degraded by either pathway. In 5' to 3' decay, transcripts are decapped by the Dcp1p/Dcp2p decapping enzyme complex, and then digested exonucleolytically by the 5' to 3' exoribonuclease, Xrn1p.[47,54-58] In 3' to 5' decay, transcripts are further deadenlylated and then degraded by a ten-subunit 3' to 5' exonuclease complex, the exosome.[59]

Poly(A) shortening includes at least two steps, an initial nuclear trimming of the pre-mRNA poly(A) tail by poly(A) nuclease (PAN) and its more extensive removal in the cytoplasm by the Ccr4p/Pop2p/Notp complex.[50,51] This eliminates most Pab1p from the mRNP, a step that appears to prompt the loss of the cap-binding translation initiation complex and the transition of the mRNP to a new state in which the binding of additional proteins enhances the enzymology of decapping and determines the subcellular site of its occurrence.[50,51] This mRNP rearrangement allows binding of the Lsm1-7p/Pat1p complex which, in turn, appears to promote interaction of the mRNP with the Dcp1p/Dcp2p complex.[60] All steps subsequent to association of the Lsm1-7p/Pat1p complex are thought to occur at a limited number of subcellular sites called P bodies.[61]

It was initially thought that degradation of nonsense-containing transcripts utilized an abbreviated normal pathway, proceeding from decapping to 5' to 3' decay without prior poly(A) shortening.[54-56,62] More recent studies indicate that the notion of a specific pathway for degrading these RNAs is too simplistic. Nonsense-containing mRNAs appear to be subject to accelerated decapping, accelerated deadenylation, and both 5' to 3' and 3' to 5' decay.[63-67] However, microarray analysis showed that the transcripts upregulated in *upf1Δ*, *nmd2Δ*, or *upf3Δ* cells displayed 70% overlap with those in *xrn1Δ* cells, thereby indicating that the substrates of the NMD pathway are preferentially degraded by the 5' to 3' pathway[68] (see chapter by He and Jacobson). Accelerated decapping is indeed deadenylation-independent, but its rate is dependent on the proximity of the nonsense codon to the mRNA 5' end.[66] Moreover, accelerated deadenylation need not be directly related to decay since a premature nonsense codon will also destabilize an mRNA in which 5'/3' interactions are established independent of a poly(A) tail.[69] These observations, as well as the diminished translational efficiencies of nonsense-containing mRNAs,[18,70] have given rise to the suggestion that these mRNAs have an altered mRNP structure that changes their accessibility to different nucleases, precludes their transition to a more stable state, or forces their localization to P bodies[65,66,71-73] (see chapter by Baker and Parker).

The existence of a premature nonsense codon within a transcript is not sufficient to promote its rapid degradation. Destabilization of yeast nonsense-containing mRNAs also requires their ongoing translation, i.e., NMD depends on recognition of the nonsense codon by the translational apparatus. This conclusion follows from observations that NMD factors and decay intermediates are localized to polysomes,[40,74-79] and that decay can be antagonized by drugs or mutations that interfere with protein synthesis[79,80] or by tRNAs that suppress termination.[81,82] Additional support for this conclusion includes experiments showing that: (a) NMD regulatory factors interact with the polypeptide release factors;[44] (see above); (b) mutations in genes encoding the NMD factors promote nonsense suppression;[18,26,34,35,41] (see above); (c) a dominant-negative form of Nmd2p is only active when localized to the cytoplasm;[36] and (d) NMD-promoted destabilization of the yeast *CPA1* mRNA is dependent on the extent of ribosome occupancy of the uORF termination codon that triggers decay.[83]

NMD Regulation by cis-Acting Elements

All functional mRNAs contain a translation termination codon, yet they are not substrates for NMD. What appears to distinguish a normal nonsense codon from one that promotes mRNA destabilization is its sequence context, or more precisely, the sequences 3' to the nonsense codon. A requirement for specific downstream elements, or DSEs, was initially suggested by experiments demonstrating that: (a) deletion of most of the *PGK1* protein coding region downstream of an early nonsense mutation reduced mRNA decay rates markedly and (b) reinsertion of a small segment of the *PGK1* coding region into the construct harboring the large deletion was sufficient to activate NMD.[9] Studies of the *GCN4* mRNA have demonstrated that a DSE can activate the NMD pathway when it is located within approximately 150 nt of the stop codon, but is not functional if traversed by translating ribosomes.[80,84] The presence of DSEs within the coding regions of wild-type mRNAs of diverse decay rates suggests that these sequences are inactive unless preceded by a termination codon. In spite of their apparently common function, DSE sequences from *PGK1* and other mRNAs have only a weak sequence consensus.[9,85-87]

One additional class of cis-acting elements appears to be capable of regulating NMD. In yeast, stabilizer elements (STEs) in the 5' leader region of the *GCN4* and *YAP1* mRNAs inactivate the NMD pathway when positioned downstream of a termination codon but still 5' of a DSE.[80,84] These STEs appear to bind the *Pub1* protein, which in turn may either antagonize activation of the decay apparatus or mimic the RNP context of a normal 3' UTR.

Models: Surveillance Complex vs Faux UTR

NMD is triggered by premature translation termination, but the features distinguishing that event from normal termination are unclear. Two predominant models for the mechanism of yeast NMD have been put forth. The first, the "surveillance complex" model,[44,88] suggests that the deposition of specific proteins (e.g., Hrp1p) during early processing events allows decay-inducing factors to maintain an mRNA association unless swept off by the ribosome in the first round(s) of translation. In the event that termination is premature, Hrp1p (or related factors) is thought to remain on the mRNA and to trigger mRNA decay by subsequent interaction with Upf1p. The manner in which Upf1p triggers decay is uncertain, but Upf1p:Dcp2p interactions[29,89] have led to speculation that Upf1p may recruit the decapping complex.[50,89,90]

The second model posits that premature termination and normal termination are biochemically different events and that, at least in yeast, NMD is triggered by the events accompanying aberrant termination at a premature nonsense codon.[66,71,72,91] In this "faux UTR" model,[71,92,93] proper termination of translation and normal rates of mRNA decay require interactions between a terminating ribosome and a specific ribonucleoprotein particle (RNP) structure or set of factors localized 3' to the stop codon. The downstream element (DSE) is thought to promote aberrant termination because it lacks a regulatory factor (or factors) normally present on a legitimate 3'-UTR and aberrant termination, in turn, is thought to allow binding of the Upf/Nmd factors and trigger mRNA decay. As described below, recent experiments using yeast both refute restriction of NMD to an early round of translation, as typifies exon junction complex-mediated NMD in mammalian cells (see chapter by Maquat), and provide strong support for the concepts of the faux UTR model.

Translation Termination at Premature Nonsense Codons Is Aberrant

The possibility that premature termination differed from normal termination was recently addressed by a series of experiments utilizing in vitro translation in yeast cell-free

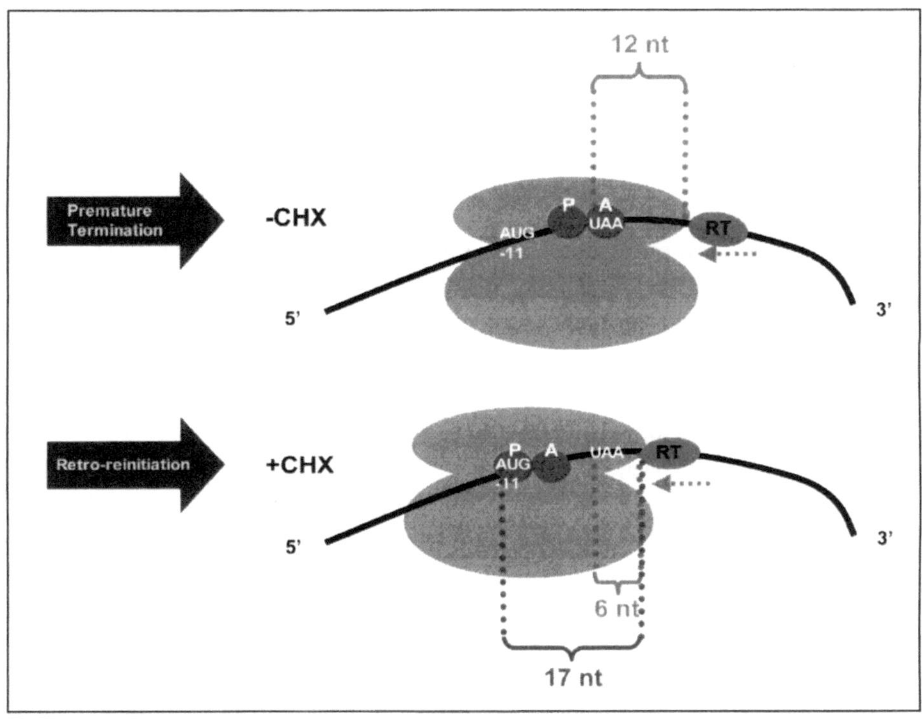

Figure 1. Positioning of ribosomes on nonsense-containing *can-100/LUC* mRNA. Toeprint analyses were used to localize ribosomes on reporter mRNAs harboring premature termination codons.[93] Top) Premature termination codon toeprints are detected at the normal +12 nt position during translation of nonsense-containing mRNAs in wild-type extracts in the absence of cycloheximide (CHX). Bottom) The aberrant +6 nt toeprints appear subsequent to CHX-chasing of the premature terminator +12 nt toeprints and correspond to ribosomes stalled with their P sites on an upstream AUG codon and ready for reinitiation of translation. RT, reverse transcriptase.

extracts combined with analyses of ribosome positioning by toeprinting, i.e., primer extension inhibition.[93] In these experiments, ribosomes encountering a premature termination codon failed to be released and yielded a +12 nt toeprint signal corresponding to the stalling of their A sites on a nonsense codon (Fig. 1, top). Similar experiments failed to detect any toeprint signals from normal termination codons, suggesting that ribosomes terminating prematurely are released much less efficiently than those encountering normal terminators.[93] Consistent with this conclusion, toeprint bands were seen at normal terminators only when Sup45p (eRF1) was nonfunctional. The addition of the elongation inhibitor cycloheximide (CHX) to translation reactions failed to elicit toeprints from normal terminators in wild-type extracts, but did allow detection of additional toeprints in close proximity to the early nonsense codons. These unanticipated bands reflected 80S ribosomes whose P sites were centered on AUG codons, protecting 16-18 nt 3' of those codons. For example, a +6 nt toeprint (relative to the termination codon) was actually attributable to ribosomes associated with an AUG codon 11 nt upstream of the premature stop[93] (see Fig. 1, bottom). These toeprint analyses, and other experiments, showed that ribosomes encountering premature termination codons (but not normal termination codons) fail to be released and instead scan backwards from the premature terminator and reinitiate translation. Additional experiments showed that ribosomes exiting premature termination codons could also scan downstream and reinitiate translation, but had a propensity for upstream scanning. Regardless of the site of post-termination reinitiation, premature stop codons had to be recognized by Sup45p and peptide hydrolysis had to be triggered prior to any reinitiation event because extracts of sup45-2 cells (possessing defective eRF1[43]) only yielded +12 nt toeprints, regardless of the presence or absence of CHX.[93]

The relevance of NMD to the retroreinitiation toeprints detected in the vicinity of premature termination codons was established by experiments showing that these toeprints are markedly diminished in extracts devoid of Upf1p or Nmd2p.[93] These observations indicated that inactivation of NMD factors may preclude or perturb events that regulate ribosome stability and movements at a premature stop codon. Further, flanking a premature nonsense codon with a normal 3'-UTR, creating a mimic of wild-type mRNA, also eliminated the AUG toeprints characteristic of aberrant termination.[93] Collectively, these experiments showed that retroreinitiation is a consequence of an aberrant termination event and raised the question of why a 3'-UTR created by a premature termination codon should differ from that of a wild-type mRNA.

3'-UTR Mimicry Reestablishes Normal Rates of mRNA Decay

The faux UTR model suggests that the downstream element (DSE) thought to be a key cis-acting regulator of NMD[9] promotes mRNA decay because it lacks a termination regulatory factor (or factors) normally present on a legitimate 3'-UTR, a hypothesis that may also explain why deletions that eliminate most coding sequences downstream of premature terminators stabilize mRNAs that would otherwise be substrates for NMD.[9,70] This inadequacy could occur because translation to the normal end of a coding region remodels an mRNP in some alternative manner[72] or because proximity to the poly(A) tail (and its bound Pab1p) has a qualitative and/or quantitative influence on the nature of proteins associating with the UTR.[71]

To test whether the proximity of Pab1p and termination codons is germane to NMD, the in vivo stability of nonsense-containing mRNAs possessing MS2 coat protein binding sites adjacent to premature terminators was assessed in yeast cells expressing MS2-Pab1p fusion proteins.[93] These experiments showed that mimicking a normal 3'-UTR by tethering Pab1p stabilizes nonsense-containing mRNAs in vivo.[93] Furthermore, partial stabilization of nonsense-containing transcripts was engendered by tethered fragments of Pab1p, suggesting that Pab1p-mediated mRNA stabilization might be attributable to protein:protein interactions characteristic of its respective domains.[51] Consistent with this possibility, immunoprecipitation of Pab1p-tethered mRNPs revealed the presence of Sup35p, but did not lead to the recovery of the translation initiation factors eIF4G or eIF4E.[93] Coimmunoprecipitation of Sup35p, a termination factor thought to play a role in the stability of conventional mRNAs,[22] suggested a role for this protein in the tethered-Pab1p-mediated reversal of NMD. This appears to be the case since tethered Sup35p also stabilized *PGK1* nonsense transcripts, albeit to a lesser extent than tethered Pab1p.[93] Similar experiments, in which Sup45p was the tethered component, failed to stabilize the same mRNA significantly, indicating that regulatory aspects of termination played a key role in antagonizing NMD and that Pab1p's role in this process most likely reflected its function as a scaffold for post-transcriptional regulators.[51]

A Higher Resolution Faux UTR Model for NMD

A key concept of the surveillance complex model,[44] and, in the case of mammalian cells, the pioneer round model,[94] is that the triggering of NMD occurs early during the translational lifetime of an mRNA. Subsequent to such early rounds of translation, factors essential for mRNA destabilization are thought to be removed from transcripts by the elongation activity of ribosomes. However, recent experiments utilizing yeast have shown that nonsense-containing mRNAs are available for NMD at every round of translation.[18,83,92] Furthermore, new results on multiple fronts appear to be consistent with the basic tenet of the faux UTR model, namely that not all termination events are equivalent and that, at least in yeast, NMD is triggered by a ribosome's failure to terminate adjacent to a properly configured 3'-UTR. As suggested by the faux UTR model, proper termination of translation and normal rates of mRNA decay are likely to require interactions between a terminating ribosome and a specific RNP structure or set of factors localized 3' to the stop codon. Failure to terminate in this manner appears to slow ribosome release and facilitate reinitiation. The latter event must only be a symptom of aberrant termination, not a cause of NMD, because elimination of reinitiation site AUGs does not alter the decay rate of nonsense-containing mRNAs.[93]

The model shown in Figure 2 addresses these observations and serves as a theoretical basis for many new questions that need to be answered. In normal termination (Fig. 2, left), we speculate, as have others,[19,20,22,23] that an interaction between Sup35p and Pab1p enhances the efficiency of termination and ensures efficient ribosome release and recycling (possibly to the 5' end of the same mRNA). While the Upf/Nmd factors are known to be able to interact with Sup35p,[45] this interaction is thought to be precluded by proximal Pab1p, possibly because Pab1p's affinity for Sup35p may exceed that of the Upf/Nmd factors, or because Pab1p is also

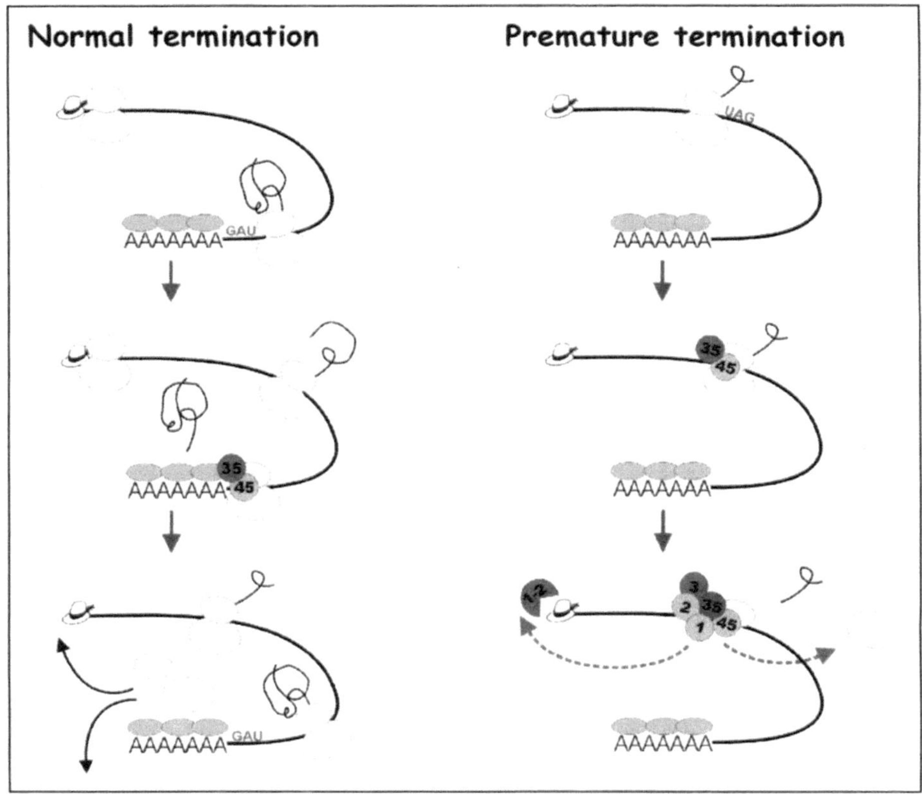

Figure 2. Proposed mechanistic differences between normal and premature termination. Normal termination is thought to be highly efficient and involve interactions between Pab1p (green ovals) and ribosome-associated Sup35p. Premature termination is thought to be inefficient, and the absence of proximal Pab1p is hypothesized to facilitate binding of the Upf/Nmd factors to the termination complex, thereby leading to peptide hydrolysis, 60S ribosomal subunit release, and Upf1p recruitment/stimulation of the Dcp1p/Dcp2p decapping complex. 1, Upf1p; 2, Nmd2p; 3, Upf3p; 35, Sup35p; 45, Sup45p. A color version of this figure is available online at www.Eurekah.com.

mediating other interactions that facilitate the termination process. Normal termination thus depends on Pab1p, Sup35p, Sup45p, proximity to a proper 3'-UTR, and, most likely, additional factors whose role may include facilitating events beyond peptide hydrolysis, e.g., ribosome recycling in cis.

In premature termination (Fig. 2, right), the absence of proximal Pab1p is thought to reduce the efficiency of termination and to allow for binding of the Upf/Nmd factors. Preferential binding of these factors to NMD substrates takes into account the extremely low abundance of Upf1p, Nmd2p, and Upf3p in yeast cells[41] and also recognizes that the only time these factors are found in association with mammalian mRNPs is when those mRNPs contain nuclear PABP, not cytoplasmic PABP[94] (only the latter PABP is homologous to yeast Pab1p).[51] Association of the Upf/Nmd factors with the termination complex may involve initial binding of Nmd2p and Upf3p to Sup35p, which, in turn, may facilitate Upf1p binding to Sup35p and Sup45p[45] (details not shown in Fig. 2). Ultimately, binding of the Upf/Nmd factors is thought to enhance the efficiency of peptide hydrolysis, since their absence promotes nonsense suppression,[35,41] and to promote the dissociation of only the 60S ribosomal subunit via the RNA helicase activity of Upf1p. In subsequent steps, the Upf/Nmd factors manifest their role in mRNA decay. The N-terminal Zn^{++}-finger domain of Upf1p is thought to recruit the Dcp1p/Dcp2p complex and promote decapping (subsequent 5' to 3' decay is not shown) and weaker interactions between the Upf/Nmd factors and components of other mRNA decay factors (e.g., the deadenylase and the exosome) are thought to promote other modes of decay (not shown).

References

1. Chin K, Pyle AM. Branch-point attack in group II introns is a highly reversible transesterification, providing a potential proofreading mechanism for 5'-splice site selection. RNA 1995; 1:391-406.
2. Freist W, Sternbach H, Cramer F. Phenylalanyl-tRNA synthetase from yeast and its discrimination of 19 amino acids in aminoacylation of tRNA(Phe)-C-C-A and tRNA(Phe)-C-C- A(3'NH2). Eur J Biochem 1996; 240(3):526-531.
3. Ibba M, Soll D. Quality control mechanisms during translation. Science 1999; 286(5446):1893-1897.
4. Jeon C, Agarwal K. Fidelity of RNA polymerase II transcription controlled by elongation factor TFIIS. Proc Natl Acad Sci USA 1996; 93:13677-13682.
5. Yarus M. Proofreading, NTPases and translation: Successful increase in specificity. Trends Biochem Sci 1992; 17(5):171-174.
6. Yoshizawa S, Fourmy D, Puglisi JD. Recognition of the codon-anticodon helix by ribosomal RNA. Science 1999; 285(5434):1722-1725.
7. Kozak M. Recognition of AUG and alternative initiator codons is augmented by G in position +4 but is not generally affected by the nucleotides in positions +5 and +6. EMBO J 1997; 16(9):2482-2492.
8. Salas-Marco J, Bedwell DM. GTP hydrolysis by eRF3 facilitates stop codon decoding during eukaryotic translation termination. Mol Cell Biol 2004; 24(17):7769-7778.
9. Peltz SW, Brown AH, Jacobson A. mRNA destabilization triggered by premature translational termination depends on at least three cis-acting sequence elements and one trans-acting factor. Genes Dev 1993; 7(9):1737-1754.
10. Peltz SW, He F, Welch E et al. Nonsense-mediated mRNA decay in yeast. Prog Nucleic Acid Res Mol Biol 1994; 47:271-298.
11. Pulak R, Anderson P. mRNA surveillance by the Caenorhabditis elegans smg genes. Genes Dev 1993; 7(10):1885-1897.
12. Kisselev L, Ehrenberg M, Frolova L. Termination of translation: Interplay of mRNA, rRNAs and release factors? EMBO J 2003; 22(2):175-182.
13. Nakamura Y, Ito K, Ehrenberg M. Mimicry grasps reality in translation termination. Cell 2000; 101(4):349-352.
14. Nakamura Y, Ito K. Making sense of mimic in translation termination. Trends Biochem Sci 2003; 28(2):99-105.
15. Frolova L, Le Goff X, Rasmussen HH et al. A highly conserved eukaryotic protein family possessing properties of polypeptide chain release factor. Nature 1994; 372(6507):701-703.

16. Stansfield I, Jones KM, Kushnirov VV et al. The products of the SUP45 (eRF1) and SUP35 genes interact to mediate translation termination in Saccharomyces cerevisiae. EMBO J 1995; 14(17):4365-4373.
17. Zhouravleva G, Frolova L, Le Goff X et al. Termination of translation in eukaryotes is governed by two interacting polypeptide chain release factors, eRF1 and eRF3. EMBO J 1995; 14(16):4065-4072.
18. Keeling KM, Lanier J, Du M et al. Leaky termination at premature stop codons antagonizes nonsense-mediated mRNA decay in S. cerevisiae. RNA 2004; 10(4):691-703.
19. Hoshino S, Imai M, Kobayashi T et al. The eukaryotic polypeptide chain releasing factor (eRF3/GSPT) carrying the translation termination signal to the 3'-poly(A) tail of mRNA. Direct association of erf3/GSPT with polyadenylate-binding protein. J Biol Chem 1999; 274(24):16677-16680.
20. Cosson B, Couturier A, Chabelskaya S et al. Poly(A)-binding protein acts in translation termination via eukaryotic release factor 3 interaction and does not influence PSI(+) propagation. Mol Cell Biol 2002; 22(10):3301-3315.
21. Cosson B, Berkova N, Couturier A et al. Poly(A)-binding protein and eRF3 are associated in vivo in human and Xenopus cells. Biol Cell 2002; 94(4-5):205-216.
22. Hosoda N, Kobayashi T, Uchida N et al. Translation termination factor eRF3 mediates mRNA decay through the regulation of deadenylation. J Biol Chem 2003; 278(40):38287-38291.
23. Uchida N, Hoshino S, Imataka H et al. A novel role of the mammalian GSPT/eRF3 associating with poly(A)-binding protein in cap/poly(A)-dependent translation. J Biol Chem 2002; 277(52):50286-50292.
24. He F, Brown AH, Jacobson A. Upf1p, Nmd2p, and Upf3p are interacting components of the yeast nonsense-mediated mRNA decay pathway. Mol Cell Biol 1997; 17(3):1580-1594.
25. Leeds P, Peltz SW, Jacobson A et al. The product of the yeast UPF1 gene is required for rapid turnover of mRNAs containing a premature translational termination codon. Genes Dev 1991; 5(12A):2303-2314.
26. Leeds P, Wood JM, Lee BS et al. Gene products that promote mRNA turnover in Saccharomyces cerevisiae. Mol Cell Biol 1992; 12(5):2165-2177.
27. Lee SI, Umen JG, Varmus HE. A genetic screen identifies cellular factors involved in retroviral -1 frameshifting. Proc Natl Acad Sci USA 1995; 92(14):6587-6591.
28. Cui Y, Hagan KW, Zhang S et al. Identification and characterization of genes that are required for the accelerated degradation of mRNAs containing a premature translational termination codon. Genes Dev 1995; 9(4):423-436.
29. He F, Jacobson A. Identification of a novel component of the nonsense-mediated mRNA decay pathway by use of an interacting protein screen. Genes Dev 1995; 9(4):437-454.
30. Altamura N, Groudinsky O, Dujardin G et al. NAM7 nuclear gene encodes a novel member of a family of helicases with a Zn-ligand motif and is involved in mitochondrial functions in Saccharomyces cerevisiae. J Mol Biol 1992; 224(3):575-587.
31. Koonin E. A new group of putative RNA helicases. Trends Biochem Sci 1992; 17:495-497.
32. Bhattacharya A, Czaplinski K, Trifillis P et al. Characterization of the biochemical properties of the human Upf1 gene product that is involved in nonsense-mediated mRNA decay. RNA 2000; 6(9):1226-1235.
33. Czaplinski K, Weng Y, Hagan KW et al. Purification and characterization of the Upf1 protein: A factor involved in translation and mRNA degradation. RNA 1995; 1(6):610-623.
34. Weng Y, Czaplinski K, Peltz SW. Genetic and biochemical characterization of mutations in the ATPase and helicase regions of the Upf1 protein. Mol Cell Biol 1996; 16(10):5477-5490.
35. Weng Y, Czaplinski K, Peltz SW. Identification and characterization of mutations in the UPF1 gene that affect nonsense suppression and the formation of the Upf protein complex but not mRNA turnover. Mol Cell Biol 1996; 16(10):5491-5506.
36. He F, Brown AH, Jacobson A. Interaction between Nmd2p and Upf1p is required for activity but not for dominant-negative inhibition of the nonsense-mediated mRNA decay pathway in yeast. RNA 1996; 2(2):153-170.
37. Kadlec J, Izaurralde E, Cusack S. The structural basis for the interaction between nonsense-mediated mRNA decay factors UPF2 and UPF3. Nat Struct Mol Biol 2004; 11(4):330-337.
38. Mendell JT, Medghalchi SM, Lake RG et al. Novel Upf2p orthologues suggest a functional link between translation initiation and nonsense surveillance complexes. Mol Cell Biol 2000; 20(23):8944-8957.
39. Lee BS, Culbertson MR. Identification of an additional gene required for eukaryotic nonsense mRNA turnover. Proc Natl Acad Sci USA 1995; 92(22):10354-10358.

40. Shirley RL, Lelivelt MJ, Schenkman LR et al. A factor required for nonsense-mediated mRNA decay in yeast is exported from the nucleus to the cytoplasm by a nuclear export signal sequence. J Cell Sci 1998; 111(Pt 21):3129-3143.
41. Maderazo AB, He F, Mangus DA et al. Upf1p control of nonsense mRNA translation is regulated by Nmd2p and Upf3p. Mol Cell Biol 2000; 20(13):4591-4603.
42. Stahl G, Bidou L, Hatin I et al. The case against the involvement of the NMD proteins in programmed frameshifting. RNA 2000; 6(12):1687-1688.
43. Stansfield I, Kushnirov VV, Jones KM et al. A conditional-lethal translation termination defect in a sup45 mutant of the yeast Saccharomyces cerevisiae. Eur JBiochem 1997; 245(3):557-563.
44. Czaplinski K, Ruiz-Echevarria MJ, Paushkin SV et al. The surveillance complex interacts with the translation release factors to enhance termination and degrade aberrant mRNAs. Genes Dev 1998; 12(11):1665-1677.
45. Wang W, Czaplinski K, Rao Y et al. The role of Upf proteins in modulating the translation read-through of nonsense-containing transcripts. EMBO J 2001; 20(4):880-890.
46. Weng Y, Czaplinski K, Peltz SW. ATP is a cofactor of the Upf1 protein that modulates its translation termination and RNA binding activities. RNA 1998; 4(2):205-214.
47. Muhlrad D, Decker CJ, Parker R. Turnover mechanisms of the stable yeast PGK1 mRNA. Mol Cell Biol 1995; 15(4):2145-2156.
48. Cao D, Parker R. Computational modeling of eukaryotic mRNA turnover. RNA 2001; 7(9):1192-1212.
49. Jacobs Anderson JS, Parker R. The 3' to 5' degradation of yeast mRNAs is a general mechanism for mRNA turnover that requires the SKI2 DEVH box protein and 3' to 5' exonucleases of the exosome complex. EMBO J 1998; 17(5):1497-1506.
50. Parker R, Song H. The enzymes and control of eukaryotic mRNA turnover. Nat Struct Mol Biol 2004; 11(2):121-127.
51. Mangus DA, Evans MC, Jacobson A. Poly(A)-binding proteins: Multifunctional scaffolds for the post-transcriptional control of gene expression. Genome Biol 2003; 4(7):223.221-223.214.
52. Decker CJ, Parker R. A turnover pathway for both stable and unstable mRNAs in yeast: Evidence for a requirement for deadenylation. Genes Dev 1993; 7(8):1632-1643.
53. Beelman CA, Parker R. Degradation of mRNA in eukaryotes. Cell 1995; 81(2):179-183.
54. Muhlrad D, Parker R. Premature translational termination triggers mRNA decapping. Nature 1994; 370(6490):578-581.
55. Dunckley T, Parker R. The DCP2 protein is required for mRNA decapping in Saccharomyces cerevisiae and contains a functional MutT motif. EMBO J 1999; 18(19):5411-5422.
56. Beelman CA, Stevens A, Caponigro G et al. An essential component of the decapping enzyme required for normal rates of mRNA turnover. Nature 1996; 382(6592):642-646.
57. Steiger M, Carr-Schmid A, Schwartz DC et al. Analysis of recombinant yeast decapping enzyme. RNA 2003; 9(2):231-238.
58. 58. Hsu CL, Stevens A. Yeast cells lacking 5'—>3' exoribonuclease 1 contain mRNA species that are poly(A) deficient and partially lack the 5' cap structure. Mol Cell Biol 1993; 13(8):4826-4835.
59. Mitchell P, Tollervey D. An NMD pathway in yeast involving accelerated deadenylation and exosome-mediated 3'—>5' degradation. Mol Cell 2003; 11(5):1405-1413.
60. Tharun S, Parker R. Targeting an mRNA for decapping: Displacement of translation factors and association of the Lsm1p-7p complex on deadenylated yeast mRNAs. Mol Cell 2001; 8(5):1075-1083.
61. Sheth U, Parker R. Decapping and decay of messenger RNA occur in cytoplasmic processing bodies. Science 2003; 300(5620):805-808.
62. Shyu AB, Belasco JG, Greenberg ME. Two distinct destabilizing elements in the c-fos message trigger deadenylation as a first step in rapid mRNA decay. Genes Dev 1991; 5(2):221-231.
63. Lejeune F, Li X, Maquat LE. Nonsense-mediated mRNA decay in mammalian cells involves decapping, deadenylating, and exonucleolytic activities. Mol Cell 2003; 12(3):675-687.
64. Chen CY, Shyu AB. Rapid deadenylation triggered by a nonsense codon precedes decay of the RNA body in a mammalian cytoplasmic nonsense-mediated decay pathway. Mol Cell Biol 2003; 23(14):4805-4813.
65. He F, Li X, Spatrick P et al. Genome-wide analysis of mRNAs regulated by the nonsense-mediated and 5' to 3' mRNA decay pathways in yeast. Mol Cell 2003; 12(6):1439-1452.
66. Cao D, Parker R. Computational modeling and experimental analysis of nonsense-mediated decay in yeast. Cell 2003; 113(4):533-545.
67. Mitchell P, Tollervey D. An NMD pathway in yeast involving accelerated deadenylation and exosome-mediated 3'—>5' degradation. Mol Cell 2003; 11(5):1405-1413.

68. He F, Li X, Spatrick P et al. Genome-wide analysis of mRNAs regulated by the nonsense-mediated and 5' to 3' mRNA decay pathways in yeast. Mol Cell 2003; 12(6):1439-1452.
69. Coller JM, Gray NK, Wickens MP. mRNA stabilization by poly(A) binding protein is independent of poly(A) and requires translation. Genes Dev 1998; 12(20):3226-3235.
70. Muhlrad D, Parker R. Recognition of yeast mRNAs as "nonsense containing" leads to both inhibition of mRNA translation and mRNA degradation: Implications for the control of mRNA decapping. Mol Biol Cell 1999; 10(11):3971-3978.
71. Jacobson A, Peltz SW. Destabilization of nonsense-containing transcripts in Saccharomyces cerevisiae. In: Sonenberg N, Hershey JWB, Mathews MB, eds. Translational Control. Second ed. Cold Spring Harbor: Cold Spring Harbor Laboratory Press, 2000:827-847.
72. Hilleren P, Parker R. mRNA surveillance in eukaryotes: Kinetic proofreading of proper translation termination as assessed by mRNP domain organization? RNA 1999; 5(6):711-719.
73. Teixeira D, Sheth U, Valencia-Sanchez MA et al. Processing bodies require RNA for assembly and contain nontranslating mRNAs. RNA 2005; 11(4):371-382.
74. Atkin AL, Schenkman LR, Eastham M et al. Relationship between yeast polyribosomes and Upf proteins required for nonsense mRNA decay. J Biol Chem 1997; 272(35):22163-22172.
75. Atkin AL, Altamura N, Leeds P et al. The majority of yeast UPF1 colocalizes with polyribosomes in the cytoplasm. Mol Biol Cell 1995; 6(5):611-625.
76. Mangus DA, Jacobson A. Linking mRNA turnover and translation: Assessing the polyribosomal association of mRNA decay factors and degradative intermediates. Methods 1999; 17(1):28-37.
77. Peltz SW, Trotta C, He F et al. Identification of the cis-acting sequences and trans-acting factors involved in nonsense-mediated mRNA decay. In: M. Tuite JMA, Sherman F, ed. Protein Synthesis and Targetting in Yeast. Vol H71. Springer-Verlag, 1993:1-10.
78. Welch EM, Jacobson A. An internal open reading frame triggers nonsense-mediated decay of the yeast SPT10 mRNA. EMBO J 1999; 18(21):6134-6145.
79. Zhang S, Welch EM, Hogan K et al. Polysome-associated mRNAs are substrates for the nonsense-mediated mRNA decay pathway in Saccharomyces cerevisiae. RNA 1997; 3(3):234-244.
80. Ruiz-Echevarria MJ, Gonzalez CI, Peltz SW. Identifying the right stop: Determining how the surveillance complex recognizes and degrades an aberrant mRNA. EMBO J 1998; 17(2):575-589.
81. Gozalbo D, Hohmann S. Nonsense suppressors partially revert the decrease of the mRNA level of a nonsense mutant allele in yeast. Curr Genet 1990; 17(1):77-79.
82. Losson R, Lacroute F. Interference of nonsense mutations with eukaryotic messenger RNA stability. Proc Natl Acad Sci USA 1979; 76(10):5134-5137.
83. Gaba A, Jacobson A, Sachs MS. Ribosome occupancy of the yeast CPA1 upstream open reading frame termination codon modulates nonsense-mediated mRNA decay. Mol Cell, (in press).
84. Ruiz-Echevarria MJ, Peltz SW. The RNA binding protein Pub1 modulates the stability of transcripts containing upstream open reading frames. Cell 2000; 101(7):741-751.
85. Hagan KW, Ruiz-Echevarria MJ, Quan Y et al. Characterization of cis-acting sequences and decay intermediates involved in nonsense-mediated mRNA turnover. Mol Cell Biol 1995; 15(2):809-823.
86. Yun DF, Sherman F. Initiation of translation can occur only in a restricted region of the CYC1 mRNA of Saccharomyces cerevisiae. Mol Cell Biol 1995; 15(2):1021-1033.
87. Zhang S, Ruiz-Echevarria MJ, Quan Y et al. Identification and characterization of a sequence motif involved in nonsense-mediated mRNA decay. Mol Cell Biol 1995; 15(4):2231-2244.
88. Gonzalez CI, Ruiz-Echevarria MJ, Vasudevan S et al. The yeast hnRNP-like protein Hrp1/Nab4 marks a transcript for nonsense- mediated mRNA decay. Mol Cell 2000; 5(3):489-499.
89. Lykke-Andersen J. Identification of a human decapping complex associated with hUpf proteins in nonsense-mediated decay. Mol Cell Biol 2002; 22(23):8114-8121.
90. Parker KC, Patterson D, Williamson B et al. Depth of proteome issues: A yeast ICAT reagent study. Mol Cell Proteomics 2004.
91. Amrani N, Ganesan R, Kervestin S et al. A faux 3'-UTR promotes aberrant termination and triggers nonsense-mediated mRNA decay. Nature 2004.
92. Maderazo AB, Belk JP, He F et al. Nonsense-containing mRNAs that accumulate in the absence of a functional nonsense-mediated mRNA decay pathway are destabilized rapidly upon its restitution. Mol Cell Biol 2003; 23(3):842-851.
93. Amrani N, Ganesan R, Kervestin S et al. A faux 3'-UTR promotes aberrant termination and triggers nonsense-mediated mRNA decay. Nature 2004; 432(7013):112-118.
94. Ishigaki Y, Li X, Serin G et al. Evidence for a pioneer round of mRNA translation: mRNAs subject to nonsense-mediated decay in mammalian cells are bound by CBP80 and CBP20. Cell 2001; 106(5):607-617.

CHAPTER 3

Endogenous Substrates of the Yeast NMD Pathway

Feng He and Allan Jacobson*

Abstract

The existence of a pathway that promotes rapid decay of nonsense-containing mRNAs raises the question of whether the substrates of this pathway are restricted to aberrant mRNAs. Additional substrates were sought by identifying mRNAs that were selectively stabilized in strains harboring mutations in one or more of the genes encoding factors essential for nonsense-mediated mRNA decay (NMD). These studies have shown that, in addition to standard mRNAs with prematurely terminated open reading frames (ORFs), the substrates of the NMD pathway include: inefficiently spliced pre-mRNAs that enter the cytoplasm with their introns intact, mRNAs in which the ribosome has bypassed the initiator AUG and commenced translation further downstream, some mRNAs containing upstream (u)ORFs, mRNAs subject to frameshifting, bicistronic mRNAs, transcripts of pseudogenes and transposable elements, and mRNAs with abnormal extensions of their 3'-untranslated regions (UTRs). Some of these substrates, e.g., the standard nonsense-containing mRNAs, the intron-containing pre-mRNAs, and the mRNAs derived from pseudogenes can all be considered to be targets of a quality control system seeking to reduce the generation of potentially deleterious polypeptide fragments. However, the existence of other classes of substrates, e.g., mRNAs with uORFs or extended 3'-UTRs, indicates that NMD has additional regulatory capabilities that include the dissociation of aberrant translation termination complexes and the modulation of specific biochemical pathways.

Historical Perspective

Targeting of Transcripts beyond Those Derived from Nonsense-Containing Alleles

The yeast, *Saccharomyces cerevisiae*, has proven to be an ideal model system for the characterization of mRNA decay pathways, including those restricted to atypical transcripts. Early experiments showing that nonsense mutations in the yeast *URA1, URA3, HIS4,* and *LEU2* genes accelerated the decay rates of the mRNAs transcribed from these genes led to the general recognition of the mRNA-destabilizing effects of premature termination codons.[1-3] When the *UPF1* gene was shown to regulate this phenomenon[3-5] an immediate and obvious question was whether the substrate specificity of its activity was limited to nonsense-containing mRNAs or extended to a broader range of potential targets. It seemed unlikely that the sole function of Upf1 protein (Upf1p) was anticipatory, i.e., dedicated to the rapid elimination of transcripts from nonsense alleles, and a survey of the abundance and half-lives of numerous mRNAs in wild-type and

*Corresponding Author: Allan Jacobson—Department of Molecular Genetics and Microbiology, University of Massachusetts Medical School, Worcester, Massachusetts 01655-0122, U.S.A. Email: allan.jacobson@umassmed.edu

Nonsense-Mediated mRNA Decay, edited by Lynne E. Maquat. ©2006 Eurekah.com.

upf1Δ strains supported this conclusion. Although the levels of most mRNAs were unaffected by the status of Upf1p activity, the *URA3* mRNA was shown to increase in abundance in *upf1Δ* cells. This altered level was shown to be attributable to the stabilization of the *PPR1* mRNA, a transcript which contains an upstream open reading frame (uORF) and encodes the transcriptional activator of the URA3 gene.[3,5-7] At the same time, it was shown that several pre-mRNAs whose splicing was either inefficient or regulated entered the cytoplasm with their introns intact and became substrates for Upf1p-dependent decay when ribosomes encountered intron-localized in-frame termination codons.[8] Termination prior to the completion of a "full" ORF thus emerged as a likely common element for transcripts subject to *UPF1*-orchestrated rapid decay and the resulting pathway was dubbed nonsense-mediated mRNA decay, or NMD.[5,6,8,9]

NMD Factors as Translational Suppressors

Subsequent to the characterization of Upf1p, additional regulators of the NMD pathway were sought. The *NMD2/UPF2* and *UPF3* genes, obtained by two-hybrid screening or by complementation of additional *upf* strains, were shown to control the stabilities of the same mRNAs as the *UPF1* gene[10-13] and, by virtue of the nonadditive effects of their combined inactivation on the abundance of endogenous substrates[12] (Fig. 1), to function in the same, or closely related, regulatory event. In turn, the identification of all three *UPF/NMD* genes clarified the nature of a diverse array of translational suppressors. Screens for extragenic omnipotent nonsense suppressors, allosuppressors of nonsense-suppressing tRNAs, regulators of frameshifting, and suppressors of upstream initiation codons had, for example, actually identified alleles of the *UPF/NMD* genes.[10-12,14-19] Characterization of the true nature of these suppressors not only extended the list of nonsense-containing mRNAs subject to NMD (e.g., the *can1-100, his5-2, ilv1-2, leu2-1, lys1-1, ura4-1, lys2-1, ade2-1, met8-1, trp3-1, trp1-1, ade3-26,* and *ilv1-1* mRNAs[20-24]), but also reinforced the notion that uORF-containing mRNAs comprised a class of NMD substrates and indicated that additional substrates included mRNAs subject to translational frameshifting.

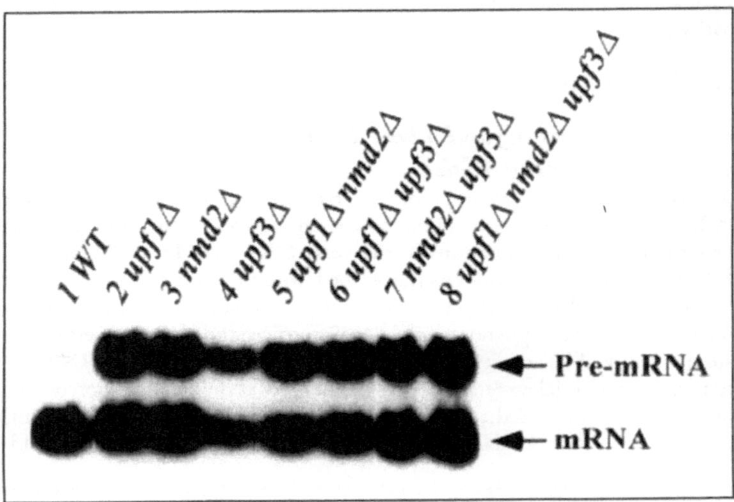

Figures 1. Single or multiple deletions of *UPF1, NMD2,* and *UPF3* genes have identical NMD phenotypes. Total RNA was isolated from *upf1Δ*, *nmd2Δ*, or *upf3Δ* strains, or from strains containing multiple NMD-inactivating mutations, and analyzed by northern blotting using a probe for the *CYH2* transcripts. Pre-mRNA identifies the intron-containing transcript that is an NMD substrate and mRNA identifies the mature transcript. Reprinted from He et al, 1997.[12]

The "Normal" mRNA Conundrum

As the search for endogenous substrates of the NMD pathway broadened, most evidence continued to suggest that premature translation termination could be a common determinant of substrate status. This was a straightforward interpretation of the results with transcripts of nonsense alleles, transcripts that retained introns, and transcripts with uORFs, and plausible for transcripts in which a frameshifting event precluded early termination with only a small fraction of traversing ribosomes. By this reasoning, mRNAs that accumulated in NMD-deficient strains without undergoing a change in half-life (e.g., *CTF13* mRNA[25]) must be indirect NMD substrates, e.g., subject to a change in transcription rate by virtue of the stabilization of another mRNA that encoded a direct regulator of the transcript in question. Although there are alternative explanations for this phenomenon,[26] a logical corollary of the premature termination hypothesis was that conventional mRNAs, i.e., those with single, simple ORFs, would not be capable of being degraded by the NMD pathway. Hence, it was initially a surprise that one such transcript, the *SPT10* mRNA, was subject to NMD.[27] However, analyses of *SPT10-PGK1* chimeric transcripts established that the *SPT10* mRNA is shunted into the NMD pathway by the leaky scanning of an initiating ribosome. In the course of normal translation ribosomes appear to occasionally scan past the contextually suboptimal *SPT10* initiator AUG, initiate at the next AUG, terminate at an out-of-frame nonsense codon, and trigger NMD. Support for this interpretation was obtained by using initiation codon context, and availability of downstream out-of-frame initiation and termination codons as criteria to identify other yeast mRNAs that behaved similarly to *SPT10* mRNA.[27]

Another apparent exception to the rule that mRNAs with "normal" ORFs could not be NMD substrates was the observation that transcripts with extended 3'-UTRs were also targets of NMD.[28-30] While the mutation generating these aberrant transcripts did not give rise to premature termination codons, it may well have altered the manner in which the normal termination codon was recognized (see below). In short, analyses of limited numbers of transcripts strongly suggested that NMD substrates might be limited to transcripts undergoing premature termination or its functional equivalent. Definitive evaluation of this hypothesis required an analysis of all the mRNAs stabilized in NMD-deficient cells.

NMD Substrate Identification by Genome-Wide Expression Profiling

Background

In yeast, wild-type mRNAs are degraded through two general and functionally redundant mechanisms: the deadenylation-dependent 5' to 3' pathway and the exosome-mediated 3' to 5' pathway.[31-34] In both of these pathways, the initial decay event is the shortening of the poly(A) tail to an oligo(A) length of 10 to 12 nucleotides.[31,35,36] After poly(A) shortening, transcripts can be degraded by either the 5' to 3' pathway or the 3' to 5' decay pathway. In the 5' to 3' decay pathway, transcripts are decapped by the Dcp1p/Dcp2p decapping enzyme complex, and then digested exonucleolytically by the 5' to 3' exoribonuclease, Xrn1p.[31,37-41] In the 3' to 5' decay pathway, transcripts are further deadenylated and then degraded by a ten-subunit 3' to 5' exonuclease complex, the exosome. It was initially thought that degradation of nonsense-containing transcripts proceeded from decapping to 5' to 3' decay without prior poly(A) shortening.[38,39,42] However, other studies have suggested that these RNAs can also be subject to accelerated poly(A) shortening and 3' to 5' decay.[26,43]

By analyzing most of an organism's transcripts, microarray technology provides a means to not only identify the substrates of all of the cellular mRNA decay pathways, but also to resolve mechanistic questions such as whether the 5' to 3' or 3' to 5' pathway predominates in NMD. Lelivelt and Culbertson[44] initiated the application of microarray analysis to the regulatory components of the NMD pathway, but did not succeed in elaborating novel classes of NMD substrates or in identifying the global set of RNAs regulated by NMD. Subsequent studies by He et al[45] analyzed genome-wide expression profiles of yeast strains containing single deletions of the *UPF1, NMD2, UPF3, DCP1,* and *XRN1* genes and double deletions of the *XRN1* gene in combination with a deletion of the *UPF1, NMD2,* or *UPF3* gene. These experiments defined the core set of transcripts regulated by NMD, identified several novel structural classes of substrates of the NMD pathway, and revealed that substrates of this pathway are also degraded by the 5' to 3' decay pathway even in the absence of functional NMD.

Genome-Wide Identification of Transcripts Regulated by Upf1p, Nmd2p, and Upf3p

He et al[45] utilized Affymetrix microarrays to identify transcripts altered two-fold or more in abundance in ***upf1Δ, nmd2Δ, upf3Δ, dcp1Δ,*** and ***xrn1Δ*** strains. These microarrays assessed the expression levels of 7839 yeast genes. Included in this total were 6086 protein-coding genes (or ORFs), 625 other genetic elements (e.g., tRNA, rRNA, snRNA, and scRNA genes, as well as transposable elements and their long terminal repeats [LTRs]), and 1128 genes designated as "other ORFs" because they were nonannotated, putative ORFs suggested by serial analysis of gene expression (SAGE) or the yeast genome sequencing project. In addition, the microarrays measured the expression of genes encoded by the mitochondrial genome and genes present on episomal plasmid DNA. These experiments showed that deletions of *UPF1, NMD2,* or *UPF3* have identical effects (strongly supporting the notion of a common function), leading to 2 to 60-fold increases of 746 transcripts (approximately 10% of the transcriptome). Deletions of the *XRN1* or *DCP1* genes affected the expression of 16-19% of the genes in the yeast genome and the expression profile of the three ***upf/nmd*** mutant strains overlapped, yet was distinct from those of the *xrn1Δ,* and *dcp1Δ* strains.

The expression data are summarized in the hierarchical clustering analysis of Figure 2. The 746 transcripts up-regulated in the NMD-deficient strains include 545 encoded by protein-coding genes, 51 encoded by other genetic elements, and 150 encoded by other ORFs (Table 1). Nineteen additional transcripts that were down-regulated in NMD-deficient cells included 6 encoded by protein-coding genes, 8 encoded by other genetic elements, and 5 encoded by other ORFs. The observed commonality of transcript phenotypes in ***upf1Δ, nmd2Δ,*** and ***upf3Δ*** strains supported the notion that Nmd2p and Upf3p are regulators of the activity of Upf1p[22,46,47] and provided a global confirmation of previously observed effects with a more limited set of mRNAs.[44,46,47]

Structural Classes of NMD-Regulated Transcripts

The transcripts up-regulated in ***nmdΔ*** cells (i.e., cells lacking Upf1p, Nmd2p, or Upf3p) showed distributions skewed towards low abundance when compared to all detectable cellular transcripts (Fig. 3), but exhibited no other simple common features.[45] They did, however, fit into distinct structural categories. These included four known classes of NMD substrates, namely mRNAs encoded by genes harboring nonsense mutations, pre-mRNAs that retained their introns in the cytoplasm, mRNAs containing uORFs, and mRNAs subject to leaky scanning. In addition, four novel structural classes of NMD substrates were identfied (Table 2). These included: mRNAs directing +1 frameshifting, bicistronic mRNAs, mRNAs encoded by pseudogenes, and transcripts encoded by transposable elements or their LTR sequences.[45] While distinct structural classes of transcripts could be identified, they were all compatible with the notion that degradation of the individual RNAs involved recognition of a translation termination codon that is either premature or in atypical context.

The RNAs containing uORFs and those encoded by transposable elements or their LTRs appear to embody the two major classes of NMD substrates.[45] The uORF-containing mRNAs, many of which encode transcription factors,[48] may be indicative of evolutionary coopting of the NMD pathway for regulatory advantage. Indeed, the encoded transcription factors may also be the primary source of indirect effects of ***nmdΔ*** mutations. Given the prevalence of transposable elements in eukaryotic genomes, these results also suggest that at least one important function of the NMD pathway is the degradation of transcripts generated from

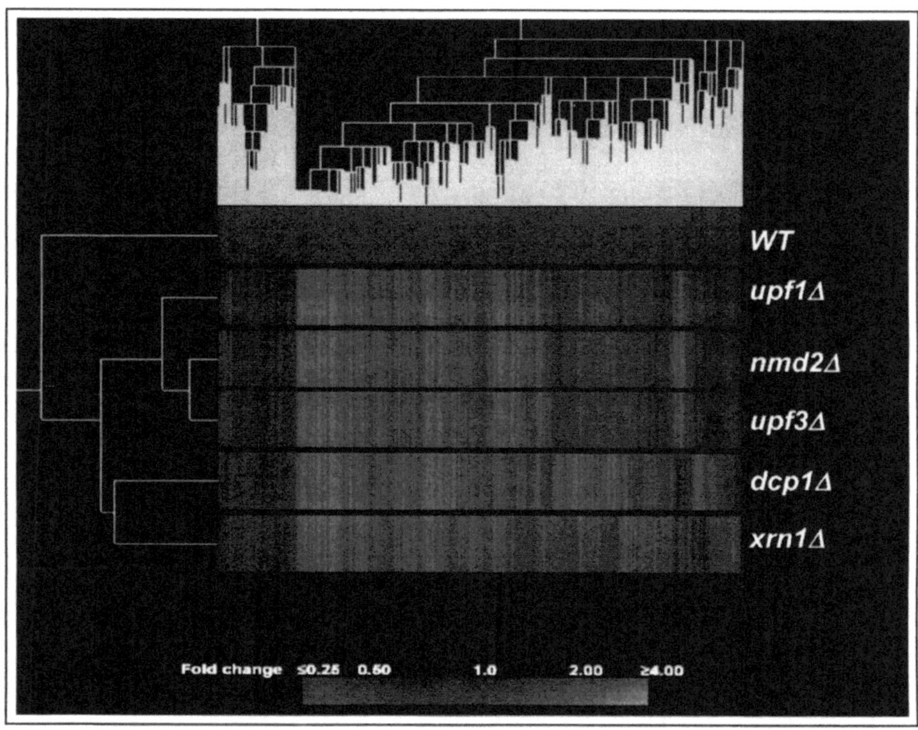

Figures 2. Expression patterns of five strains defective in either NMD or 5' to 3' mRNA decay. Twenty-six expression profiling experiments were carried out for the wild-type (WT), *upf1Δ*, *nmd2Δ*, *upf3Δ*, *dcp1Δ*, and *xrn1Δ* strains. A set of 2152 transcripts represented by 2271 probe sets exhibited more than 2-fold increases or decreases in their levels of expression and a change P value ≤ 0.05 in at least one of the mutant strains. These 2152 transcripts were clustered by using a hierarchical clustering algorithm (GeneSpring) in two-dimensions according to gene and experiment vectors. In this analysis, the average difference (i.e., expression level) of each transcript in the mutant strains was normalized to that in the wild-type strain. The resulting expression ratios from at least four independent replicates were averaged, transformed to base two logarithms, and clustered using the standard correlation coefficient as a distance metric. Genes are plotted along the horizontal axis, with the gene cluster tree above. Strains are plotted along the vertical axis, with the experiment cluster tree on the left side of the plot. Reprinted from He et al,[45] ©2003, with permission from Elsevier. A color version of this figure is available online at www.Eurekah.com.

these genetic elements. When combined with earlier observations on the substrate status of mRNAs with unprocessed introns and other sources of premature nonsense codons, the detection of transposable element and pseudogene RNAs in the collection of core transcripts supports the notion that the NMD pathway plays an important role in the quality control of gene expression by eliminating RNAs capable of giving rise to potentially deleterious translation products.[30,49,50]

Functional Classes and Genomic Distribution of NMD Substrates

The ORF-coding transcripts that were up-regulated in *nmdΔ* strains could be assigned to twelve distinct functional classes, including metabolism, energy generation, cell fate, cell cycle and DNA processing, cell rescue and defense, cellular transport and transport mechanisms, control of cellular organization, regulation of interaction with cellular environment,

Table 1. Deletions of* UPF1, NMD2, UPF3, DCP1, *or* XRN1 *genes result in differential expression of a subset of yeast genes

	Increases				Decreases				Both
Strain	ORFs	Other Elements	Other ORFs	Sub-Total	ORFs	Other Elements	Other ORFs	Sub-Total	Total
upf1Δ	576	56	186	818	9	5	5	19	837
	562	51	157	770	9	5	5	19	789
nmd2Δ	661	58	197	916	15	16	19	50	966
	646	53	170	869	15	16	19	50	917
upf3Δ	682	56	199	937	17	21	15	53	990
	668	51	171	890	17	20	15	52	940
nmdΔ	560	56	176	792	6	8	5	19	811
pool	545	51	150	746	6	8	5	19	765
dcp1Δ	948	60	225	1233	52	16	27	95	1328
	930	55	195	1180	52	14	25	91	1269
xrn1Δ	996	67	217	1280	94	117	26	237	1517
	984	63	191	1238	94	112	25	231	1467

High-density oligonucleotide DNA arrays were used to analyze the genome-wide expression profiles of the wild-type and *upf1Δ*, *nmd2Δ*, *upf3Δ*, *dcp1Δ*, and *xrn1Δ* strains. The expression level of each transcript in the wild-type strain was compared to that of each mutant strain. Transcripts whose level of expression increased or decreased more than 2-fold with a change P value ≤ 0.05 were identified. The number of probe sets as well as the number of transcripts in each RNA category is presented (upper and lower numbers, respectively). ORFs indicate transcripts encoded by protein-coding genes; other elements indicate transcripts encoded by genes of small RNAs, tRNAs, rRNA as well as by transposable elements and their LTRs; and *other* ORFs indicate transcripts encoded by genes of the mitochondrial genome and plasmid DNA as well as the putative, non-annotated ORFs identified by serial analysis of gene expression (SAGE) and the yeast genome sequence project. *nmdΔ pool* refers to the combined data of the *upf1Δ*, *nmd2Δ*, and *upf3Δ* strains. Reprinted from He et al,[45] ©2003, with permission from Elsevier.

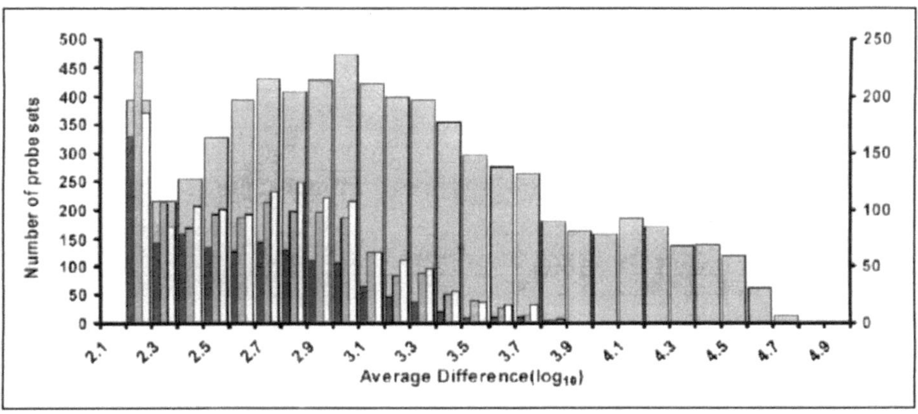

Figures 3. Transcripts up-regulated in *nmd*Δ, *dcp1*Δ, or *xrn1*Δ strains exhibit intrinsically low abundance in wild-type cells. Histograms are used to display the normal distributions of abundance (i.e., average difference) for the 7049 probe sets considered to be expressed in at least one of the six yeast strains analyzed (blue bars), the 792 probe sets up-regulated in the *nmd*Δ strains (pink bars), the 1233 probe sets up-regulated in the *dcp1*Δ strain (cyan bars), and the 1280 probe sets up-regulated in the *xrn1*Δ strain (yellow bars). The horizontal axis represents the log_{10} transformed average difference values. The vertical axes represent the frequency of occurrence, with the left axis used for the data from all 7049 probe sets and the right axis utilized for the up-regulated probe sets. In all cases, the average difference value for each transcript was the average of the values from five independent replicates carried out for the wild-type strain. Reprinted from He et al,[45] ©2003, with permission from Elsevier. A color version of this figure is available online at www.Eurekah.com.

transcription, protein synthesis, protein fate, and transport facilitation.[45] Since the 545 ORFs up-regulated in *nmd*Δ strains have a genome-wide average frequency distribution of 8.0% (545 out of 6335 ORF genes in the yeast genome), only two of the functional categories (cell rescue and defense [15.1%], and transport facilitation [16.0%]) are over-represented by NMD-regulated RNAs. In contrast, genes involved in protein synthesis appear to be under-represented, having a frequency of 1.7%.[45]

The NMD pathway may play a significant regulatory role for several specific processes in yeast cells because genes having products involved in the same cellular process or pathway showed coordinate up-regulation in *nmd*Δ strains. For example, inactivation of NMD led to increases in the expression levels of: six genes involved in telomere maintenance (*EST1, EST2, EST3, STN1, YKU70,* and *TEL1*), seven genes involved in the thiamine biosynthesis (*THI4, THI6, THI11, THI20, THI21, THI22,* and *YOR192C*), ten genes involved in pre-mRNA splicing (*BRR1, CDC40, CUS1, DBR1, HAH49, MSL1, PPR3, PPR46, SAD1,* and *SPP2),* three genes encoding products containing covalent phosphopantetheine as a prosthetic group (*PPT2, LYS5 and FAA2*), twelve genes involved in peroxisomal function (*CAT2, DCI1, DAL7, ECI1, FAA2, FOX2, IDP3, PEX4, PEX6, PEX10, PXA1,* and *PXA2*), fourteen genes involved in nitrogen metabolism (*AMD2, MEP2, GDH2, NIT2, SEO1, ARG81, UGA3, GZF3, DUR1,2, DAL3, DAL2, DUR3 YJR149W,* and *YOL153C*), and 11 genes involved in DNA repair (*APN2, RAD1, DNL4, RAD57, REV3, NTG2, MEC3, YKU70, MLH3, THI4,* and *RNR3*).[45] In addition, an independent analysis of the data noted that a significant number of up-regulated genes were involved in amino acid homeostasis.[51] In short, the NMD pathway appears to have important regulatory capabilities beyond mRNA surveillance.

The ORFs up-regulated in NMD-deficient strains are distributed over sixteen chromosomes with a random distribution pattern in most chromosomal regions (Fig. 4). However, up-regulated ORFs located in at least fourteen telomere proximal regions show a nonrandom,

Table 2. Relative expression levels of representative transcripts from different classes of NMD substrates

RNA Type	Gene Name	Probe Set	*upf1Δ*	*nmd2Δ*	*upf3Δ*	*dcp1Δ*	*xrn1Δ*
Nonsense-containing	*YEL063C (CAN1)*	5771_at	9.7	12.1	11.0	4.9	6.5
	YOR128C (ADE2)	8420_at	42.2	41.7	53.7	15.0	35.1
	YDR007W (TRP1)	6468_at	18.0	14.5	14.3	21.1	12.9
Pre-mRNAs	*YJR021C (MER2)*	10998_at	10.1	10.1	13.2	19.8	12.4
	YDL115C	6616_at	7.2	10.0	11.1	4.7	3.5
	YGL251C (MER3)	5262_at	5.7	4.5	5.6	4.3	5.0
uORF-containing	*YOR302W (CPA1)*	8236_at	5.0	6.8	5.7	3.4	4.0
	YOR303W (CPA1)	8237_at	3.8	5.5	5.6	3.3	4.1
	YLR233C (EST1)	10080_at	14.0	10.5	14.0	9.9	8.8
	YJL023C (PET130)	11044_at	7.3	4.5	5.1	7.7	9.6
	YOL055C (THI20)	8596_at	8.0	7.8	7.9	8.6	7.0
Leaky scanners	*YCR007C*	3453_at	2.4	2.8	2.5	4.8	4.2
	YHL050C	3187_s_at	2.9	5.2	3.2	1.1	1.3
	YIL107C (PFK26)	4204_at	2.8	3.1	2.7	2.5	2.1
+1 frameshifting	*YIL009C-A (EST3)*	4117_at	3.9	3.6	3.7	2.6	2.3
	YILWTY3-1	3137_s_at	6.5	3.5	3.4	8.6	4.0
Polycistronic	*YLR315W*	10030_at	3.0	2.9	2.7	2.9	2.5
	YLR317W	9986_at	3.1	4.4	3.5	3.3	3.3
	YMR181C	9451_at	2.9	3.4	2.7	6.0	1.8
	YMR182C (RGM1)	9452_at	4.4	5.5	4.4	2.8	1.7
Pseudogenes	*YCR099C*	6794_at	3.7	3.7	3.5	8.1	5.3
	YCR100C	6795_at	4.8	6.1	4.2	6.5	5.5
	YIL164C (NIT1)	4238_at	16.3	20.7	21.5	16.2	14.9
	YIL165C	4237_at	9.6	9.0	8.9	11.4	10.4
	YIR043C	4078_i_at	12.5	17.6	16.4	4.0	6.8
Transposable elements	*YDLWTAU1*	3475_f_at	7.0	5.1	5.1	15.2	9.5
	YJLWTY4-1	3900_f_at	4.4	3.5	3.2	5.9	3.8
	YJLWTY4-1	3899_s_at	8.5	4.0	4.4	8.7	4.3
	YLRCDELTA27	3730_at	4.5	6.6	5.9	10.0	5.7
	YNLWSIGMA2	3661_at	3.7	2.9	2.8	5.2	4.2

The up-regulation of different structural classes of RNAs was monitored by microarray analysis, as in Figure 2 and Table 1. The systematic gene names, probe set names, and the average fold transcript increases in *upf1Δ*, *nmd2Δ*, *upf3Δ*, *dcp1Δ*, and *xrn1Δ* strains are shown. Reprinted from He et al,[45]

clustered distribution pattern. Eighty of the 223 ORFs (35.9 %) located in the 20-kb telomere regions showed up-regulation in NMD-deficient strains whereas only 465 out of 5863 (7.9%) ORFs located in nontelomere regions showed comparable up-regulation. These results indicate that the NMD pathway directly or indirectly controls the expression of genes near yeast telomeres, a conclusion consistent with the observation that six genes involved in telomere maintenance showed up-regulation in *nmdΔ* strains (see above) and previous demonstrations that yeast strains defective in NMD have short telomeres and derepressed expression of genes normally silenced in telomeric regions.[52,53]

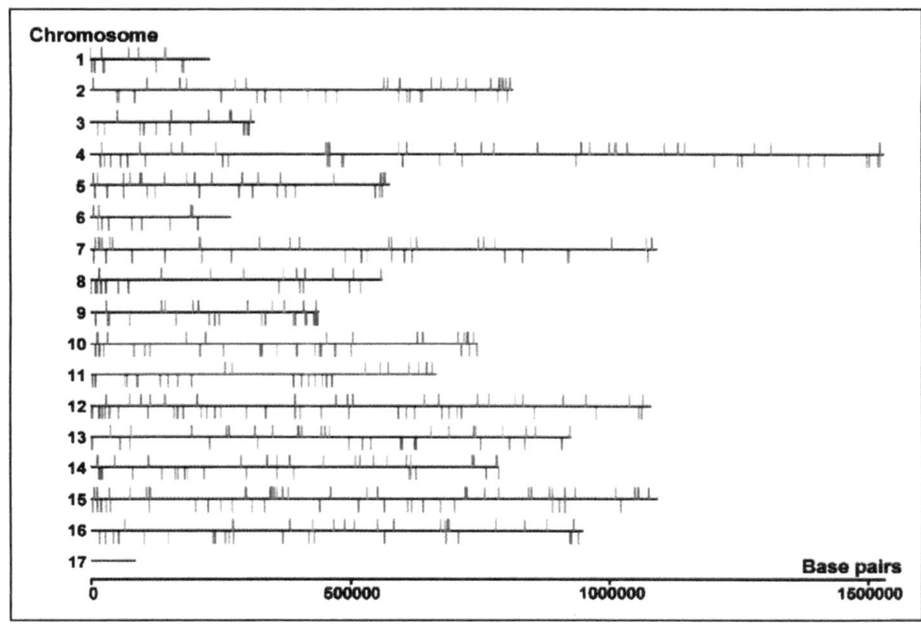

Figure 4. Genome-wide map of the physical locations of genes up-regulated in *nmdΔ* strains. The 545 protein-coding genes up-regulated by inactivation of the NMD pathway were mapped to their respective chromosomal loci. Horizontal green lines represent chromosomes and vertical red bars represent genes. Reprinted from He et al,[45] ©2003, with permission from Elsevier. A color version of this figure is available online at www.Eurekah.com.

Direct Targets of the Pathway: Identification Using Inducible NMD

Since expression profiling experiments only measure the relative steady-state levels of each transcript, the changes observed in NMD-deficient strains could result from direct effects on mRNA decay or from indirect effects on other processes that influence mRNA accumulation. To distinguish these possibilities, and to further define the substrates of the NMD pathway, we utilized an inducible system[54] coupled with microarray analysis to evaluate the effect of reactivating NMD on global mRNA accumulation in NMD-deficient cells. In this inducible NMD system, the expression of *NMD2* is under the control of the *GAL1* promoter. Yeast cells lacking endogenous Nmd2p, but harboring the *GAL1-NMD2* allele, are inactive in NMD when grown in raffinose-containing media, but fully functional in NMD 20 minutes after the addition of galactose to their growth media. During a 60-minute time course, galactose-induced expression of *NMD2* caused the down-regulation of 591 transcripts (F. He, P. Spatrick, C. Li, A. Lim, and A. Jacobson, manuscript in preparation). Comparisons of this set of transcripts to those up-regulated in NMD-deficient strains revealed that the vast majority also showed increased steady-state levels when NMD was inactivated and encompassed all of the classes of NMD substrates. Hence, most transcripts down-regulated upon reactivation of NMD are direct targets of the NMD pathway. These experiments also demonstrate that, at least in yeast, substrates of the NMD pathway are available for NMD at each round of translation and are thus not limited to newly synthesized or newly exported mRNAs, i.e., those involved in a pioneer round of translation.[55] These observations reinforce the notion of NMD in yeast as a cytoplasmic pathway and call into question models which suggest that NMD is triggered by the failure of elongating ribosomes to dissociate decay-inducing protein complexes deposited on the mRNA during early processing events, as data indicate is the case for mammals (see chapter by Maquat).[49]

Substrates of the NMD Pathway are Degraded by a 5' to 3' Mechanism Even in NMD-Deficient Cells

It was initially thought that degradation of nonsense-containing yeast transcripts proceeded from decapping to 5' to 3' decay without prior poly(A) shortening.[38,39,42] However, more recent studies suggest that these RNAs can also be subject to accelerated poly(A) shortening and 3' to 5' decay.[26,43] The latter pathway must be a minor activity because microarray analyses showed that 77% of the transcripts up-regulated by inactivation of the NMD pathway were also up-regulated in *xrn1Δ* cells.[45] Hence, when NMD is active, its substrates are mainly degraded by the 5' to 3' pathway. Interestingly, further analyses showed that deletion of the *XRN1* gene alone, or double deletions of the *XRN1* gene in combination with the *UPF1*, *NMD2*, or *UPF3* genes, affected the expression of almost the same set of mRNAs.[45] This implies that the substrates of the NMD pathway are also degraded by the 5' to 3' decay pathway even in the absence of functional NMD. If these RNAs are also capable of being degraded by the 3' to 5' pathway when NMD is active,[43] it would suggest that nonsense codon recognition triggers availability for rapid decay rather than acceleration of a predetermined decay pathway. This, in turn, indicates that premature termination leads an mRNP to exit the conventional cycle of translation and decay and enter a pool in which RNA degradation is rapid but indiscriminate.

Further Attributes of NMD Substrates

Destabilization of Nonsense-Containing mRNAs Requires Their Translation

The mere presence of a premature nonsense codon within a transcript is not sufficient to promote its degradation by NMD. Destabilization of yeast nonsense-containing mRNAs also requires their translation, i.e., NMD depends on recognition of the nonsense codon by the translational apparatus. This conclusion follows from observations that NMD factors and decay intermediates are localized to polysomes[56-61] and that decay can be antagonized by drugs or mutations that interfere with protein synthesis[27,61,62] or by tRNAs that suppress termination.[1,63] Additional support for this conclusion includes experiments showing that: (a) Upf/Nmd factors interact with the polypeptide release factors Sup35p (eRF3) and Sup45p (eRF1),[64] (b) mutations in genes encoding these factors (e.g., the *UPF1* gene) not only lead to the stabilization of nonsense-containing mRNAs, but also promote nonsense suppression,[4,22,65-67] and (c) NMD-promoted destabilization of the yeast *CPA1* mRNA is dependent on the extent of ribosome occupancy of the termination codon that triggers decay (A. Gaba, M. Sachs, and A. Jacobson, in press). The requirement for translatability implies that events and factors that alter mRNA translatability should also affect NMD. Since metazoan NMD has been thought to be dependent on mRNA association with specific components of the exon junction complex (EJC),[68,69] and since these factors can also influence the efficiency with which an mRNA is translated,[70,71] it is possible that the apparent NMD role of the EJC is related to its ability to ensure efficient mRNA translation.

Regulation of Substrate Status by cis-Acting Elements: The Importance of Termination Codon Context

All functional mRNAs contain a translation termination codon, yet they are not substrates for NMD. What appears to distinguish a normal nonsense codon from one that promotes mRNA destabilization is its sequence context, or more precisely, the sequences 3' to the nonsense codon. A requirement for specific downstream elements, or DSEs, was initially suggested by experiments demonstrating that: (a) deletion of most of the *PGK1* protein coding region downstream of an early nonsense mutation reduced mRNA decay rates markedly and (b) reinsertion of a small segment of the *PGK1* coding region into the construct harboring the large deletion was sufficient to activate NMD.[5] Studies of the *GCN4* mRNA have demonstrated that a DSE can activate the NMD pathway when it is located within approximately 150 nt of the stop codon, but is not functional if traversed by translating ribosomes.[62,72] The presence of DSEs within the coding regions of wild-type mRNAs of diverse decay rates suggests that these sequences are inactive unless preceded by an upstream termination codon. In spite of their apparently common function, DSE sequences from *PGK1* and other mRNAs have only a weak sequence consensus.[5,73-75]

An additional type of cis-acting element appears to be capable of regulating NMD. In yeast, stabilizer elements (STEs) in the 5' leader region of the *GCN4* and *YAP1* mRNAs inactivate the NMD pathway when positioned downstream of a termination codon but still 5' of a DSE.[62,72] These STEs appear to bind Pub1p which, in turn may either antagonize activation of the decay apparatus or mimic the RNP context of a normal 3' -UTR.

The two predominant models for the mechanism of NMD propose different explanations for the role of these cis-acting elements. The surveillance complex model,[64,76] and its metazoan descendants,[55,68] suggests that the deposition of specific proteins (including Hrp1p in yeast) during early processing events allows decay-inducing factors to maintain an mRNA association unless swept off by the ribosome in an early round of translation. In the event that termination is premature, the bound factors are thought to remain on the mRNA and to trigger mRNA decay by subsequent interaction with the Upf1p. The manner in which Upf1p triggers decay is uncertain, but Upf1p:Dcp2p interactions[11,77] have led to speculation that Upf1p may recruit the decapping complex.[77-79] The second model posits that premature termination and normal termination are biochemically different events and that, at least in yeast, NMD is triggered by the events accompanying aberrant termination at a premature nonsense codon.[26,49,80,81] In this *faux* UTR model,[49,54] proper termination of translation and normal rates of mRNA decay require interactions between a terminating ribosome and a specific RNP structure or set of factors localized 3' to the stop codon. The DSE is thought to promote aberrant termination because it lacks a regulatory factor (or factors) normally present on a legitimate 3'-UTR and aberrant termination, in turn, is thought to allow binding of the Upf/Nmd factors and trigger mRNA decay.

As noted above, our experiments with inducible NMD indicate that nonsense-containing mRNAs in yeast are always available for NMD and are not restricted to decay in an early round of translation. These observations, and analyses of termination efficiency in vitro, have provided strong support for the concepts of the *faux* UTR model (see chapter by Amrani and Jacobson). The crux of the *faux* UTR model is that mRNA decay is activated by the intrinsically aberrant nature of certain termination events, including those which are premature.[81] One mRNA feature thought to be important for the maintenance of normal termination is proximity of the termination codon to the mRNA poly(A) tail and its associated poly(A)-binding protein and other factors.[81] Such dependence on a properly configured 3'-UTR provides additional insight into the determination of NMD substrate status. mRNAs with extended 3'-UTRs,[29] for example, are likely to be substrates for NMD because their termination codon are no longer in their normal, i.e., proper, context.

Concluding Remarks

The destabilizing effects of premature nonsense codons were originally recognized more than 25 years ago.[1] What was once thought to represent a trivial deprotection of mRNA by release of ribosomes has now come to illustrate the complex crosstalk between the pathways of mRNA decay and translation.[49,82] While the most obvious substrate for NMD is a transcript derived from a gene in which a mutation has given rise to a premature nonsense codon, the fate of the mRNA would be no different if its nonsense-containing status was attributable to errors in transcription, pre-mRNA splicing, or RNA editing, or caused by failure of the ribosome to maintain the normal reading frame. In addition to ridding the cell of such "classic" aberrant transcripts NMD also targets a large number of other RNAs, including intron-containing pre-mRNAs that enter the cytoplasm, uORF-containing mRNAs, bicistronic mRNAs, transcripts of pseudogenes and transposable elements, mRNAs subject to frameshifting or leaky scanning, and mRNAs with abnormal extensions of their 3'-UTRs. Rapid elimination of all of these transcripts certainly minimizes the accumulation of potentially toxic "junk" in the gene expression pathway, but there appear to be broader functions for NMD. Since, premature termination is aberrant,[81] the NMD pathway must also play a role in liberating and recycling ribosomes and other components of the translation apparatus from termination complexes that would otherwise be inefficiently dissociated. Moreover, NMD must also be viewed as a potential regulatory circuit for some normal mRNAs.[25,27,45,52] While there is preliminary evidence that the latter regulatory circuits may have been evolutionarily conserved,[51] it remains to be determined whether selective coopting of the NMD pathway has been maintained throughout the eukaryotes.

Acknowledgements

Grant support to A. Jacobson from the U.S. National Institutes of Health (GM27757 and HD48137) is gratefully acknowledged.

References

1. Losson R, Lacroute F. Interference of nonsense mutations with eukaryotic messenger RNA stability. Proc Natl Acad Sci USA 1979; 76(10):5134-5137.
2. Pelsy F, Lacroute F. Effect of ochre nonsense mutations on yeast URA1 mRNA stability. Curr Genet 1984; 8:277-282.
3. Leeds P, Peltz SW, Jacobson A et al. The product of the yeast UPF1 gene is required for rapid turnover of mRNAs containing a premature translational termination codon. Genes Dev 1991; 5(12A):2303-2314.
4. Leeds P, Wood JM, Lee BS et al. Gene products that promote mRNA turnover in Saccharomyces cerevisiae. Mol Cell Biol 1992; 12(5):2165-2177.
5. Peltz SW, Brown AH, Jacobson A. mRNA destabilization triggered by premature translational termination depends on at least three cis-acting sequence elements and one trans-acting factor. Genes Dev 1993; 7(9):1737-1754.
6. Peltz SW, He F, Welch E et al. Nonsense-mediated mRNA decay in yeast. Prog Nucleic Acid Res Mol Biol 1994; 47:271-298.
7. Kebaara B, Nazarenus T, Taylor R et al. The Upf-dependent decay of wild-type PPR1 mRNA depends on its 5'-UTR and first 92 ORF nucleotides. Nucleic Acids Res 2003; 31(12):3157-3165.
8. He F, Peltz SW, Donahue JL et al. Stabilization and ribosome association of unspliced pre-mRNAs in a yeast upf1⁻ mutant. Proc Natl Acad Sci USA 1993; 90(15):7034-7038.
9. Peltz SW, Jacobson A. mRNA Turnover in Saccharomyces cerevisiae. In: Belasco J, Brawerman G, eds. Control of Messenger RNA Stability. New York: Academic Press, 1993.
10. Cui Y, Hagan KW, Zhang S et al. Identification and characterization of genes that are required for the accelerated degradation of mRNAs containing a premature translational termination codon. Genes Dev 1995; 9(4):423-436.
11. He F, Jacobson A. Identification of a novel component of the nonsense-mediated mRNA decay pathway by use of an interacting protein screen. Genes Dev 1995; 9(4):437-454.
12. He F, Brown AH, Jacobson A. Upf1p, Nmd2p, and Upf3p are interacting components of the yeast nonsense-mediated mRNA decay pathway. Mol Cell Biol 1997; 17(3):1580-1594.
13. Lee BS, Culbertson MR. Identification of an additional gene required for eukaryotic nonsense mRNA turnover. Proc Natl Acad Sci USA 1995; 92(22):10354-10358.
14. Cui Y, Dinman JD, Peltz SW. Mof4-1 is an allele of the UPF1/IFS2 gene which affects both mRNA turnover and -1 ribosomal frameshifting efficiency. EMBO J 1996; 15(20):5726-5736.
15. Dinman JD, Wickner RB. Translational maintenance of frame: Mutants of Saccharomyces cerevisiae with altered -1 ribosomal frameshifting efficiencies. Genetics 1994; 136(1):75-86.
16. Culbertson MR, Underbrink KM, Fink GR. Frameshift suppression Saccharomyces cerevisiae. Genetic properties of group II suppressors. Genetics 1980; 95(4):833-853.
17. Hampsey M, Na JG, Pinto I et al. Extragenic suppressors of a translation initiation defect in the cyc1 gene of Saccharomyces cerevisiae. Biochimie 1991; 73(12):1445-1455.
18. Lee SI, Umen JG, Varmus HE. A genetic screen identifies cellular factors involved in retroviral -1 frameshifting. Proc Natl Acad Sci USA 1995; 92(14):6587-6591.

19. Pinto I, Na JG, Sherman F et al. cis- and trans-acting suppressors of a translation initiation defect at the cyc1 locus of Saccharomyces cerevisiae. Genetics 1992; 132(1):97-112.
20. Song JM, Liebman SW. Allosuppressors that enhance the efficiency of omnipotent suppressors in Saccharomyces cerevisiae. Genetics 1987; 115(3):451-460.
21. Crouzet M, Tuite MF. Genetic control of translational fidelity in yeast: Molecular cloning and analysis of the allosuppressor gene SAL3. Mol Gen Genet 1987; 210(3):581-583.
22. Maderazo AB, He F, Mangus DA et al. Upf1p control of nonsense mRNA translation is regulated by Nmd2p and Upf3p. Mol Cell Biol 2000; 20(13):4591-4603.
23. Ono B, Ishino-Arao Y, Tanaka M et al. Recessive nonsense suppressors in Saccharomyces cerevisiae: Action spectra, complementation groups and map positions. Genetics 1986; 114(2):363-374.
24. Ono BI, Tanaka M, Kominami M et al. Recessive UAA suppressors of the yeast Saccharomyces cerevisiae. Genetics 1982; 102(4):653-664.
25. Dahlseid JN, Puziss J, Shirley RL et al. Accumulation of mRNA coding for the ctf13p kinetochore subunit of Saccharomyces cerevisiae depends on the same factors that promote rapid decay of nonsense mRNAs. Genetics 1998; 150(3):1019-1035.
26. Cao D, Parker R. Computational modeling and experimental analysis of nonsense-mediated decay in yeast. Cell 2003; 113(4):533-545.
27. Welch EM, Jacobson A. An internal open reading frame triggers nonsense-mediated decay of the yeast SPT10 mRNA. EMBO J 1999; 18(21):6134-6145.
28. Das B, Guo Z, Russo P et al. The role of nuclear cap binding protein Cbc1p of yeast in mRNA termination and degradation. Mol Cell Biol 2000; 20(8):2827-2838.
29. Muhlrad D, Parker R. Aberrant mRNAs with extended 3' UTRs are substrates for rapid degradation by mRNA surveillance. RNA 1999; 5(10):1299-1307.
30. Pulak R, Anderson P. mRNA surveillance by the Caenorhabditis elegans smg genes. Genes Dev 1993; 7(10):1885-1897.
31. Muhlrad D, Decker CJ, Parker R. Turnover mechanisms of the stable yeast PGK1 mRNA. Mol Cell Biol 1995; 15(4):2145-2156.
32. Cao D, Parker R. Computational modeling of eukaryotic mRNA turnover. RNA 2001; 7(9):1192-1212.
33. Jacobs Anderson JS, Parker R. The 3' to 5' degradation of yeast mRNAs is a general mechanism for mRNA turnover that requires the SKI2 DEVH box protein and 3' to 5' exonucleases of the exosome complex. EMBO J 1998; 17(5):1497-1506.
34. Mangus D, Evans M, Jacobson A. Poly(A)-binding proteins: Multifunctional scaffolds for the post-transcriptional control of gene expression. Genome Biol 2003; 4(7):223.
35. Decker CJ, Parker R. A turnover pathway for both stable and unstable mRNAs in yeast: Evidence for a requirement for deadenylation. Genes Dev 1993; 7(8):1632-1643.
36. Beelman CA, Parker R. Degradation of mRNA in eukaryotes. Cell 1995; 81(2):179-183.
37. Muhlrad D, Decker CJ, Parker R. Deadenylation of the unstable mRNA encoded by the yeast MFA2 gene leads to decapping followed by 5'—>3' digestion of the transcript. Genes Dev 1994; 8(7):855-866.
38. Dunckley T, Parker R. The DCP2 protein is required for mRNA decapping in Saccharomyces cerevisiae and contains a functional MutT motif. EMBO J 1999; 18(19):5411-5422.
39. Beelman CA, Stevens A, Caponigro G et al. An essential component of the decapping enzyme required for normal rates of mRNA turnover. Nature 1996; 382(6592):642-646.
40. Steiger M, Carr-Schmid A, Schwartz DC et al. Analysis of recombinant yeast decapping enzyme. RNA 2003; 9(2):231-238.
41. Hsu CL, Stevens A. Yeast cells lacking 5'—>3' exoribonuclease 1 contain mRNA species that are poly(A) deficient and partially lack the 5' cap structure. Mol Cell Biol 1993; 13(8):4826-4835.
42. Muhlrad D, Parker R. Premature translational termination triggers mRNA decapping. Nature 1994; 370(6490):578-581.
43. Mitchell P, Tollervey D. An NMD pathway in yeast involving accelerated deadenylation and exosome-mediated 3'—>5' degradation. Mol Cell 2003; 11(5):1405-1413.
44. Lelivelt MJ, Culbertson MR. Yeast Upf proteins required for RNA surveillance affect global expression of the yeast transcriptome. Mol Cell Biol 1999; 19(10):6710-6719.
45. He F, Li X, Spatrick P et al. Genome-wide analysis of mRNAs regulated by the nonsense-mediated and 5' to 3' mRNA decay pathways in yeast. Mol Cell 2003; 12(6):1439-1452.
46. He F, Brown AH, Jacobson A. Upf1p, Nmd2p, and Upf3p are interacting components of the yeast nonsense-mediated mRNA decay pathway. Mol Cell Biol 1997; 17(3):1580-1594.

47. He F, Jacobson A. Upf1p, Nmd2p, and Upf3p regulate the decapping and exonucleolytic degradation of both nonsense-containing mRNAs and wild-type mRNAs. Mol Cell Biol 2001; 21(5):1515-1530.
48. Vilela C, Linz B, Rodrigues-Pousada C et al. The yeast transcription factor genes YAP1 and YAP2 are subject to differential control at the levels of both translation and mRNA stability. Nucleic Acids Res 1998; 26(5):1150-1159.
49. Jacobson A, Peltz SW. Destabilization of nonsense-containing transcripts in Saccharomyces cerevisiae. In: Sonenberg N, Hershey JWB, Mathews MB, eds. Translational Control of Gene Expression. 2nd ed. Cold Spring Harbor: Cold Spring Harbor Laboratory Press, 2000:827-847.
50. Hentze MW, Kulozik AE. A perfect message: RNA surveillance and nonsense-mediated decay. Cell 1999; 96(3):307-310.
51. Mendell JT, Sharifi NA, Meyers JL et al. Nonsense surveillance regulates expression of diverse classes of mammalian transcripts and mutes genomic noise. Nature Genet 2004; 36(10):1073-1078.
52. Lew JE, Enomoto S, Berman J. Telomere length regulation and telomeric chromatin require the nonsense-mediated mRNA decay pathway. Mol Cell Biol 1998; 18(10):6121-6130.
53. Dahlseid JN, Lew-Smith J, Lelivelt MJ et al. mRNAs encoding telomerase components and regulators are controlled by UPF genes in Saccharomyces cerevisiae. Eukaryot Cell 2003; 2(1):134-142.
54. Maderazo AB, Belk JP, He F et al. Nonsense-containing mRNAs that accumulate in the absence of a functional nonsense-mediated mRNA decay pathway are destabilized rapidly upon its restitution. Mol Cell Biol 2003; 23(3):842-851.
55. Ishigaki Y, Li X, Serin G et al. Evidence for a pioneer round of mRNA translation: mRNAs subject to nonsense-mediated decay in mammalian cells are bound by CBP80 and CBP20. Cell 2001; 106(5):607-617.
56. Atkin AL, Schenkman LR, Eastham M et al. Relationship between yeast polyribosomes and Upf proteins required for nonsense mRNA decay. J Biol Chem 1997; 272(35):22163-22172.
57. Atkin AL, Altamura N, Leeds P et al. The majority of yeast UPF1 colocalizes with polyribosomes in the cytoplasm. Mol Biol Cell 1995; 6(5):611-625.
58. Mangus DA, Jacobson A. Linking mRNA turnover and translation: Assessing the polyribosomal association of mRNA decay factors and degradative intermediates. Methods 1999; 17(1):28-37.
59. Peltz SW, Trotta C, He F et al. Identification of the cis-acting sequences and trans-acting factors involved in nonsense-mediated mRNA decay. In: Tuite M, JM A, Sherman F, eds. Protein Synthesis and Targetting in Yeast. Vol H71. Springer-Verlag, 1993:1-10.
60. Shirley RL, Lelivelt MJ, Schenkman LR et al. A factor required for nonsense-mediated mRNA decay in yeast is exported from the nucleus to the cytoplasm by a nuclear export signal sequence. J Cell Sci 1998; 111(Pt 21):3129-3143.
61. Zhang S, Welch EM, Hogan K et al. Polysome-associated mRNAs are substrates for the nonsense-mediated mRNA decay pathway in Saccharomyces cerevisiae. RNA 1997; 3(3):234-244.
62. Ruiz-Echevarria MJ, Gonzalez CI, Peltz SW. Identifying the right stop: Determining how the surveillance complex recognizes and degrades an aberrant mRNA. EMBO J 1998; 17(2):575-589.
63. Gozalbo D, Hohmann S. Nonsense suppressors partially revert the decrease of the mRNA level of a nonsense mutant allele in yeast. Curr Genet 1990; 17(1):77-79.
64. Czaplinski K, Ruiz-Echevarria MJ, Paushkin SV et al. The surveillance complex interacts with the translation release factors to enhance termination and degrade aberrant mRNAs. Genes Dev 1998; 12(11):1665-1677.
65. Weng Y, Czaplinski K, Peltz SW. Genetic and biochemical characterization of mutations in the ATPase and helicase regions of the Upf1 protein. Mol Cell Biol 1996; 16(10):5477-5490.
66. Weng Y, Czaplinski K, Peltz SW. Identification and characterization of mutations in the UPF1 gene that affect nonsense suppression and the formation of the Upf protein complex but not mRNA turnover. Mol Cell Biol 1996; 16(10):5491-5506.
67. Salas-Marco J, Bedwell DM. GTP hydrolysis by eRF3 facilitates stop codon decoding during eukaryotic translation termination. Mol Cell Biol 2004; 24:7769-7778.
68. Maquat LE. Nonsense-mediated mRNA decay: Splicing, translation, and mRNP dynamics. Nature Rev Mol Cell Biol 2004; 5:89-99.
69. Lykke-Andersen J. Making structural sense of nonsense-mediated decay. Nature Struct Mol Biol 2004; 11(4):305-306.
70. Nott A, Le Hir H, Moore MJ. Splicing enhances translation in mammalian cells: An additional function of the exon junction complex. Genes Dev 2004; 18(2):210-222.
71. Wiegand HL, Lu S, Cullen BR. Exon junction complexes mediate the enhancing effect of splicing on mRNA expression. Proc Natl Acad Sci USA 2003; 100(20):11327-11332.

72. Ruiz-Echevarria MJ, Peltz SW. The RNA binding protein Pub1 modulates the stability of transcripts containing upstream open reading frames. Cell 2000; 101(7):741-751.
73. Hagan KW, Ruiz-Echevarria MJ, Quan Y et al. Characterization of cis-acting sequences and decay intermediates involved in nonsense-mediated mRNA turnover. Mol Cell Biol 1995; 15(2):809-823.
74. Yun DF, Sherman F. Initiation of translation can occur only in a restricted region of the CYC1 mRNA of Saccharomyces cerevisiae. Mol Cell Biol 1995; 15(2):1021-1033.
75. Zhang S, Ruiz-Echevarria MJ, Quan Y et al. Identification and characterization of a sequence motif involved in nonsense-mediated mRNA decay. Mol Cell Biol 1995; 15(4):2231-2244.
76. Gonzalez CI, Ruiz-Echevarria MJ, Vasudevan S et al. The yeast hnRNP-like protein Hrp1/Nab4 marks a transcript for nonsense- mediated mRNA decay. Mol Cell 2000; 5(3):489-499.
77. Lykke-Andersen J. Identification of a human decapping complex associated with hUpf proteins in nonsense-mediated decay. Mol Cell Biol 2002; 22:8114-8121.
78. Jacobson A. Regulation of mRNA decay: Decapping goes solo. Molec Cell 2004; 15:1-2.
79. Parker R, Song H. The enzymes and control of eukaryotic mRNA turnover. Nature Struct Mol Biol 2004; 11:121-127.
80. Hilleren P, Parker R. mRNA surveillance in eukaryotes: Kinetic proofreading of proper translation termination as assessed by mRNP domain organization? RNA 1999; 5(6):711-719.
81. Amrani N, Ganesan R, Kervestin S et al. A faux 3'-UTR promotes aberrant termination and triggers nonsense-mediated mRNA decay. Nature 2004; 432(7013):112-118.
82. Jacobson A, Peltz SW. Interrelationships of the pathways of mRNA decay and translation in eukaryotic cells. Annu Rev Biochem 1996; 65:693-739.

Section II
Mammals

Chapter 4

NMD in Mammalian Cells:

A History

Lynne E. Maquat*

Abstract

Premature termination codons (PTCs) that are caused by either frameshift or nonsense mutations have long been known to decrease mRNA half-life by a process that is called nonsense-mediated mRNA decay (NMD). We now understand that NMD in mammalian cells is generally triggered when a UAA, UGA or UAG nonsense codon resides more than ~25 nucleotides upstream of a post-splicing exon junction complex (EJC) of proteins. Constituents of the EJC include the Upf NMD factors. These factors function in part to recruit proteins that degrade mRNA from both 5' and 3' ends. NMD appears to require nonsense codon recognition during a pioneer round of translation. This round of translation involves newly synthesized mRNA that is bound by the mostly nuclear cap binding protein (CBP) heterodimer CBP80-CBP20. As mRNA matures, CBP80-CBP20 is replaced by eukaryotic translation initiation factor (eIF)4E, which is the mostly cytoplasmic cap binding protein. By the time CBP80-CBP20 has been replaced by eIF4E, the majority if not all of the EJCs have been removed and the mRNA is immune to NMD. NMD down-regulates mRNAs that produce truncated and potentially deleterious proteins as a type of quality control or mRNA surveillance. NMD also down-regulates particular physiologic mRNAs that encode functional proteins as a means of controlling proper gene expression. The essential nature of both regulatory aspects of NMD most likely explains why NMD is required for mammalian development.

In the Beginning

It first became evident that a PTC could reduce the half-life of mammalian mRNA by studying newly synthesized β-globin transcripts in nucleated cells of bone marrow aspirates from patients with the hemolytic anemia β°-thalassemia.[1,2] Until then, the only indication that a PTC could alter RNA half-life in eukaryotes derived from studies of *Saccharomyces cerevisiae*. Approach-to-steady-state labeling experiments demonstrated that each of two PTCs within the URA3 gene reduced the half-life of polyadenylated *URA3* RNA.[3] Subsequently, important insight into the mechanism of NMD in mammalian cells derived from systematic studies of transcripts that were associated with inherited diseases in addition to β°-thalassemia, most notably triosephosphate isomerase (TPI) deficiency. Mechanistic insights also arose from studies of immunoglobulin (Ig) and T-cell receptor (TCR)β RNAs. These RNAs derive from genes that have the remarkable capacity to undergo somatic-cell rearrangement and hypermutation, both of which often result in the generation of a PTC (see chapter by Gudikote and Wilkinson).

*Lynne E. Maquat—Department of Biochemistry and Biophysics, School of Medicine and Dentistry, 601 Elmwood Avenue, Box 712, University of Rochester, Rochester, New York 14642, U.S.A. Email: lynne_maquat@urmc.rochester.edu

Nonsense-Mediated mRNA Decay, edited by Lynne E. Maquat. ©2006 Eurekah.com.

NMD Targets Newly Synthesized mRNA: The Role of Intron Position

Translation most often terminates within the last exon of an mRNA, and translation termination within TPI mRNA is no exception. Analyses of single and compensating pairs of frameshift mutations that individually shortened but in combination restored the translational reading frame demonstrated that the abundance of TPI mRNA is reduced when translation terminates sufficiently upstream of the last exon-exon junction.[4,5] At first, it was difficult to explain how a PTC could be influenced by an exon-exon junction or, as we put it at that time, the distance of the PTC "within the penultimate exon relative to the final intron".[5] The final intron could be substituted with another spliceable intron without compromising the extent of NMD, suggesting that NMD required the process of nuclear splicing rather than a specific intron sequence.[6]

Subsequently, the 50-55 nucleotide rule was established. This rule stipulates that a nonsense codon elicits NMD provided that it resides at least 50-55 nucleotides upstream of an intron (Fig. 1),[7] and it was based on several lines of evidence. As one line of evidence, moving the last intron of TPI RNA was found to move the boundary between PTCs that did and did not elicit NMD when the PTC resided within the penultimate exon before and after the move.[6,8] For example, moving the last intron downstream by nine nucleotides moved the boundary downstream by nine nucleotides. As another line of evidence, inserting nucleotides immediately upstream of the last intron of either TPI or β-globin RNA was also

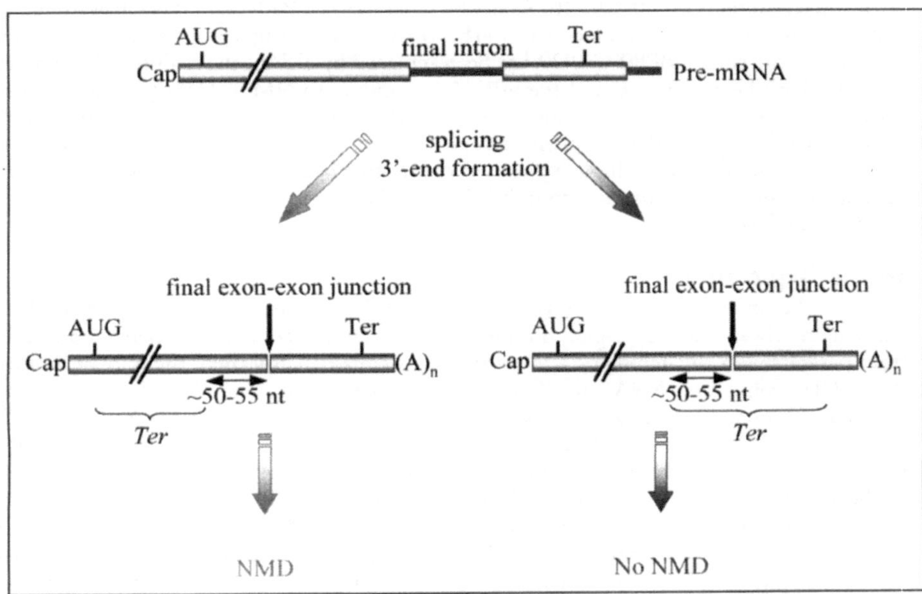

Figure 1. The ~50-55 nucleotide rule. Only the final intron within a generic mammalian pre-mRNA and, as a consequence, the final exon-exon junction within the product mRNA is shown. NMD generally results when translation terminates at a nonsense codon that is located more than ~50-55 nucleotides (nt) upstream of a splicing-generated exon-exon junction. In contrast, NMD does not occur when translation terminates at a nonsense codon that is located either less than ~50-55 nucleotides upstream of an exon-exon junction or downstream of such a junction. AUG denotes the translation initiation codon. The 5'-most (italicized) Ter signifies a nonsense codon that is premature. The 3'-most (nonitalicized) Ter denotes the normal termination codon, which usually resides within the final exon.

found to move the boundary between the two types of PTCs upstream by the same number of nucleotides.[6,8,9] As yet another line of evidence, for 98% of physiologic transcripts in which the normal termination codon is followed by one or more introns, the distance between the termination codon and the last intron is less than 50 nucleotides.[7] Since physiologic transcripts generally would not be expected to be NMD targets, a rule specifying that NMD does not occur when an intron resides less than 50 nucleotides downstream of a nonsense codon supports this expectation.

Despite the role of intron position in NMD, PTC recognition likely involves translationally active cytoplasmic ribosomes. For example, not only compensating frameshift mutations but also suppressor tRNA, both of which restore the site of translation termination to normal, were found to inhibit NMD.[4,10] Furthermore, introducing a highly structured 5' untranslated region (0UTR) that blocks translation initiation also inhibited NMD.[10]

To make matters more perplexing, PTCs were found to elicit "nucleus-associated" NMD. For example, PTCs reduced the half-life of TPI mRNA that copurified with nuclei without reducing the half-life of TPI mRNA in the cytoplasm.[6,11,12] Earlier studies had demonstrated that PTCs within dihydrofolate reductase (DHFR) transcripts also reduced the overall cellular abundance of DHFR mRNA without affecting its cytoplasmic half-life.[13] The effect of a PTC was not only specific to the allele within which it resides but also specific to the splicing isoform within which it resides, indicating that a PTC acts in cis but not in trans to down-regulate only the spliced mRNA within which it resides.[4,11] These findings ran counter to the proposal that translation in the cytoplasm feeds back to degrade mRNA in the nucleoplasm.[14]

Even today, the cellular site of nucleus-associated NMD remains uncertain. Proposals for nuclear translation and the possibility of NMD occurring within the nucleoplasm have been raised but subsequently dismissed by most but not all scientists.[15,16] Currently, a favorite model posits that nucleus-associated NMD takes place during mRNA export, at a point where mRNA copurifies with nuclei but can be translated by cytoplasmic ribosomes (Fig. 2). In fact, at least one mRNA that is large enough to be visually monitored using electron microscopy is invariably exported 5'-end-first from the nucleus to the cytoplasm so as to be translated by cytoplasmic ribosomes before its 3' end has completely passed through the nuclear pore complex.[17]

Nucleus-associated NMD has also been reported for many other transcripts in addition to TPI mRNA. These transcripts include the major urinary protein mRNA,[11] β-globin mRNA in nonerythroid cells,[18] adenine phosphoribosyltransferase mRNA,[19] and TCRβ mRNA.[20] The absence of NMD in the cytoplasm is not because mRNA fails to associate with polysomes.[21] Therefore, for those mRNAs that are subject to nucleus-associated NMD, we rationalized that "cytoplasmic mRNA is either no longer associated with factors that are required for nonsense decay or is associated with factors that antagonize decay".[14]

PTC recognition occurs after splicing. For example, we demonstrated using TPI mRNA[22] and others subsequently demonstrated using TCRβ mRNA[23] that PTCs that spanned two exons, and thus could be recognized by ribosomes only after splicing, were competent to elicit NMD.[22] Consistent with recognition after splicing, PTCs within the Igμ or TPI gene did not detectably affect the rate of gene transcription.[12,24] Furthermore, PTCs within the TPI gene did not detectably affect the level of any of the six introns within TPI pre-mRNA.[12] The failure of PTCs to affect splicing rates was more recently reported for DHFR and TCRβ genes.[25] A specific PTC can certainly influence splicing by altering a cis-acting affector of splicing, such as an exonic splicing enhancer. However, it remains controversial that a PTC can also influence splicing by altering the translational reading frame (see chapter by Zhang and Krainer).

While most PTC-containing mRNAs in mammalian cells are targeted for nucleus-associated NMD, relatively few are targeted for cytoplasmic NMD (Fig. 2). One example is provided by glutathione peroxidase (GPx)1 mRNA.[26,27] Nevertheless, the cytoplasmic NMD of GPx1 mRNA still follows the 50-55 nucleotide rule.[28]

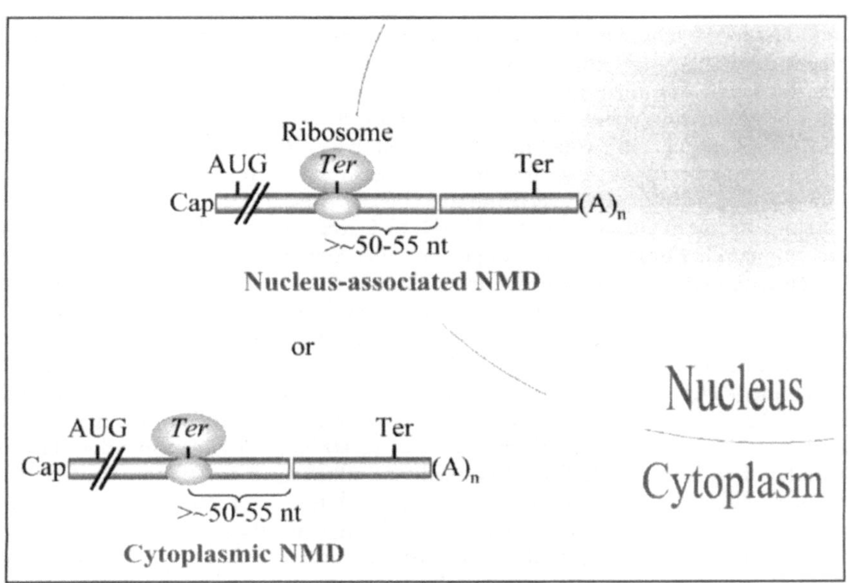

Figure 2. NMD is nucleus-associated or cytoplasmic. The majority of mRNAs that are targeted for NMD in mammalian cells are subject to nucleus-associated NMD. Since translation is not known to occur within the nucleus,[15] nucleus-associated NMD has been proposed to take place during mRNA export. In such a way, an mRNA that copurifies with nuclei could be translated by cytoplasmic ribosomes. In support of this proposal, a large mRNA of the insect *Chironomus tentans* is known to undergo concomitant export and translation.[17] Relatively few mRNAs that are targeted for NMD in mammalian cells are known to be subject to cytoplasmic NMD. Determinants of the cellular site of NMD have yet to be defined. However, it is reasonable to assume they are cis-acting mRNA sequences. Notably, the extent to which nucleus-associated and cytoplasmic NMD occurs in processing bodies is currently unknown (see chapter by Singh and Lykke-Andersen). AUG denotes the translation initiation codon. The 5'-most (italicized) Ter signifies a nonsense codon that resides more than ~50-55 nucleotides upstream of a splicing-generated exon-exon junction. The 3'-most (nonitalicized) Ter denotes the normal termination codon.

NMD Generally Requires a Post-Splicing Exon Junction Complex Downstream of the Site of Translation Termination

We had proposed that the role of intron position in NMD reflects a splicing-dependent "mark" that translating ribosomes must approximate or traverse in order to confer mRNA stability".[14] We now know the so-called "mark" is a complex of proteins called the exon junction complex (EJC). Evidence that pre-mRNA splicing alters the composition of proteins that bind to the exon-exon junctions of product mRNA was first demonstrated by splicing and cross-linking in vitro-generated RNAs in nuclear extracts.[29] Characterizations by a number of labs demonstrated that the remarkably stable EJC, which is deposited ~20-24 nucleotides upstream of mRNA exon-exon junctions,[30] generally consists of at least 13 proteins (Table 1). These proteins, which have a variety of functions, include the NMD factors Upf3 or Upf3X (also called Upf3a or Upf3b), Upf2 and, ultimately, Upf1 (reviewed in refs. 31,32; Fig. 3).

Each Upf factor was identified and named after its ortholog in *S. cerevisiae* (see chapter by Singh and Lykke-Andersen).[33-37] By tethering a Upf factor downstream of a termination codon and either (i) using small interfering (si)RNA to down-regulate another Upf factor

Table 1. Proteins involved in NMD (Shaded area indicates stable EJC components)

Protein	Feature(s)
Upf1	Transiently interacts with EJC[57] (see chapter by Singh and Lykke-Andersen); Group 1 helicase;[76-78] ATPase;[76-78] interacts with CBP80,[63] eRF1[79] and eRF3[79] (see text); last of the Upf proteins to join the EJC;[35,38] phosphorylation cycle necessary for NMD (see chapter by Yamashita et al); also functions in other translation-dependent mRNA decay pathways such as Staufen1-mediated mRNA decay (see chapter by Kim and Maquat) or stem-loop binding protein-mediated decay of histone H2A mRNA at the end of S phase (see chapter by Kaygun and Marzluff)
Upf2	EJC component[30,58,80] (see chapter by Singh and Lykke-Andersen); interacts with both Upf3 and Upf3X;[37,58] functions in NMD after Upf3 or Upf3X but before Upf1[35,38]
Upf3 (Upf3a)	EJC component[58,80,81] (see chapter by Singh and Lykke-Andersen); may function comparably to Upf3X but with a more predominant role in NMD since it interacts with the EJC component Y14 more strongly than Upf3X;[81] functions before Upf2 and Upf1[35,38]
Upf3X (Upf3b)	EJC component[58,80,81] (see chapter by Singh and Lykke-Andersen); may function comparably to but less predominantly than Upf3 in NMD[81,82]
Y14	EJC component;[30,58,80,83] forms heterodimer with Magoh;[84] interacts with either Upf3 or Upf3X[30,58,80,83]
Magoh	EJC component;[58,80] forms heterodimer with Y14;[84] interacts with either Upf3 or Upf3X[58,80]
RNPS1	EJC component[30,58,80]
Barentsz (MLN51)	EJC component;[85] interacts with eIF4AIII and Y14-Magoh[85]
PYM	EJC component;[86] interacts with Y14/Magoh[86]
eIF4AIII	EJC component;[87-90] DEAD box helicase;[91] interacts with Barentsz and Y14-Magoh;[88-90] forms "platform" of EJC[90]
SRm160	EJC component[29,30,58,80]
Pnn/DRS	EJC component[92]
REF/Aly	EJC component;[30,58,80] also functions in mRNA export[93]
UAP56	EJC component;[58,80] also functions in mRNA export[93]
TAP/NFX1:p15	EJC component;[30,58,80] also functions in mRNA export[30,58,80,83]
CBP80-CBP20	Major nuclear cap-binding proteins;[94,95] CBP80 interacts with Upf1 so as to promote interaction between Upf1 and Upf2[63]
Smg1	Phosphoinositide 3-kinase-related protein kinase that phosphorylates Upf1 and other proteins (see chapters by Yamashita et al and Abraham and Oliveira)
Smg5, Smg6, Smg7	Function in Upf1 dephosphorylation (see chapters by Singh and Lykke-Andersen, and Yamashita et al); Smg7 provides link between identification of a nonsense codon as the type that elicits mRNA decay and decay process[46]

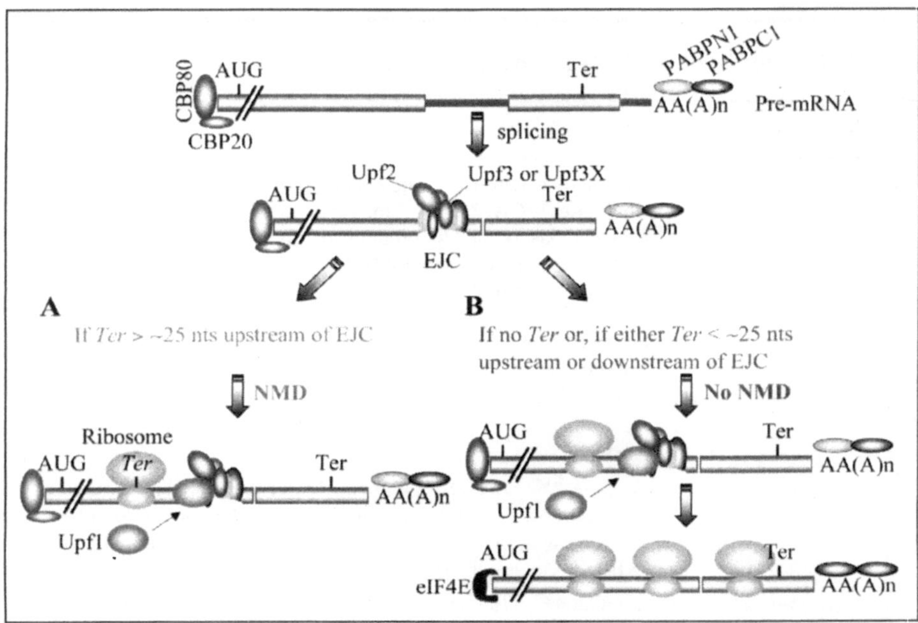

Figure 3. NMD targets CBP80-CBP20-bound newly synthesized mRNA but does not detectably target eIF4E-bound steady-state mRNA. Pre-mRNA, which consists of exons (boxes) and introns (lines between boxes), exists in the nucleus bound by cap binding protein (CBP)80 and CBP20 at the 5' cap and, after 3'-end formation, the major nuclear poly(A)-binding protein (PABP)N1 and, generally, the major cytoplasmic poly(A) binding protein PABPC1 at the 3' poly(A) tail (N. Hosoda, D. Bear, B. Reinert, T. Howard and L.E. Maquat, unpububished data). The process of pre-mRNA splicing deposits an exon junction complex (EJC) of proteins ~20-24 nucleotides upstream of each exon-exon junction. This EJC consists minimally of proteins involved in pre-mRNA splicing, such as RNPS1, SRm160 and UAP56, Pnn/DRS; proteins involved in mRNA export, such as Y14, REF/Aly and Magoh; and proteins of incompletely understood function, such as eIF4AIII, Barentsz/MLN51 and probably PYM. The EJC also consists of the NMD factors Upf3 or 3X, Upf2 and, ultimately, Upf1. The resulting mRNA and associated proteins constitute the "pioneer" translation initiation complex. A) NMD is triggered when translation terminates more than ~25 nucleotides upstream of an EJC. While the nature of the trigger is unknown, it may involve the interaction of Upf1 first with eukaryotic release factors and subsequently with EJC-associated Upf2. Notably, NMD is not generally 100% efficient. mRNA that escapes NMD is remodeled to the steady-state translation initiation complex (not shown), which is immune to NMD. The 5'-most (italicized) Ter signifies a nonsense codon that resides more than ~50-55 nucleotides upstream of a splicing-generated exon-exon junction. The 3'-most (nonitalicized) Ter denotes the normal termination codon. B) If translation terminates less than ~25 nucleotides upstream of an EJC or downstream of an EJC, as it generally does normally, then NMD is not triggered and CBP80-CBP20-bound mRNA is remodeled to eIF4E-bound mRNA. During remodeling, the EJCs and associated Upf factors are removed, and PABPN1 is replaced by PABPC1. Whether translation or other processes are required for remodeling is unclear. However, translation has been implicated in removal of the Y14 component of the EJC.

or (ii) coexpressing nontetherable Upf1 that was mutated within the helicase domain, the order of Upf factor function in NMD was determined to be first Upf3 or Upf3X, then Upf2 and, finally, Upf1.[35,38] Upf1 has recently been shown to interact with eukaryotic translation release factor (eRF)1 and eRF3 (F. Lejeune and L.E. Maquat, unpublished data; G. Singh and J. Lykke-Andersen, personal communication), both of which function in translation termination. Considering that NMD generally requires translation termination upstream of an EJC, Upf1 may interact with eRFs before, simultaneously with or after EJC-associated Upf2 (Fig. 3). If translation terminates less than ~25 nucleotides upstream of the final EJC (i.e., less than ~50-55 nucleotides upstream of the final exon-exon junction) or downstream of the final EJC, then all EJCs are thought to be removed by translating ribosomes (Fig. 3).[39]

In addition to the Upf factors, NMD requires Smg factors that were named after their orthologs in *Caenorhabditis elegans*. Each of these factors modifies the phosphorylation status of Upf1 (see chapters by Yamashita et al, and Abraham and Oliveira). Smg1 is a phosphoinositide 3-kinase-related protein kinase that phosphorylates Upf1.[40-42] Smg5, Smg6 and Smg7 promote Upf1 dephosphorylation.[43-46] Interestingly, tethering Smg7 to mRNA elicits mRNA decay even when there is no termination codon upstream of the tethering site.[46] It was concluded that Smg7 participates at a late step in the NMD pathway, possibly after nonsense codon recognition. Consistent with this conclusion, Smg7 functions in NMD after Smg5 and Smg 6.[46]

The importance of EJCs to NMD is reflected by the failure of PTCs within transcripts that derive from genes that are naturally intronless to elicit NMD. As examples, PTCs within histone H4, heat shock protein 70 and melanocortin 4-receptor transcripts fail to elicit NMD.[47,48] Inserting an intron sufficiently downstream of a PTC within histone H4 RNA leads to NMD.[47] Therefore, the immunity of this and probably other transcripts that derive from naturally intronless genes is simply due to the lack of an EJC rather than to an RNA sequence that inhibits NMD.

However, data suggest that there are mechanisms of NMD that do not involve an EJC. For example, TCRβ transcripts provide an exception to the 50-55 nucleotide rule. PTCs that are situated as few as 8-10 nucleotides upstream of the final exon-exon junction elicit some degree of NMD, suggesting that some component of messenger ribonucleoprotein particles (mRNPs) other than the junction function in NMD.[23,49] (see chapter by Gudikote and Wilkinson).

Other examples of what appear to be EJC-independent NMD were revealed when the last intron of the β-globin or TPI gene was deleted. As noted above, moving the position of the last intron within either β-globin or TPI RNA comparably moves the boundary between PTCs that do and do not elicit NMD and that reside within the penultimate exon.[6,8,9] This result indicates that a PTC within the penultimate exon of either RNA elicits NMD depending on the position of that PTC relative to the last EJC. However, for those PTCs within the penultimate exon that normally elicit NMD, deleting the last intron of either gene fails to eliminate NMD.[6,8,9] This finding suggests that some component of mRNP functionally substitutes for the last EJC when the last EJC is not formed, and we have called such functional substitution a "failsafe" mechanism.[8,9] The failsafe mechanism is less efficient than EJC-dependent NMD and may be related to the EJC-independent mechanism of *S. cerevisiae* that involves an abnormally long or "faux" 3' UTR (ref. 31; see chapter by Amrani and Jacobson). However, the failsafe mechanism in mammalian cells differs from EJC-independent NMD in yeast by requiring splicing upstream of the PTC, since PTC-containing mRNAs that derive from intronless TPI or β-globin genes are immune to NMD.[8,9]

Additional exceptions to the 50-55 nucleotide rule are apparent with the finding that PTCs within the first exon of β-globin mRNA that are situated close to the initiation codon fail to elicit NMD.[50] Failure of PTCs within the first exon of β-globin mRNA to elicit NMD is not due to translation reinitiation at downstream AUG codons, which is known to inhibit NMD when it occurs before the next EJC,[51] or to effects on splicing.[52] Nor do PTCs that reside close to the initiation codon always fail to elicit NMD, as exemplified using TPI mRNA.[51] As another example of an exception to the 50-55 nucleotide rule, editing of a CAA codon to a UAA nonsense codon within apolipoprotein B transcripts fails to elicit NMD. Failure depends on the presence of an editing complex that must exist a particular distance downstream of the edited codon (ref. 53; Y. Ishigaki and L.E. Maquat, unpublished data).

It is worth noting that not all UGA codons direct translation termination. For example, GPx1 mRNA, which encodes a selenoprotein, is targeted for NMD when the UGA codon within its translational reading frame directs translation termination.[26,27] However, GPx1 mRNA is not targeted for NMD when this UGA codon directs incorporation of the 21st amino acid selenocysteine.[26,27]

NMD is linked to pre-mRNA metabolism not only because it generally depends on a post-splicing EJC but also in more mysterious ways (reviewed in ref. 32). For example, over-expressing of SF2/ASF, SC35 or other members of the SR protein family of splicing factors was found to augment the NMD of β-globin mRNA by an unknown mechanism.[54]

By processes that are not understood, the efficiency of NMD can be modified by sequences other than an EJC. 5' PTCs have long been known to reduce the level of Igμ RNA more dramatically than 3' PTCs.[55] Work undertaken more recently has demonstrated that cis-acting sequences within TCRβ RNA increase the efficiency of NMD to almost 100% when they are situated upstream of a PTC (see chapter by Gudikote and Wilkinson).[56] These sequences include part of an exon and the flanking intron,[56] and they appear to promote efficient splicing.

Evidence That NMD Is Restricted to CBP80-Bound mRNA and Does Not Detectably Target eIF4E-Bound mRNA: The Pioneer Translation Initiation Complex

In agreement with our kinetic studies indicating that NMD is restricted to newly synthesized mRNA, we have found that both nucleus-associated and cytoplasmic NMD target mRNA that is bound by the major nuclear cap binding protein (CBP) heterodimer CBP80-CBP20.[57] Before this finding, only mRNA bound by eukaryotic translation initiation factor (eIF)4E, which is the major cytoplasmic cap binding protein, was thought to be translated. Consistent with the importance of the EJC to NMD, CBP80-CBP20-bound mRNA is detectably associated with the Upf factors and other EJC constituents whereas eIF4E-bound mRNA is not.[57,58] eIF4E, unlike CBP80-CBP20, fails to detectably associate with de novo-synthesized caps.[58] These findings led us to conclude that mRNA is immune to NMD by the time eIF4E replaces CBB80-CBP20 and forms the steady-state initiation complex. Thus, NMD is a consequence of nonsense codon recognition during a "pioneer round" of translation.

The efficiency with which the pioneer translation initiation complex undergoes NMD, which is rarely if ever 100%, influences the efficiency with which this complex is remodeled to the steady-state translation initiation complex, which produces the bulk of cellular proteins. Translation is a very complicated process that involves many translation initiation factors. It follows that there are more similarities than differences in the ways the pioneer and steady-state complexes initiate translation. For example, translation of both complexes involves eIF2, eIF3, eIF4AI, poly(A) binding protein (PABP)C1 (formerly called PABP1; N. Hosoda, D. Bear, F. Lejeune, B. Reinert, T. Howard, and L.E. Maquat, unpublished data), and eIF4G.[59,60]

While the translation of CBP80-CBP20-bound mRNA is likely to occur in many organisms, mammalian cells may be unique by restricting NMD to CBP80-CBP20-bound mRNA.[61] In *S. cerevisiae*, either CBP80-CBP20 (the yeast orthologs of which are Cbc1p-Cbc2p) or eIF4E can bind to de novo-synthesized caps.[62] Accordingly, NMD targets both CBP80-CBP20-mediated and eIF4E-mediated pioneer rounds of translation.[62] NMD in *S. cerevisiae* also differs from NMD in mammalian cells by taking place not only during the pioneer round of translation but also during steady-state translation (see chapter by He and Jacobson). Not only does NMD in mammalian cells appear to be restricted to CBP80-CBP20-bound mRNA, but CBP80 promotes the binding of Upf1 to Upf2.[63] Therefore, CBP80 not only coexists on mRNA that contains EJCs, but it facilitates EJC function in NMD.

Evidence That NMD Degrades mRNA from Both 5' and 3' Ends

Once a termination codon has been recognized as abnormal, mRNA is degraded by the same activities that mediate general mRNA decay. Results from a number of studies indicate that NMD involves decapping followed by 5'-to-3' exonucleolytic decay as well as deadenylation followed by 3'-to-5' exonucleolytic decay (Fig. 4). First, down-regulating the decapping enzyme Dcp2, the poly(A) ribonuclease PARN or the PM/Scl100 component of the exosome, which mediates 3'-to-5' exonucleolytic decay, inhibits both nucleus-associated and cytoplasmic NMD.[64] Furthermore, down-regulating Dcp2 or the 5'-to-3' exonuclease Xrn1 inhibits the reduction in mRNA abundance that results from tethering Smg7 downstream of a termination codon.[46] Second, antibodies to each Upf protein coimmunopurify Dcp2, Xrn1, the Rat1 (also called Xrn2) 5'-to-3' exonuclease, PARN, PM/Scl100, and Rrp4 and Rrp41 components of the exosome.[46,64,65] Additionally, human Upf1 interacts with Dcp1 in yeast two-hybrid analyses.[66] Results of yeast two-hybrid analyses also indicate that human Upf2 interacts with Lsm1, which is required for decapping, as well as Dcp2, Rat1 and PM/Scl100.[66] Third, measurements of poly(A) tail lengths indicate that NMD augments deadenylation.[67] In addition to decay from either mRNA end, studies of PTC-containing β-globin mRNA in human cells indicate that NMD can also involve incomplete deadenylation followed by decapping.[68]

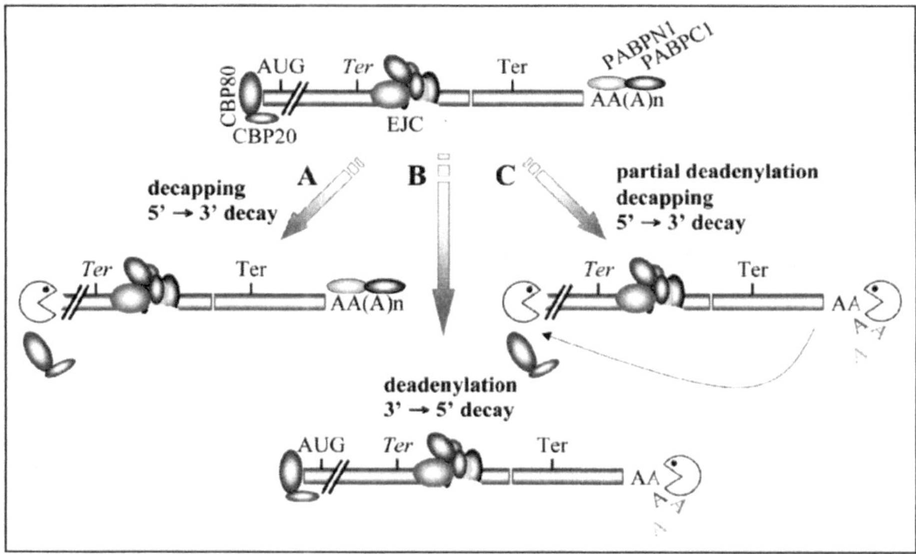

Figure 4. NMD involves mRNA decay from both 5' and 3' ends. NMD occurs when translation terminates on CBP80-CBP20-bound mRNA at nonsense codon that resides more than 50-55 nucleotides upstream of an exon-exon junction that is bound by an exon junction complex (EJC). Data indicate that NMD initiates at either mRNA end. A) NMD of the mRNA 5' end involves decapping followed by 5'-to-3' exonucleolytic decay. B) NMD of the mRNA 3' end involves deadenylation followed by 3'-to-5' exonucleolytic decay. C) Notably, NMD can also involve partial deadenylation followed by decapping and, presumably, 5'-to-3' exonucleolytic decay. Currently, it is not possible to determine the relative efficiencies of 5'-to-3' and 3'-to-5' decay. The 5'-most (italicized) Ter signifies a nonsense codon that resides more than ~50-55 nucleotides upstream of a splicing-generated exon-exon junction. The 3'-most (nonitalicized) Ter denotes the normal termination codon.

The relative rates at which mRNA is degraded from each end is currently unknown. Also unknown are the relative rates at which deadenylation is followed by 3'-to-5' exonucleolytic decay or by decapping and then 5'-to-3' exonucleolytic decay.

The decay of PTC-containing β-globin mRNA in mouse erythroid cells appears to be in a class by itself. While decay may involve the pathways described above, studies of the erythroid tissues of mice transgenic for one of several β°-thasassemic β-globin alleles or mouse erythroleukemic cells stably transfected with one of several of these alleles reveal readily detectable β-globin mRNA decay intermediates. These intermediates are polyadenylated but lack regions of the mRNA 5' end.[69-72] Data indicate that the decay intermediates are the result of endonucleolytic cleavage, preferentially at UG dinucleotides, by a polysome-associated activity that is similar to the activity of *Xenopus* PMR1.[72,73] The new 5' ends subsequently undergo capping and are apparently resistant to further 5'-to-3' decay.[70] Endonucleolytic cleavage generally characterizes mRNA-specific pathways. Since NMD is not an mRNA-specific pathway, the PMR1-like activity that degrades β-globin mRNA in erythroid cells, while dependent on the recognition of a PTC during translation, may be superimposed upon or may supercede the exonucleolytic pathways that typify NMD.

The Importance of NMD

The critical nature of NMD is exemplified by the findings that mouse embryos that are inactive in NMD resorb shortly after implantation.[74] Additionally, blastocysts isolated 3.5 days post-coitum that are inactive in NMD undergo apoptosis in culture after a brief growth period.[74] The inviability of NMD-deficient embryos and cells probably reflects the combined failure to properly regulate natural substrates and eliminate transcripts that are generated in error (see chapters by Sharifi and Dietz, and Soergel et al).

Future Frontiers

Studies of mammalian-cell NMD have revealed a number of unforeseen RNA metabolic pathways. These include mechanistic connections between pre-mRNA splicing in the nucleoplasm and mRNA decay in the cytoplasm, and the translation of CBP80-CBP20-bound mRNA. Studies of mammalian-cell NMD have also elucidated the structure and function of RNA-protein particles both before and after splicing. Nevertheless, there is much yet to learn.

The following questions remain unanswered. What is the cellular site of nucleus-associated NMD? What differentiates mRNAs that are targeted for nucleus-associated NMD from mRNAs that are targeted for cytoplasmic NMD? Are all EJCs uniform in composition, and do all EJCs function comparably in NMD? What is the mechanism of EJC-independent NMD? How do factors that function in NMD orchestrate the series of events that begin with translation initiation and end with mRNA decay? What triggers remodeling of the pioneer translation initiation complex to the steady-state translation initiation complex, and is there is an obligate order to the steps that comprise remodeling? Does NMD occur in cytoplasmic processing bodies, which are apparently sites at which general mRNAs are degraded from the 5' end (see chapter by Singh and Lykke-Andersen)? How do studies demonstrating that PTC-containing human β-globin in mouse erythroid cells are targeted for endonucleolytic relate to studies demonstrating that NMD generally involves decay from mRNA 5' and 3' ends? Does NMD exist largely to eliminate mistakes in gene expression or primarily to achieve proper levels of productive gene expression (see chapters by Sharifi and Dietz, and Soergel et al). What is the evolutionary explanation that Upf and Smg factors function in other metabolic pathways (see chapters by Abraham and Oliveira, Kim and Maquat, Kaygun and Marzluff, Azzalin et al, and Zhang and Krainer).[75] Furthermore, how does function in multiple pathways impact each pathway?

Acknowledgements

I thank Yoon Ki Kim for generating figures and members of the lab, especially Holly Kuzmiak, for comments on the manuscript. I apologize for not referencing all relevant papers in the very large body of literature that pertains to NMD. L.E. Maquat was supported by NIH R01 DK33938 and GM059614.

References

1. Maquat LE, Kinniburgh AJ, Rachmilewitz EA et al. Unstable beta-globin mRNA in mRNA-deficient beta-o-thalassemia. Cell 1981; 27:543-553.
2. Kinniburgh AJ, Maquat LE, Schedl T et al. mRNA-deficient beta zero-thalassemia results from a single nucleotide deletion. Nucleic Acids Res 1982; 10:5421-5427.
3. Losson R, Lacroute F. Interference of nonsense mutations with eukaryotic messenger RNA stability. Proc Natl Acad Sci USA 1979; 76:5134-5137.
4. Daar IO, Maquat LE. Premature translation termination mediates triosephosphate isomerase mRNA degradation. Mol Cell Biol 1988; 8:802-813.
5. Cheng J, Fogel-Petrovic M, Maquat LE. Translation to near the distal end of the penultimate exon is required for normal levels of spliced triosephosphate isomerase mRNA. Mol Cell Biol 1990; 10:5215-5225.
6. Cheng J, Belgrader P, Zhou X et al. Introns are cis effectors of the nonsense-codon-mediated reduction in nuclear mRNA abundance. Mol Cell Biol 1994; 14:6317-6325.
7. Nagy E, Maquat LE. A rule for termination-codon position within intron-containing genes: When nonsense affects RNA abundance. Trends Biochem Sci 1998; 23:198-199.
8. Zhang J, Sun X, Qian Y et al. Intron function in the nonsense-mediated decay of beta-globin mRNA: Indications that pre-mRNA splicing in the nucleus can influence mRNA translation in the cytoplasm. RNA 1998; 4:801-815.
9. Zhang J, Sun X, Qian Y et al. At least one intron is required for the nonsense-mediated decay of triosephosphate isomerase mRNA: A possible link between nuclear splicing and cytoplasmic translation. Mol Cell Biol 1998; 18:5272-5283.
10. Belgrader P, Cheng J, Maquat LE. Evidence to implicate translation by ribosomes in the mechanism by which nonsense codons reduce the nuclear level of human triosephosphate isomerase mRNA. Proc Natl Acad Sci USA 1993; 90:482-486.
11. Belgrader P, Maquat LE. Nonsense but not missense mutations can decrease the abundance of nuclear mRNA for the mouse major urinary protein, while both types of mutations can facilitate exon skipping. Mol Cell Biol 1994; 14:6326-6336.
12. Cheng J, Maquat LE. Nonsense codons can reduce the abundance of nuclear mRNA without affecting the abundance of pre-mRNA or the half-life of cytoplasmic mRNA. Mol Cell Biol 1993; 13:1892-1902.
13. Urlaub G, Mitchell PJ, Ciudad CJ et al. Nonsense mutations in the dihydrofolate reductase gene affect RNA processing. Mol Cell Biol 1989; 9:2868-2880.
14. Maquat LE. When cells stop making sense: Effects of nonsense codons on RNA metabolism in vertebrate cells. RNA 1995; 1:453-465.
15. Dahlberg JE, Lund E, Goodwin EB. Nuclear translation: What is the evidence? RNA 2003; 9:1-8.
16. Iborra FJ, Jackson DA, Cook PR. The case for nuclear translation. J Cell Sci 2004; 117:5713-5720.
17. Daneholt B. Assembly and transport of a premessenger RNP particle. Proc Natl Acad Sci USA 2001; 98:7012-7017.
18. Kugler W, Enssle J, Hentze MW et al. Nuclear degradation of nonsense mutated beta-globin mRNA: A post-transcriptional mechanism to protect heterozygotes from severe clinical manifestations of beta-thalassemia? Nucleic Acids Res 1995; 23:413-418.
19. Kessler O, Chasin LA. Effects of nonsense mutations on nuclear and cytoplasmic adenine phosphoribosyltransferase RNA. Mol Cell Biol 1996; 16:4426-4435.
20. Li S, Leonard D, Wilkinson MF. T cell receptor (TCR) mini-gene mRNA expression regulated by nonsense codons: A nuclear-associated translation-like mechanism. J Exp Med 1997; 185:985-992.
21. Stephenson LS, Maquat LE. Cytoplasmic mRNA for human triosephosphate isomerase is immune to nonsense-mediated decay despite forming polysomes. Biochimie 1996; 78:1043-1047.
22. Zhang J, Maquat LE. Evidence that the decay of nucleus-associated nonsense mRNA for human triosephosphate isomerase involves nonsense codon recognition after splicing. RNA 1996; 2:235-243.
23. Wang J, Hamilton JI, Carter MS et al. Alternatively spliced TCR mRNA induced by disruption of reading frame. Science 2002; 297:108-110.

24. Jack HM, Berg J, Wabl M. Translation affects immunoglobulin mRNA stability. Eur J Immunol 1989; 19:843-847.
25. Lytle JR, Steitz JA. Premature termination codons do not affect the rate of splicing of neighboring introns. RNA 2004; 10:657-668.
26. Weiss SL, Sunde RA. Cis-acting elements are required for selenium regulation of glutathione peroxidase-1 mRNA levels. RNA 1998; 4:816-827.
27. Moriarty PM, Reddy CC, Maquat LE. Selenium deficiency reduces the abundance of mRNA for Se-dependent glutathione peroxidase 1 by a UGA-dependent mechanism likely to be nonsense codon-mediated decay of cytoplasmic mRNA. Mol Cell Biol 1998; 18:2932-2939.
28. Sun X, Moriarty PM, Maquat LE. Nonsense-mediated decay of glutathione peroxidase 1 mRNA in the cytoplasm depends on intron position. EMBO J 2000; 19:4734-4744.
29. Le Hir H, Moore MJ, Maquat LE. Pre-mRNA splicing alters mRNP composition: Evidence for stable association of proteins at exon-exon junctions. Genes Dev 2000; 14:1098-1108.
30. Le Hir H, Izaurralde E, Maquat LE et al. The spliceosome deposits multiple proteins 20-24 nucleotides upstream of mRNA exon-exon junctions. EMBO J 2000; 19:6860-6869.
31. Amrani N, Ganesan R, Kervestin S et al. A faux 3'-UTR promotes aberrant termination and triggers nonsense-mediated mRNA decay. Nature 2004; 432:112-118.
32. Lejeune F, Maquat LE. Mechanistic links between nonsense-mediated mRNA decay and pre-mRNA splicing in mammalian cells. Curr Opin Cell Biol 2005; 17:309-315.
33. Perlick HA, Medghalchi SM, Spencer FA et al. Mammalian orthologues of a yeast regulator of nonsense transcript stability. Proc Natl Acad Sci USA 1996; 93:10928-10932.
34. Applequist SE, Selg M, Raman C et al. Cloning and characterization of hUPF1, a human homolog of the Saccharomyces cerevisiae nonsense mRNA-reducing UPF1 protein. Nucleic Acids Res 1997; 25:814-821.
35. Lykke-Andersen J, Shu MD, Steitz JA. Human Upf proteins target an mRNA for nonsense-mediated decay when bound downstream of a termination codon. Cell 2000; 103:1121-1131.
36. Mendell JT, Medghalchi SM, Lake RG et al. Novel Upf2p orthologues suggest a functional link between translation initiation and nonsense surveillance complexes. Mol Cell Biol 2000; 20:8944-8957.
37. Serin G, Gersappe A, Black JD et al. Identification and characterization of human orthologues to Saccharomyces cerevisiae Upf2 protein and Upf3 protein (Caenorhabditis elegans SMG-4). Mol Cell Biol 2001; 21:209-223.
38. Kim YK, Furic L, Desgroseillers L et al. Mammalian Staufen1 recruits Upf1 to specific mRNA 3'UTRs so as to elicit mRNA decay. Cell 2005; 120:195-208.
39. Dostie J, Dreyfuss G. Translation is required to remove Y14 from mRNAs in the cytoplasm. Curr Biol 2002; 12:1060-1067.
40. Pal M, Ishigaki Y, Nagy E et al. Evidence that phosphorylation of human Upfl protein varies with intracellular location and is mediated by a wortmannin-sensitive and rapamycin-sensitive PI 3-kinase-related kinase signaling pathway. RNA 2001; 7:5-15.
41. Denning G, Jamieson L, Maquat LE et al. Cloning of a novel phosphatidylinositol kinase-related kinase: Characterization of the human SMG-1 RNA surveillance protein. J Biol Chem 2001; 276:22709-22714.
42. Yamashita A, Ohnishi T, Kashima I et al. Human SMG-1, a novel phosphatidylinositol 3-kinase-related protein kinase, associates with components of the mRNA surveillance complex and is involved in the regulation of nonsense-mediated mRNA decay. Genes Dev 2001; 15:2215-2228.
43. Chiu SY, Serin G, Ohara O et al. Characterization of human Smg5/7a: A protein with similarities to Caenorhabditis elegans SMG5 and SMG7 that functions in the dephosphorylation of Upfl. RNA 2003; 9:77-87.
44. Gatfield D, Unterholzner L, Ciccarelli FD et al. Nonsense-mediated mRNA decay in Drosophila: At the intersection of the yeast and mammalian pathways. EMBO J 2003; 22:3960-3970.
45. Ohnishi T, Yamashita A, Kashima I et al. Phosphorylation of hUPF1 induces formation of mRNA surveillance complexes containing hSMG-5 and hSMG-7. Mol Cell 2003; 12:1187-1200.
46. Unterholzner L, Izaurralde E. SMG7 acts as a molecular link between mRNA surveillance and mRNA decay. Mol Cell 2004; 16:587-596.
47. Maquat LE, Li X. Mammalian heat shock p70 and histone H4 transcripts, which derive from naturally intronless genes, are immune to nonsense-mediated decay. RNA 2001; 7:445-456.
48. Brocke KS, Neu-Yilik G, Gehring NH et al. The human intronless melanocortin 4-receptor gene is NMD insensitive. Hum Mol Genet 2002; 11:331-335.
49. Carter MS, Li S, Wilkinson MF. A splicing-dependent regulatory mechanism that detects translation signals. EMBO J 1996; 15:5965-5975.

50. Inacio A, Silva AL, Pinto J et al. Nonsense mutations in close proximity to the initiation codon fail to trigger full nonsense-mediated mRNA decay. J Biol Chem 2004; 279:32170-32180.
51. Zhang J, Maquat LE. Evidence that translation reinitiation abrogates nonsense-mediated mRNA decay in mammalian cells. EMBO J 1997; 16:826-833.
52. Danckwardt S, Neu-Yilik G, Thermann R et al. Abnormally spliced beta-globin mRNAs: A single point mutation generates transcripts sensitive and insensitive to nonsense-mediated mRNA decay. Blood 2002; 99:1811-1816.
53. Chester A, Somasekaram A, Tzimina M et al. The apolipoprotein B mRNA editing complex performs a multifunctional cycle and suppresses nonsense-mediated decay. EMBO J 2003; 22:3971-3982.
54. Zhang Z, Krainer AR. Involvement of SR proteins in mRNA surveillance. Mol Cell 2004; 16:597-607.
55. Baumann B, Potash MJ, Kohler G. Consequences of frameshift mutations at the immunoglobulin heavy chain locus of the mouse. EMBO J 1985; 4:351-359.
56. Gudikote JP, Wilkinson MF. T-cell receptor sequences that elicit strong down-regulation of premature termination codon-bearing transcripts. EMBO J 2002; 21:125-134.
57. Ishigaki Y, Li X, Serin G et al. Evidence for a pioneer round of mRNA translation: mRNAs subject to nonsense-mediated decay in mammalian cells are bound by CBP80 and CBP20. Cell 2001; 106:607-617.
58. Lejeune F, Ishigaki Y, Li X et al. The exon junction complex is detected on CBP80-bound but not eIF4E-bound mRNA in mammalian cells: Dynamics of mRNP remodeling. EMBO J 2002; 21:3536-3545.
59. Chiu SY, Lejeune F, Ranganathan AC et al. The pioneer translation initiation complex is functionally distinct from but structurally overlaps with the steady-state translation initiation complex. Genes Dev 2004; 18:745-754.
60. Lejeune F, Ranganathan AC, Maquat LE. eIF4G is required for the pioneer round of translation in mammalian cells. Nat Struct Mol Biol 2004; 11:992-1000.
61. Maquat LE. Nonsense-mediated mRNA decay : A comparative analysis of different species. Current Genomics 2004; 5:175-190.
62. Gao Q, Das B, Sherman F et al. Cap-binding protein 1-mediated and eukaryotic translation initiation factor 4E-mediated pioneer rounds of translation in yeast. Proc Natl Acad Sci USA 2005; 102:4258-4263.
63. Hosoda N, Kim YK, Lejeune F et al. CBP80 promotes the interaction of Upf1 with Upf2 during nonsense-mediated mRNA decay in mammalian cells. Nat Struct Mol Biol 2005; in press.
64. Lejeune F, Li X, Maquat LE. Nonsense-mediated mRNA decay in mammalian cells involves decapping, deadenylating, and exonucleolytic activities. Mol Cell 2003; 12:675-687.
65. Lykke-Andersen J. Identification of a human decapping complex associated with hUpf proteins in nonsense-mediated decay. Mol Cell Biol 2002; 22:8114-8121.
66. Lehner B, Sanderson CM. A protein interaction framework for human mRNA degradation. Genome Res 2004; 14:1315-1323.
67. Chen CY, Shyu AB. Rapid deadenylation triggered by a nonsense codon precedes decay of the RNA body in a mammalian cytoplasmic nonsense-mediated decay pathway. Mol Cell Biol 2003; 23:4805-4813.
68. Couttet P, Grange T. Premature termination codons enhance mRNA decapping in human cells. Nucleic Acids Res 2004; 32:488-494.
69. Lim S, Mullins JJ, Chen CM et al. Novel metabolism of several beta zero-thalassemic beta-globin mRNAs in the erythroid tissues of transgenic mice. EMBO J 1989; 8:2613-2619.
70. Lim SK, Maquat LE. Human beta-globin mRNAs that harbor a nonsense codon are degraded in murine erythroid tissues to intermediates lacking regions of exon I or exons I and II that have a cap-like structure at the 5' termini. EMBO J 1992; 11:3271-3278.
71. Lim SK, Sigmund CD, Gross KW et al. Nonsense codons in human beta-globin mRNA result in the production of mRNA degradation products. Mol Cell Biol 1992; 12:1149-1161.
72. Stevens A, Wang Y, Bremer K et al. Beta-Globin mRNA decay in erythroid cells: UG site-preferred endonucleolytic cleavage that is augmented by a premature termination codon. Proc Natl Acad Sci USA 2002; 99:12741-12746.
73. Bremer KA, Stevens A, Schoenberg DR. An endonuclease activity similar to Xenopus PMR1 catalyzes the degradation of normal and nonsense-containing human beta-globin mRNA in erythroid cells. RNA 2003; 9:1157-1167.
74. Medghalchi SM, Frischmeyer PA, Mendell JT et al. Rent1, a trans-effector of nonsense-mediated mRNA decay, is essential for mammalian embryonic viability. Hum Mol Genet 2001; 10:99-105.

75. Maquat LE. Nonsense-mediated mRNA decay: Splicing, translation and mRNP dynamics. Nat Rev Mol Cell Biol 2004; 5:89-99.
76. Weng Y, Czaplinski K, Peltz SW. Identification and characterization of mutations in the UPF1 gene that affect nonsense suppression and the formation of the Upf protein complex but not mRNA turnover. Mol Cell Biol 1996; 16:5491-5506.
77. Weng Y, Czaplinski K, Peltz SW. Genetic and biochemical characterization of mutations in the ATPase and helicase regions of the Upf1 protein. Mol Cell Biol 1996; 16:5477-5490.
78. Czaplinski K, Weng Y, Hagan KW et al. Purification and characterization of the Upf1 protein: A factor involved in translation and mRNA degradation. RNA 1995; 1:610-623.
79. Czaplinski K, Ruiz-Echevarria MJ, Paushkin SV et al. The surveillance complex interacts with the translation release factors to enhance termination and degrade aberrant mRNAs. Genes Dev 1998; 12:1665-1677.
80. Le Hir H, Gatfield D, Izaurralde E et al. The exon-exon junction complex provides a binding platform for factors involved in mRNA export and nonsense-mediated mRNA decay. EMBO J 2001; 20:4987-4997.
81. Kim VN, Kataoka N, Dreyfuss G. Role of the nonsense-mediated decay factor hUpf3 in the splicing-dependent exon-exon junction complex. Science 2001; 293:1832-1836.
82. Lykke-Andersen J, Shu MD, Steitz JA. Communication of the position of exon-exon junctions to the mRNA surveillance machinery by the protein RNPS1. Science 2001; 293:1836-1839.
83. Kataoka N, Yong J, Kim VN et al. Pre-mRNA splicing imprints mRNA in the nucleus with a novel RNA-binding protein that persists in the cytoplasm. Mol Cell 2000; 6:673-682.
84. Shi H, Xu RM. Crystal structure of the Drosophila Mago nashi-Y14 complex. Genes Dev 2003; 17:971-976.
85. Degot S, Le Hir H, Alpy F et al. Association of the breast cancer protein MLN51 with the exon junction complex via its speckle localizer and RNA binding module. J Biol Chem 2004; 279:33702-33715.
86. Bono F, Ebert J, Unterholzner L et al. Molecular insights into the interaction of PYM with the Mago-Y14 core of the exon junction complex. EMBO Rep 2004; 5:304-310.
87. Chan CC, Dostie J, Diem MD et al. eIF4A3 is a novel component of the exon junction complex. RNA 2004; 10:200-209.
88. Ferraiuolo MA, Lee CS, Ler LW et al. A nuclear translation-like factor eIF4AIII is recruited to the mRNA during splicing and functions in nonsense-mediated decay. Proc Natl Acad Sci USA 2004; 101:4118-4123.
89. Palacios IM, Gatfield D, St Johnston D et al. An eIF4AIII-containing complex required for mRNA localization and nonsense-mediated mRNA decay. Nature 2004; 427:753-757.
90. Shibuya T, Tange TO, Sonenberg N et al. eIF4AIII binds spliced mRNA in the exon junction complex and is essential for nonsense-mediated decay. Nat Struct Mol Biol 2004; 11:346-351.
91. Tanner NK, Cordin O, Banroques J et al. The Q motif: A newly identified motif in DEAD box helicases may regulate ATP binding and hydrolysis. Mol Cell 2003; 11:127-138.
92. Li C, Lin RI, Lai MC et al. Nuclear Pnn/DRS protein binds to spliced mRNPs and participates in mRNA processing and export via interaction with RNPS1. Mol Cell Biol 2003; 23:7363-7376.
93. Luo ML, Zhou Z, Magni K et al. Pre-mRNA splicing and mRNA export linked by direct interactions between UAP56 and Aly. Nature 2001; 413:644-647.
94. Izaurralde E, McGuigan C, Mattaj IW. Nuclear localization of a cap-binding protein complex. Cold Spring Harb Symp Quant Biol 1995; 60:669-675.
95. Lewis JD, Izaurralde E. The role of the cap structure in RNA processing and nuclear export. Eur J Biochem 1997; 247:461-469.

CHAPTER 5

Human Upf Proteins in NMD

Guramrit Singh and Jens Lykke-Andersen*

Abstract

The human Upf (hUpf) proteins work at the core of the nonsense-mediated mRNA decay (NMD) pathway. The three hUpf proteins, hUpf1, hUpf2 and hUpf3, form the hUpf complex, which is critical for the recognition and degradation of mRNAs containing premature termination codons (PTCs). The recognition of PTC-containing mRNAs by the hUpf complex in mammalian cells is promoted by the splicing dependent exon-junction complex (EJC), with which hUpf3 interacts. Following the recognition of PTCs, the hUpf complex is believed to disrupt mRNP structure to prevent further translation and trigger mRNA decay. Emerging evidence suggests that hSmg proteins involved in phosphorylation and dephosphorylation of hUpf1 may play a key role in delivering PTC-containing mRNAs to the mRNA decay machinery.

Introduction

In recent years the cellular machinery that identifies mRNAs with premature termination codons (PTCs) and subjects them to NMD has been characterized in several eukaryotes. The NMD machinery has the capability to discriminate PTC-containing mRNAs from normal mRNAs, and to inhibit translation and activate decay of the NMD target mRNAs. The three Upf (*Up-f*rameshift) proteins, Upf1, Upf2 and Upf3, work at the heart of this pathway in all organisms studied. The Upf proteins were first discovered in yeast,[1-3] and orthologs have subsequently been identified in other eukaryotes.[4] The conservation across species is highest for Upf1 and lower for Upf2 and Upf3.[4] However, despite this conservation, mechanistic differences may exist between various organisms in how PTCs are recognized by the NMD machinery.[5] In this chapter, we discuss the current understanding of the role of Upf proteins in the recognition of mRNAs that are targeted for NMD in humans, and how these proteins may shunt the mRNA from the translational pool to the mRNA decay machinery.

Evidence That Upf Proteins Are Involved in the Human NMD Pathway

It is now well established that human Upf (referred to as hUpf hereafter) proteins play an essential role in NMD. Initial evidence came from identification of the human ortholog to yeast Upf1p based on sequence similarity.[6,7] Several residues of yeast Upf1p that are essential for NMD were found to be conserved in hUpf1. An arginine-to-cysteine mutation at residue 844 was shown to create a dominant-negative form of hUpf1 that impairs NMD in human cells.[8] A similar mutation was earlier shown to inhibit NMD in yeast.[9] Subsequently, hUpf2 and hUpf3, the human orthologs to the other two essential yeast NMD proteins, were identified. The demonstration that they interact with hUpf1 implicated these two proteins in the human NMD pathway as well.[10-12]

*Corresponding Author: Jens Lykke-Andersen—Molecular, Cellular and Developmental Biology, 347 UCB, University of Colorado at Boulder, Boulder, Colorado 80309-0347, U.S.A. Email: jens.lykke-andersen@colorado.edu

Further evidence that hUpf proteins function in NMD was obtained by showing that each hUpf protein induces the NMD-like decay of mRNAs to which they are artificially tethered.[12] This gain-of-function manipulation of hUpf proteins triggers mRNA decay only when the proteins are tethered downstream of a translation termination codon.[12] More recently, RNA interference-mediated depletion of hUpf proteins from human cells was shown to inhibit NMD, demonstrating that these proteins are essential for NMD.[13,14]

hUpf1 Is a RNA Helicase and a Phosphoprotein

Among the three core NMD proteins, Upf1 shows the highest sequence conservation.[4] The central region of hUpf1 is 58% identical to yeast Upf1p[6] and, like yeast Upf1p, it is an ATP-dependent RNA helicase.[15] The conserved central region consists of two putative cysteine-rich zinc-finger motifs and seven group I helicase motifs (Fig. 1). The role of the zinc-finger motifs in NMD is not known. The ATPase activity of the protein resides in helicase motifs Ia and II (Fig. 1), and is linked to the essential 5' to 3' helicase activity of the protein.[15] Mutated variants of hUpf1 that lack ATPase activity due to mutation of highly conserved aspartate and glutamate residues in helicase motif II show no double-stranded RNA unwinding activity[15] and are inactive in NMD in yeast.[16]

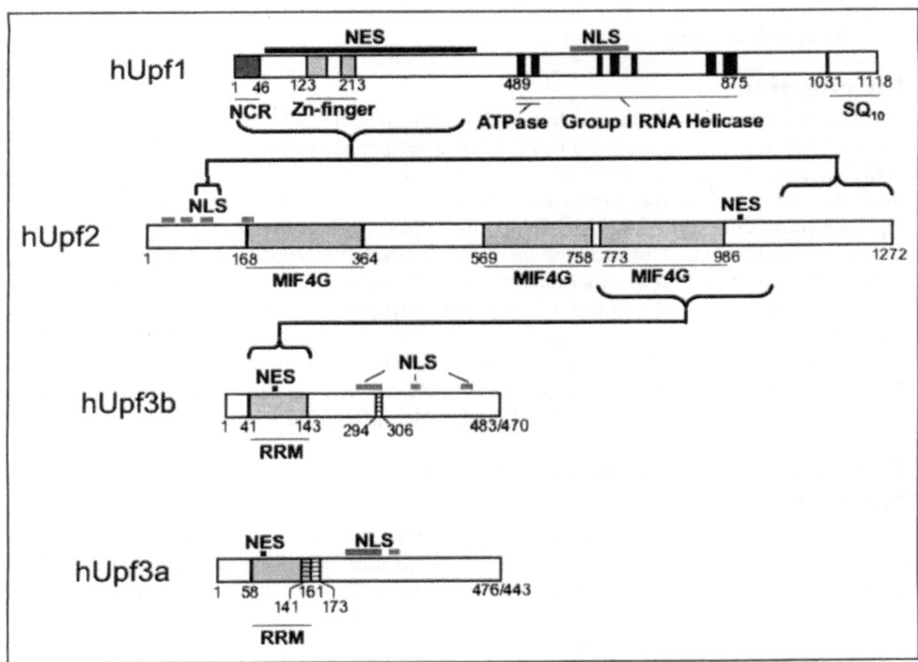

Figure 1. Schematic representations of human Upf proteins. Specific domains are shaded grey or black for each protein. A black bar on top of each representation indicates a putative nuclear export signal (NES) and a grey bar represents a putative nuclear localization signal (NLS). Regions of interactions between hUpf proteins are indicated by brackets with connecting lines. Amino-acids 141-173 corresponding to the alternatively spliced exon 4 of hUpf3a, and amino-acids 294-306 corresponding to the alternatively spliced exon 8 of hUpf3b are shown by striped rectangles. NCR: N-terminal conserved region, found in Upf1 proteins of all metazoans that have been analyzed; SQ_{10}: region containing ten SQ motifs; MIF4G: middle portion of eIF4G-like domain; RRM: RNA recognition motif.

The C-terminus of hUpf1 contains several serine/glutamine (SQ) and serine/glutamine/proline (SQP) repeats (Fig. 1). Multiple serines (S1073, S1078, S1096 and S1116) in SQ repeats are targets of phosphorylation by hSmg1, a phosphatidylinositol 3-kinase related protein kinase involved in NMD (see chapter by Yamashita et al).[17] The N-terminus of hUpf1 contains a proline-glycine rich region, and it is also rich in negatively charged residues. An N-terminal region of hUpf1 (amino acids 1-46), which is conserved among metazoans but is not found in yeast, has recently been shown to mediate interaction with Smg5, a protein that forms part of a complex responsible for dephosphorylating Upf1 in both humans and *C. elegans* (see chapters by Yamashita et al and Anderson).[18-20] Thus, the N- and C-termini of hUpf1, which are missing from yeast Upf1p, are involved in regulating the protein by a phosphorylation and dephosphorylation cycle.

hUpf1 primarily localizes to the cytoplasm.[11,12] However, evidence shows that hUpf1 shuttles in and out of the nucleus, and its export out of the nucleus is mediated by Crm1.[14] A conventional nuclear export signal (NES) or nuclear localization signal (NLS) has not been identified in hUpf1. However, specific regions of hUpf1 have been shown to possess NLS (amino acids 596-697) and NES (amino acids 55-416) function (Fig. 1).[14]

hUpf2 and hUpf3 Function in NMD

hUpf2 interacts with both hUpf1 and hUpf3.[10-12] hUpf2 contains three conserved middle of eIF4G-like (MIF4G) domains with similarity to a domain in the middle of eukaryotic translation initiation factor 4G(eIF4G) (Fig. 1). The region overlapping with the last of the three MIF4G domains is responsible for interaction with hUpf3.[10,11] The importance of the MIF4G domains in NMD has been studied in *Schizosaccharomyces pombe* where mutations in the phylogenetically conserved "FIGEL" motif of the last two MIF4G domains of Upf2 inhibit NMD.[10] In addition to the MIF4G domains, hUpf2 contains more than one putative NLS in its N-terminus. Even though the N-terminal region of hUpf2 (amino acids 1-120) can target a heterologous protein to the nucleus,[10] it is currently unknown whether hUpf2, like hUpf1, is a nucleocytoplasmic shuttling protein (Fig. 1).[10-12] However, at steady state, hUpf2 is cytoplasmic and concentrated in the perinuclear region.[11,12]

hUpf3 is the least conserved component of the hUpf proteins.[4] The human genome contains two different hUPF3 genes, encoding hUpf3a (also known as hUpf3) and hUpf3b (also known as hUpf3X since the corresponding gene maps to the X-chromosome), respectively.[11,12] hUpf3a and hUpf3b transcripts are alternatively spliced to encode two isoforms for each protein (Fig. 1). hUpf3 proteins are nucleocytoplasmic shuttling proteins that contain putative NLS and NES motifs and primarily localize in the nucleus.[11,12] The hUpf3 proteins also contain an RNA recognition motif (RRM) in the N-terminal region (Fig. 1). The hUpf3 RRM, which was originally hypothesized to constitute an RNA-binding surface,[12] lacks critical aromatic amino acid residues for RNA-binding.[21] Recently, the RRM domain of hUpf3b was shown to constitute the surface that interacts with hUpf2 (see below).[21]

Interactions between hUpf Proteins

The three yeast Upf proteins are known to form a complex in which Upf2p serves as a bridge between Upf1p and Upf3p.[2] A similar complex exists between hUpf proteins.[10-12] The regions of hUpf proteins that mediate these interactions were predicted based on mapped interaction domains of yeast Upf proteins, and tested by deletion analyses. The hUpf2-interaction domain of hUpf1 is found in the N-terminus spanning the zinc-finger domain (amino acids 1-415). This conserved N-terminal region interacts with the C-terminal region of hUpf2 (amino acids 1084-1272) (Fig. 1). However, a hUpf2 deletion protein lacking these residues shows residual weak interaction with hUpf1,[10,11] possibly mediated by an N-terminal region of hUpf2 (amino acids 94-133).[11] Thus, there are two potential hUpf1-interacting regions in hUpf2.

Deletion analysis mapped the region of hUpf2 necessary for interaction with hUpf3 to its third MIF4G domain, while the hUpf2-interacting surface of hUpf3 overlaps with its RRM motif (Fig. 1).[11,12] A recent crystal structure of this MIF4G domain of hUpf2 in complex with the RRM domain of hUpf3b provides detailed insight into how hUpf2 interacts with hUpf3b.[21] The α-helices formed by the MIF4G domain interact with the β-sheet surface of the hUpf3b RRM domain. The interaction between hUpf2 and hUpf3b is mediated by positively charged residues of the ribonucleoprotein motif 2 (RNP2) of the hUpf3b RRM domain and negatively charged residues in hUpf2. Specific mutations in these residues abolish complex formation.[21]

Even though the interaction between Upf3 and Upf2 is conserved between species, hUpf3b lacking its hUpf2-interacting domain is capable of destabilizing an mRNA when tethered downstream of a stop codon.[22] This observation suggests that hUpf3b does not require an interaction with hUpf2 to trigger tethered decay of an mRNA. However, it is unknown if this is also true in the natural NMD pathway.

Interactions between hUpf Proteins and the Translation Machinery

Translation is essential for NMD.[23-25] NMD substrates need to undergo at least one round of translation during which the PTC is recognized. Evidence suggests that this occurs during a first, so-called pioneer, round of translation (see chapter by Maquat).[26-29] However, experiments in yeast show that while NMD may normally take place during a pioneer round of translation,[30] this is not an absolute requirement, because an mRNA that has undergone several rounds of translation can be a target of NMD (see chapters by Baker and Parker, and Amrani and Jacobson).[31]

The key to understanding how the hUpf complex functions in NMD will be to understand how it communicates with the translational machinery. Translation initiation is the rate-limiting and the most tightly regulated step of translation. The translation factor eIF4G acts as a scaffold for assembly of the translation initiation complex and plays a central role in translation initiation. eIF4G interacts with the mRNA cap binding proteins, CBC (cap-binding complex)[28,32] and eIF4E[33], which predominantly localize to the nucleus and cytoplasm, respectively (Fig. 2).[29,33] Evidence in mammalian cells suggests that NMD is associated with CBC-initiated translation (see chapter by Maquat).[26,28] eIF4G helps recruit the 40S ribosomal subunit via the eIF3 complex and other associated factors.[33,34] The release factors eRF1 and eRF3 are essential for translation termination. eRF1 recognizes the stop codon in the A-site of the ribosome and catalyzes the release of the nascent peptide.[31] eRF3 may promote release of eRF1 from the ribosome and ribosome recycling via its interaction with poly(A) binding protein (PABP) C1.[33,35] PABPC1 also interacts with eIF4G at the cap to circularize the mRNA, a key feature of efficient translation in eukaryotes.[33]

Several interactions between hUpf proteins and the translation factors described above have been reported. Both yeast and human Upf1 proteins have been shown to interact with the yeast translation termination factors eRF1 and eRF3.[36] However, the role of this critical interaction in discriminating a premature from a normal termination codon is poorly understood. In yeast, Upf proteins have been shown to play a role in translation termination. Yeast strains depleted of Upf proteins are reported to show increased readthrough at termination codons,[37-39] although conflicting evidence about the role of yeast Upf proteins in translation termination exists.[39,40] A similar role for human Upf proteins has not been tested. The interplay between hUpf1 and eRFs 1 and 3 likely holds a key to understanding the mechanism of PTC-containing mRNA recognition and should be an important topic for future studies.

hUpf1 has also been found to interact with PABPC1,[41] but the significance of this interaction in NMD is unknown. Another NMD factor that may link the hUpf complex to the translation process is hUpf2. Two-hybrid assays indicate that the third middle portion of the eIF4G (MIF4G) domain of hUpf2 can interact with the translation initiation factors eIF4AI and hSui1, the latter of which is an eIF3 subunit.[10] Since the MIF4G domain of eIF4G interacts with the same factors, it has been speculated that hUpf2 competes with eIF4G for MIF4G-mediated

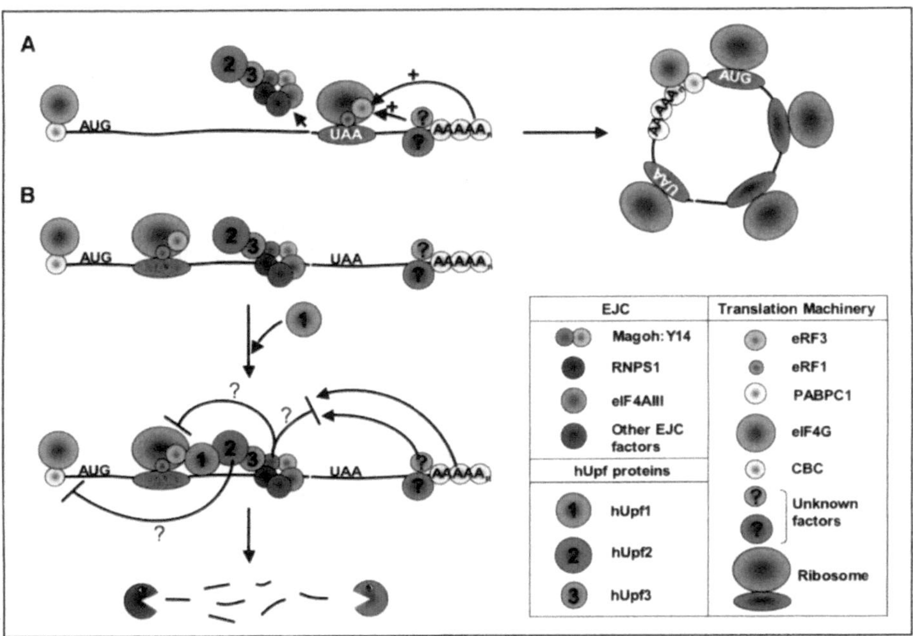

Figure 2. Models of events after translation termination on normal and PTC-containing mRNAs in mammalian cells. A) On normal mRNAs, the absence of an exon junction complex (EJC) downstream of the termination codon at the time of translation termination promotes efficient translation termination. This may require interactions (indicated by arrows) between PABPC1 and eRF3 and potentially other unknown factors as indicated. An efficient translation termination event channels normal mRNAs into the translational pool (shown on the right). B) A translation termination event at a PTC (5'-most UAA, shown in red) that is more than ~50-55 nucleotides upstream of the last exon-exon junction leads to NMD. An EJC that remains bound to the mRNA 3'-untranslated region (UTR) at the time of translation termination may promote inefficient termination, perhaps due to the inhibition (indicated by ⊥) of interactions between the termination complex and PABPC1 or other unknown factors the at 3'UTR. The recruitment of the hUpf complex triggers degradation by the mRNA decay machinery. The code for specific proteins is given on the right. The identities of mRNA decay factors (specified by pac-men), are at present only poorly defined and not shown.

interactions, so as to inhibit translation initiation on NMD-target mRNAs (Fig. 2).[10] However, there is no direct evidence to support this otherwise attractive model of hUpf2 function in NMD. It is possible that the interactions of hUpf proteins with translation initiation factors and PABPC1 could destabilize the closed-loop structure of the mRNP and thereby inhibit translation and expose the mRNA cap and poly(A) tail to decay enzymes.[10]

hUpf Proteins Are Recruited to mRNAs

A major unanswered question is how hUpf proteins are recruited to mRNAs. Intriguingly, hUpf proteins concentrate in different sub-cellular compartments: hUpf1 and hUpf2 proteins in the cytoplasm, and hUpf3 proteins primarily in the nucleus.[11,12] Nevertheless, each hUpf protein should be recruited to a target mRNA before it can undergo NMD. This observation suggests that the assembly of the hUpf complex on mRNA targets is a dynamic process.

Several lines of evidence indicate that hUpf3 proteins are recruited to mRNAs in a splicing-dependent manner. The hUpf3 proteins associate specifically with mRNAs that have undergone splicing in human cells.[12,42-44] Moreover, *Xenopus* Upf3 specifically interacts with spliced RNA in *Xenopus* oocyte extracts.[45] A hallmark of splicing in mammalian cells is the deposition of an exon-junction complex (EJC) ~20-24 nucleotides upstream of exon-exon junctions after splicing (see chapter by Maquat).[46] The interaction of hUpf3 with spliced mRNA was shown to be mediated by the EJC.[43-45] The hUpf3 proteins coimmunoprecipitate with core EJC subunits, RNPS1,[44] Y14,[43] and eIF4AIII (G. Singh and J. Lykke-Andersen, unpublished data), although it is not clear which interactions are direct.[43,47] The interaction of hUpf3b with Y14 has been studied in most detail and depends on a 14-amino acid conserved region in the C-terminal region of hUpf3b.[22] Deletion of this conserved stretch or an arginine-to-alanine mutation at residue 423 abolishes the ability of hUpf3b to form a complex with Y14 in cell extracts.[22] It is currently unknown whether hUpf3b interacts with other EJC components through the same C-terminal region.

hUpf2, like hUpf3, associates with spliced mRNA,[45] and is detected in mRNPs purified with antibodies against the nuclear cap binding protein, which also contain EJC components and hUpf3.[26,27] This suggests that hUpf2 may be recruited to the mRNA via the EJC and hUpf3. hUpf2 may also interact with mRNA directly. The crystal structure of the third MIF4G domain of hUpf2 revealed a basic patch that possesses RNA binding activity. This RNA-binding region of the MIF4G domain is not part of the hUpf3b-binding surface, and directly binds RNA in vitro with or without hUpf3b.[21] However, the significance of this RNA binding property of hUpf2 to the NMD pathway is unknown.

Unlike hUpf3 and hUpf2, hUpf1 has not been reported to specifically associate with spliced mRNAs. The cellular pool of hUpf1 evenly distributes between polysomes, a sub-polysomal fraction and a free cytoplasmic fraction.[48] How hUpf1 is recruited to mRNAs is not clear, and will be an important topic for future studies. eRF1 and eRF3,[36] hUpf2 (as a part of the EJC:hUpf3 complex) or PABPC1[41] or CBC (N. Hosoda and L.E. Maquat, personal communication) are all candidates for proteins responsible for recruiting hUpf1 to target mRNAs. In addition, hUpf1 can itself bind RNA via its RNA helicase motif. The RNA binding activity of hUpf1 is modulated by ATP, a cofactor of hUpf1.[15] The ability of hUpf1 to bind RNA is reduced in the presence of ATP, suggesting that binding of ATP to hUpf1 would cause the protein to dissociate from RNA while hydrolysis of bound ATP would allow it to bind RNA.[15,49] However, the specific role in NMD of the RNA helicase function of hUpf1 is currently unclear.

These observations taken together have led to the hypothesis that hUpf3 and hUpf2 are recruited to mRNAs through splicing via the EJC, whereas hUpf1 may be recruited by the hUpf2:hUpf3:EJC complex or alternatively via eRFs after translation termination (Fig. 2). Another possibility, which is more in line with the role of Upf proteins in translation termination in yeast, is that the hUpf proteins all enter the mRNA with the termination complex. However, these models need more experimental testing, and the question still remains how the hUpf proteins, once recruited to mRNA, help distinguish NMD targets from normal stable mRNAs.

hUpf Proteins Help Discriminate PTC-Containing from Normal mRNAs

The key step in the NMD pathway is the discrimination between a normal termination codon and a PTC that triggers. The high degree of conservation in the translation process between eukaryotes would suggest that the basic mechanism of PTC recognition is similar in all eukaryotes. Yet, only mammalian cells appear to rely on pre-mRNA splicing for recognition of PTCs. Evidence suggests that in the majority of cases, PTCs in mammalian cells are recognized when they are situated more than 50-55 nucleotides upstream of last exon-exon junction[24,50-53] in a way that depends on the EJC that is deposited upstream of the exon-exon junctions (see chapter by Maquat). By contrast, no EJC has been observed in yeast, and in flies the EJC plays no apparent role in NMD (see chapter by Behm-Ansmant and Izaurralde).[54] Evidence in yeast suggests that translation termination at a PTC is kinetically different from translation termination at a normal stop codon and depends on the nature of the downstream

3' untranslated region (UTR) (see chapter by Amrani and Jacobson).[55] Moreover, a PTC is recognized as a normal termination codon when poly(A) binding protein (Pab1p) is tethered 40-75 nucleotides downstream.[55] These data are consistent with a so-called faux 3'UTR model, in which a normal termination event takes place only in the presence of a normal 3'UTR whereas an aberrant 3'UTR triggers abnormal termination and NMD (see chapter by Amrani and Jacobson).[55-57] Although similar experiments have yet to be performed in mammalian cells, the conservation of the translational machinery and the Upf complex, and the observation that hUpf1 associates with PABPC1 and eRFs 1 and 3, suggests that mammalian NMD may function in a similar manner.

How could these observations be reconciled with the involvement of the EJC in mammalian NMD? Perhaps in mammalian cells, the presence of an EJC that remains bound to an mRNA 3'UTR after translation termination due to the presence of an exon-exon junction more than 50-55 nucleotides downstream of the PTC[50-52,58] signals an aberrant 3'UTR, which negatively influences the termination process (Fig. 2). The eRFs have been shown in both yeast and humans to associate with poly(A) binding protein.[35,59] Possibly, an EJC that remains bound to an mRNA 3'UTR after a premature translation termination event disrupts the communication between these translation factors, which in turn influences translation termination. hUpf3 may have evolved to acquire interactions with EJC components so as to link splicing with NMD in mammalian cells. Consistent with this hypothesis the Y14-binding region of hUpf3 is not conserved in yeast.[22] Clearly, more experiments are needed to understand how PTCs are recognized in mammalian cells and how this may differ from, or resemble the mechanism in other organisms.

hUpf Proteins Trigger mRNA Decay

The last step in the elimination of an mRNA that harbors a PTC is its destruction by the cellular mRNA decay machinery. In yeast, where the events in degradation of NMD substrates are best understood, decay mainly proceeds from the 5' end (see chapter by Baker and Parker). By contrast, in *Drosophila*, the decay of NMD substrates is initiated by endonucleolytic cleavage followed by exonucleolytic decay of the 5' and 3' fragments (see chapter by Behm-Ansmant and Izaurralde).[60]

The degradation of NMD substrates in human cells is poorly understood but has been found to depend on enzymes involved in decapping, deadenylation, and 5' to 3' and 3' to 5' exonucleolytic decay.[61] In addition, rapid deadenylation of NMD substrates has been detected.[62] hUpf1 has been found in complex with the human decapping proteins, hDcp1 and hDcp2,[61,63] as well as with components of the deadenylation machinery, and the 5' to 3' and 3' to 5' exonucleolytic decay machineries.[61] Preliminary evidence suggests that decay of the body of the transcript may proceed from both ends with similar kinetics.[61] However, the relative contribution of specific mRNA decay enzymes in human NMD is poorly understood. Moreover, it is unknown whether an endonucleolytic cleavage takes place, although endonucleolytic intermediates of PTC-containing β-globin mRNA have been detected in mouse erythroid cells (see chapter by Maquat).[64-68]

How does the hUpf complex, once assembled onto PTC-containing mRNAs, communicate with the mRNA decay machinery? Although little is known about this process, new evidence suggests a link between a phosphorylation-dephosphorylation cycle of hUpf1 and the delivery of PTC-containing mRNAs to the decay machinery.[5,69,70] The proteins hSmg5 and hSmg7 have been found in complexes containing phosphorylated hUpf1 and may mediate hUpf1 dephosphorylation via protein phosphatase 2A (see chapters by Yamashita et al and Anderson).[18,19] When exogenously expressed, these proteins localize to sub-cytoplasmic structures enriched in mRNA decay factors, called processing bodies.[69] Evidence in yeast and human cells suggests that processing bodies constitute cytoplasmic sites of mRNA decay.[71,72] Exogenous expression of hSmg7 also leads to detection of coexpressed hUpf1 protein in processing bodies.[69] Moreover, this hSmg7-mediated recruitment of hUpf1 to processing bodies was disrupted by mutation of residues in hSmg7 critical for binding to hUpf1 in vitro.[70] This suggests that the complex that associates with phosphorylated hUpf1 and triggers dephosphorylation may also deliver PTC-containing mRNAs to the mRNA decay machinery. However, this model needs to be further tested in future experiments.

Conclusions and Future Perspectives

The hUpf complex is essential to human NMD. However, the mechanism by which this complex helps recognize PTC-containing mRNAs and divert them from the translational pool to the mRNA decay machinery is poorly understood. Key issues that need to be addressed in the future are how the hUpf proteins, which concentrate in different sub-cellular compartments, are recruited to mRNA targets, and what events lead to recognition of a termination codon as a PTC. Experiments aimed at studying the communication between the hUpf complex and translation termination factors should be critical to understanding these processes. Another important goal is to understand how mRNA decay factors are recruited to PTC-containing mRNAs. Future research should provide insight into these issues and help to elucidate the unsolved mysteries of NMD.

Acknowledgements

This work was supported by the National Science Foundation (NSF) grant 0328888 to J. Lykke-Andersen.

References

1. Cui Y, Hagan KW, Zhang S et al. Identification and characterization of genes that are required for the accelerated degradation of mRNAs containing a premature translational termination codon. Genes Dev 1995; 9(4):423-436.
2. He F, Brown AH, Jacobson A. Upf1p, Nmd2p, and Upf3p are interacting components of the yeast nonsense-mediated mRNA decay pathway. Mol Cell Biol 1997; 17(3):1580-1594.
3. Leeds P, Peltz SW, Jacobson A et al. The product of the yeast UPF1 gene is required for rapid turnover of mRNAs containing a premature translational termination codon. Genes Dev 1991; 5(12A):2303-2314.
4. Culbertson MR, Leeds PF. Looking at mRNA decay pathways through the window of molecular evolution. Curr Opin Genet Dev 2003; 13(2):207-214.
5. Conti E, Izaurralde E. Nonsense-mediated mRNA decay: Molecular insights and mechanistic variations across species. Curr Opin Cell Biol 2005; 17(3):316-325.
6. Perlick HA, Medghalchi SM, Spencer FA et al. Mammalian orthologues of a yeast regulator of nonsense transcript stability. Proc Natl Acad Sci USA 1996; 93(20):10928-10932.
7. Applequist SE, Selg M, Raman C et al. Cloning and characterization of HUPF1, a human homolog of the Saccharomyces cerevisiae nonsense mRNA-reducing UPF1 protein. Nucleic Acids Res 1997; 25(4):814-821.
8. Sun X, Perlick HA, Dietz HC et al. A mutated human homologue to yeast Upf1 protein has a dominant-negative effect on the decay of nonsense-containing mRNAs in mammalian cells. Proc Natl Acad Sci USA 1998; 95(17):10009-10014.
9. Leeds P, Wood JM, Lee BS et al. Gene products that promote mRNA turnover in Saccharomyces cerevisiae. Mol Cell Biol 1992; 12(5):2165-2177.
10. Mendell JT, Medghalchi SM, Lake RG et al. Novel Upf2p orthologues suggest a functional link between translation initiation and nonsense surveillance complexes. Mol Cell Biol 2000; 20(23):8944-8957.
11. Serin G, Gersappe A, Black JD et al. Identification and characterization of human orthologues to Saccharomyces cerevisiae Upf2 protein and Upf3 protein (Caenorhabditis elegans SMG-4). Mol Cell Biol 2001; 21(1):209-223.
12. Lykke-Andersen J, Shu MD, Steitz JA. Human Upf proteins target an mRNA for nonsense-mediated decay when bound downstream of a termination codon. Cell 2000; 103(7):1121-1131.
13. Kim YK, Furic L, Desgroseillers L et al. Mammalian Staufen1 recruits Upf1 to specific mRNA 3'UTRs so as to elicit mRNA decay. Cell 2005; 120(2):195-208.
14. Mendell JT, ap Rhys CM, Dietz HC. Separable roles for rent1/hUpf1 in altered splicing and decay of nonsense transcripts. Science 2002; 298(5592):419-422.

15. Bhattacharya A, Czaplinski K, Trifillis P et al. Characterization of the biochemical properties of the human Upfl gene product that is involved in nonsense-mediated mRNA decay. RNA 2000; 6(9):1226-1235.
16. Weng Y, Czaplinski K, Peltz SW. Genetic and biochemical characterization of mutations in the ATPase and helicase regions of the Upfl protein. Mol Cell Biol 1996; 16(10):5477-5490.
17. Yamashita A, Ohnishi T, Kashima I et al. Human SMG-1, a novel phosphatidylinositol 3-kinase-related protein kinase, associates with components of the mRNA surveillance complex and is involved in the regulation of nonsense-mediated mRNA decay. Genes Dev 2001; 15(17):2215-2228.
18. Ohnishi T, Yamashita A, Kashima I et al. Phosphorylation of hUPF1 induces formation of mRNA surveillance complexes containing hSMG-5 and hSMG-7. Mol Cell 2003; 12(5):1187-1200.
19. Anders KR, Grimson A, Anderson P. SMG-5, required for C. elegans nonsense-mediated mRNA decay, associates with SMG-2 and protein phosphatase 2A. EMBO J 2003; 22(3):641-650.
20. Chiu SY, Serin G, Ohara O et al. Characterization of human Smg5/7a: A protein with similarities to Caenorhabditis elegans SMG5 and SMG7 that functions in the dephosphorylation of Upfl. RNA 2003; 9(1):77-87.
21. Kadlec J, Izaurralde E, Cusack S. The structural basis for the interaction between nonsense-mediated mRNA decay factors UPF2 and UPF3. Nat Struct Mol Biol 2004; 11(4):330-337.
22. Gehring NH, Neu-Yilik G, Schell T et al. Y14 and hUpf3b form an NMD-activating complex. Mol Cell 2003; 11(4):939-949.
23. Thermann R, Neu-Yilik G, Deters A et al. Binary specification of nonsense codons by splicing and cytoplasmic translation. EMBO J 1998; 17(12):3484-3494.
24. Carter MS, Doskow J, Morris P et al. A regulatory mechanism that detects premature nonsense codons in T-cell receptor transcripts in vivo is reversed by protein synthesis inhibitors in vitro. J Biol Chem 1995; 270(48):28995-29003.
25. Belgrader P, Cheng J, Maquat LE. Evidence to implicate translation by ribosomes in the mechanism by which nonsense codons reduce the nuclear level of human triosephosphate isomerase mRNA. Proc Natl Acad Sci USA 1993; 90(2):482-486.
26. Ishigaki Y, Li X, Serin G et al. Evidence for a pioneer round of mRNA translation: mRNAs subject to nonsense-mediated decay in mammalian cells are bound by CBP80 and CBP20. Cell 2001; 106(5):607-617.
27. Lejeune F, Ishigaki Y, Li X et al. The exon junction complex is detected on CBP80-bound but not eIF4E-bound mRNA in mammalian cells: Dynamics of mRNP remodeling. EMBO J 2002; 21(13):3536-3545.
28. Chiu SY, Lejeune F, Ranganathan AC et al. The pioneer translation initiation complex is functionally distinct from but structurally overlaps with the steady-state translation initiation complex. Genes Dev 2004; 18(7):745-754.
29. Lejeune F, Ranganathan AC, Maquat LE. eIF4G is required for the pioneer round of translation in mammalian cells. Nat Struct Mol Biol 2004; 11(10):992-1000.
30. Gao Q, Das B, Sherman F et al. Cap-binding protein 1-mediated and eukaryotic translation initiation factor 4E-mediated pioneer rounds of translation in yeast. Proc Natl Acad Sci USA 2005; 102(12):4258-4263.
31. Maderazo AB, Belk JP, He F et al. Nonsense-containing mRNAs that accumulate in the absence of a functional nonsense-mediated mRNA decay pathway are destabilized rapidly upon its restitution. Mol Cell Biol 2003; 23(3):842-851.
32. McKendrick L, Thompson E, Ferreira J et al. Interaction of eukaryotic translation initiation factor 4G with the nuclear cap-binding complex provides a link between nuclear and cytoplasmic functions of the m(7) guanosine cap. Mol Cell Biol 2001; 21(11):3632-3641.
33. Kapp LD, Lorsch JR. The molecular mechanics of eukaryotic translation. Annu Rev Biochem 2004; 73:657-704.
34. Pestova TV, Kolupaeva VG. The roles of individual eukaryotic translation initiation factors in ribosomal scanning and initiation codon selection. Genes Dev 2002; 16(22):2906-2922.
35. Hoshino S, Imai M, Kobayashi T et al. The eukaryotic polypeptide chain releasing factor (eRF3/GSPT) carrying the translation termination signal to the 3'-poly(A) tail of mRNA. Direct association of eRF3/GSPT with polyadenylate-binding protein. J Biol Chem 1999; 274(24):16677-16680.
36. Czaplinski K, Ruiz-Echevarria MJ, Paushkin SV et al. The surveillance complex interacts with the translation release factors to enhance termination and degrade aberrant mRNAs. Genes Dev 1998; 12(11):1665-1677.
37. Weng Y, Czaplinski K, Peltz SW. Identification and characterization of mutations in the UPF1 gene that affect nonsense suppression and the formation of the Upf protein complex but not mRNA turnover. Mol Cell Biol 1996; 16(10):5491-5506.

38. Wang W, Czaplinski K, Rao Y et al. The role of Upf proteins in modulating the translation read-through of nonsense-containing transcripts. EMBO J 2001; 20(4):880-890.
39. Keeling KM, Lanier J, Du M et al. Leaky termination at premature stop codons antagonizes nonsense-mediated mRNA decay in S. cerevisiae. RNA 2004; 10(4):691-703.
40. Harger JW, Dinman JD. Evidence against a direct role for the Upf proteins in frameshifting or nonsense codon readthrough. RNA 2004; 10(11):1721-1729.
41. Schell T, Kocher T, Wilm M et al. Complexes between the nonsense-mediated mRNA decay pathway factor human upf1 (up-frameshift protein 1) and essential nonsense-mediated mRNA decay factors in HeLa cells. Biochem J 2003; 373(Pt 3):775-783.
42. Lykke-Andersen J. mRNA quality control: Marking the message for life or death. Curr Biol 2001; 11(3):R88-91.
43. Kim VN, Kataoka N, Dreyfuss G. Role of the nonsense-mediated decay factor hUpf3 in the splicing-dependent exon-exon junction complex. Science 2001; 293(5536):1832-1836.
44. Lykke-Andersen J, Shu MD, Steitz JA. Communication of the position of exon-exon junctions to the mRNA surveillance machinery by the protein RNPS1. Science 2001; 293(5536):1836-1839.
45. Le Hir H, Gatfield D, Izaurralde E et al. The exon-exon junction complex provides a binding platform for factors involved in mRNA export and nonsense-mediated mRNA decay. EMBO J 2001; 20(17):4987-4997.
46. Le Hir H, Izaurralde E, Maquat LE et al. The spliceosome deposits multiple proteins 20-24 nucleotides upstream of mRNA exon-exon junctions. EMBO J 2000; 19(24):6860-6869.
47. Fribourg S, Gatfield D, Izaurralde E et al. A novel mode of RBD-protein recognition in the Y14-Mago complex. Nat Struct Biol 2003; 10(6):433-439.
48. Pal M, Ishigaki Y, Nagy E et al. Evidence that phosphorylation of human Upfl protein varies with intracellular location and is mediated by a wortmannin-sensitive and rapamycin-sensitive PI 3-kinase-related kinase signaling pathway. RNA 2001; 7(1):5-15.
49. Weng Y, Czaplinski K, Peltz SW. ATP is a cofactor of the Upfl protein that modulates its translation termination and RNA binding activities. RNA 1998;4(2):205-214.
50. Nagy E, Maquat LE. A rule for termination-codon position within intron-containing genes: When nonsense affects RNA abundance. Trends Biochem Sci 1998; 23(6):198-199.
51. Sun X, Moriarty PM, Maquat LE. Nonsense-mediated decay of glutathione peroxidase 1 mRNA in the cytoplasm depends on intron position. EMBO J 2000; 19(17):4734-4744.
52. Zhang J, Sun X, Qian Y et al. Intron function in the nonsense-mediated decay of beta-globin mRNA: Indications that pre-mRNA splicing in the nucleus can influence mRNA translation in the cytoplasm. RNA 1998; 4(7):801-815.
53. Cheng J, Belgrader P, Zhou X et al. Introns are cis effectors of the nonsense-codon-mediated reduction in nuclear mRNA abundance. Mol Cell Biol 1994; 14(9):6317-6325.
54. Gatfield D, Unterholzner L, Ciccarelli FD et al. Nonsense-mediated mRNA decay in Drosophila: At the intersection of the yeast and mammalian pathways. EMBO J 2003; 22(15):3960-3970.
55. Amrani N, Ganesan R, Kervestin S et al. A faux 3'-UTR promotes aberrant termination and triggers nonsense-mediated mRNA decay. Nature 2004; 432(7013):112-118.
56. Hilleren P, Parker R. mRNA surveillance in eukaryotes: Kinetic proofreading of proper translation termination as assessed by mRNP domain organization? RNA 1999; 5(6):711-719.
57. Hilleren P, Parker R. Mechanisms of mRNA surveillance in eukaryotes. Annu Rev Genet 1999; 33:229-260.
58. Zhang J, Sun X, Qian Y et al. At least one intron is required for the nonsense-mediated decay of triosephosphate isomerase mRNA: A possible link between nuclear splicing and cytoplasmic translation. Mol Cell Biol 1998; 18(9):5272-5283.
59. Hosoda N, Kobayashi T, Uchida N et al. Translation termination factor eRF3 mediates mRNA decay through the regulation of deadenylation. J Biol Chem 2003; 278(40):38287-38291.
60. Gatfield D, Izaurralde E. Nonsense-mediated messenger RNA decay is initiated by endonucleolytic cleavage in Drosophila. Nature 2004; 429(6991):575-578.
61. Lejeune F, Li X, Maquat LE. Nonsense-mediated mRNA decay in mammalian cells involves decapping, deadenylating, and exonucleolytic activities. Mol Cell 2003; 12(3):675-687.
62. Chen CY, Shyu AB. Rapid deadenylation triggered by a nonsense codon precedes decay of the RNA body in a mammalian cytoplasmic nonsense-mediated decay pathway. Mol Cell Biol 2003; 23(14):4805-4813.
63. Lykke-Andersen J. Identification of a human decapping complex associated with hUpf proteins in nonsense-mediated decay. Mol Cell Biol 2002; 22(23):8114-8121.
64. Bremer KA, Stevens A, Schoenberg DR. An endonuclease activity similar to Xenopus PMR1 catalyzes the degradation of normal and nonsense-containing human beta-globin mRNA in erythroid cells. RNA 2003; 9(9):1157-1167.

65. Stevens A, Wang Y, Bremer K et al. Beta -Globin mRNA decay in erythroid cells: UG site-preferred endonucleolytic cleavage that is augmented by a premature termination codon. Proc Natl Acad Sci USA 2002; 99(20):12741-12746.
66. Lim S, Mullins JJ, Chen CM et al. Novel metabolism of several beta zero-thalassemic beta-globin mRNAs in the erythroid tissues of transgenic mice. EMBO J 1989; 8(9):2613-2619.
67. Lim SK, Sigmund CD, Gross KW et al. Nonsense codons in human beta-globin mRNA result in the production of mRNA degradation products. Mol Cell Biol 1992; 12(3):1149-1161.
68. Lim SK, Maquat LE. Human beta-globin mRNAs that harbor a nonsense codon are degraded in murine erythroid tissues to intermediates lacking regions of exon I or exons I and II that have a cap-like structure at the 5' termini. EMBO J 1992; 11(9):3271-3278.
69. Unterholzner L, Izaurralde E. SMG7 acts as a molecular link between mRNA surveillance and mRNA decay. Mol Cell 2004; 16(4):587-596.
70. Fukuhara N, Ebert J, Unterholzner L et al. SMG7 is a 14-3-3-like adaptor in the nonsense-mediated mRNA decay pathway. Mol Cell 2005; 17(4):537-547.
71. Sheth U, Parker R. Decapping and decay of messenger RNA occur in cytoplasmic processing bodies. Science 2003; 300(5620):805-808.
72. Cougot N, Babajko S, Seraphin B. Cytoplasmic foci are sites of mRNA decay in human cells. J Cell Biol 2004; 165(1):31-40.

Chapter 6

NMD and the Immune System

Jayanthi P. Gudikote and Miles F. Wilkinson*

Abstract

The immune system is a surveillance mechanism that recognizes improper substances outside host cells. Signature components of microorganisms, including their proteins and unique modifications of their nucleic acids, are recognized by immune system cells and immune system molecules that affect microorganism destruction. Counterpart surveillance mechanisms exist that recognize improper substances inside host cells. One of these is nonsense-mediated mRNA decay (NMD), which destroys mRNAs harboring premature termination codons (PTCs). Such PTC-bearing transcripts are potentially dangerous, as some of them encode truncated proteins that have dominant-negative or deleterious gain-of-function activity.

In this chapter, we will focus on the relationship between the immune system and NMD. We will discuss how molecules essential for immune function—immunoglobulin (Ig) and T-cell receptor (TCR) transcripts—are frequent targets of NMD and why this may be crucial for the survival of B and T lymphocytes. We will also detail discoveries about the nature of the NMD mechanism that acts on Ig and TCR transcripts and how this mechanism appears to be different from the one that acts on most other transcripts. We will also detail other immune system transcripts that may be targeted for destruction by NMD. Lastly, we will discuss the possible physiological relevance of NMD to the immune system.

Ig and TCR Gene Rearrangements Frequently Generate Nonsense Codons

The two white-blood-cell types that recognize foreign substances in an antigen-specific manner are B and T lymphocytes (usually called B and T cells, respectively). Each B cell carries on its surface a unique receptor called Ig (or antibody) that allows it to recognize a particular foreign molecule (or antigen) from viruses, bacteria, or parasites. The recognition of an antigen by a naïve B cell allows it to proliferate into armies of B cells that mature to become plasma cells. Each of these plasma cells secretes copious amounts of a single type of antibody with a unique antigen-binding specificity that activates mechanisms that destroy the antibody-bound pathogen. Like B cells, T cells have antigen-specific receptors, or TCRs, on their surfaces. Each T-cell clone that expands upon antigen recognition has a unique TCR. Unlike Ig, the TCR is always surface bound, it is never secreted. Its exclusive function is to activate T cells when they come in contact with foreign antigens on the surface of host cells. This recognition event drives T cells to proliferate in response to the antigen and, in some cases, to kill the cells presenting the foreign antigen on their surfaces.

*Corresponding Author: Miles F. Wilkinson—Department of Immunology, The University of Texas M.D. Anderson Cancer Center, Houston, Texas 77030, U.S.A.
Email: mwilkins@mdanderson.org

Nonsense-Mediated mRNA Decay, edited by Lynne E. Maquat. ©2006 Eurekah.com.

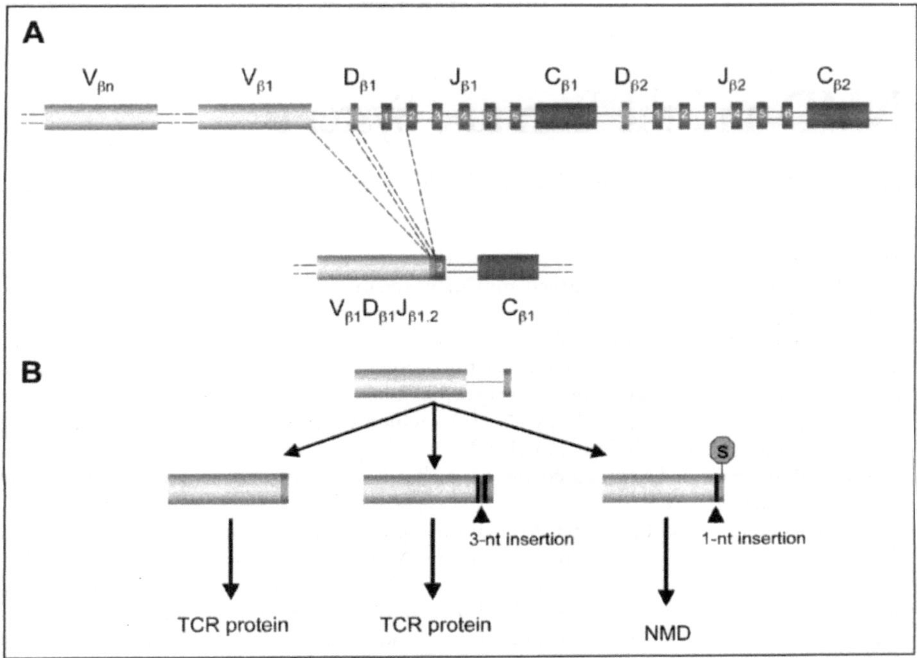

Figure 1. TCR and Ig gene rearrangements frequently generate nonsense codons. A) Diagram of the TCRβ gene, including the V, D, and J segments that undergo programmed rearrangement. Recombinase enzymes specific to the immune system join these segments in a manner that brings together only a single V, D, and J segment upstream of the C (constant) region in a given lymphocyte. B) To achieve additional diversity during programmed rearrangements, the terminal transferase enzyme adds extra nucleotides between the V, D, and J segments. A functional receptor is made only when nucleotides in multiples of 3 (such as the three in the middle example) are added between segments (only the V-D junction is shown). If the number of nucleotides added is not a multiple of 3 (such as the one in the example on the right), this will cause a frameshift that generates a downstream PTC (octagon). The rearrangement and nucleotide insertion events shown here for the TCRβ gene also occur in the other TCR and Ig genes, although some of these genes do not contain D elements.

A major problem in modern biology was to identify the mechanism responsible for generating the literally millions of different Ig and TCR molecules expressed in different B- and T-cell clones, respectively. In the 1970s, this mechanism was shown to be a "random genetic generator" that recombines Ig- and TCR-gene segments in different combinations during lymphocyte development.[1] This gene rearrangement process is highly regulated and transient to ensure that a given mature lymphocyte stably expresses only a single type of receptor.

The units of diversity in the Ig- and TCR- genes are the many variable (V) and joining (J) segments that they contain (Fig. 1A). Recombinase enzymes specific to the immune system join these segments in a manner that brings together only a single V segment and a single J segment in a given lymphocyte. Some Ig and TCR genes also contain diversity (D) segments that are recombined in the same way to provide even more receptor diversity. Because many V, D, and J segments exist, this recombination mechanism generates a vast array of unique VJ and VDJ exons encoding receptors with different binding specificities.

Additional diversity is achieved by nucleotides added between the V, D, and J segments during programmed rearrangement (Fig. 1B). One means by which this occurs is the addition of nontemplate-directed nucleotides (N nucleotides) to the junctions of the segments by the enzyme terminal transferase. A second source is coding nucleotides (P nucleotides), which are transferred from the complementary strand to the coding strand.

While N and P nucleotides greatly increase lymphocyte receptor diversity, they also have the unfortunate consequence of frequently generating frameshift mutations and thereby causing the formation of PTCs. Only when nucleotides in multiples of 3 (e.g., three or six nucleotides) are added between V, D, and J segments can the reading frame be maintained such that a functional receptor can be made (Fig. 1B). Two thirds of the time, a "nonproductive" rearrangement occurs when the number of nucleotides inserted at a junction is not a multiple of 3. This leads to the production of PTC-bearing transcripts encoding truncated, nonfunctional Ig or TCR molecules.[2,3] These truncated receptors could have a dominant-negative effect by interfering with the processing or function of the normal full-length protein encoded by the other allele. In addition, truncated Ig and TCR proteins may not fold properly and may thereby induce a stress response in the endoplasmic reticulum of lymphocytes.[4]

NMD and Immune Receptors

Because NMD specifically recognizes and destroys transcripts harboring PTCs, it provides a perfect solution to remove the "problem" transcripts generated from nonproductively rearranged Ig and TCR genes. By rapidly degrading these aberrant transcripts, NMD largely prevents the translation of truncated, potentially deleterious receptor molecules in lymphocytes.

The first hint that such a mechanism operated came 20 years ago from the work of Nobel Prize-winner Georges Kohler. He found that B-cell hybridomas containing nonproductively rearranged Ig heavy chain (Igμ) genes expressed lower levels of Igμ mRNA than did hybridomas containing in-frame Igμ genes.[5] This finding was subsequently confirmed by others.[6,7]

The identification of a similar mechanism in T cells had to wait until another major mystery in immunology was solved: the identification of the elusive TCR. Sequencing of the first TCR cDNAs in the 1980s revealed that the vast majority were in frame, which was a puzzle, as two thirds of TCR gene rearrangements are out of frame. Most immunologists ignored this puzzle, choosing to focus instead on their good fortune that most of their cloned TCR cDNAs were in frame, allowing them to quickly generate plasmid constructs to study the biochemical and biological functions of TCRs.

One of the authors of this chapter, Miles Wilkinson, entered the scene by serendipity, having discovered that the mouse T-cell clone SL12.4 gives rise to high levels of precursor TCRβ mRNA but only trace levels of mature mRNA.[8,9] Thinking that TCRβ splicing in this cell clone might be negatively regulated by an unstable repressor protein, the protein synthesis inhibitor cycloheximide was tested for the ability to upregulate the level of TCRβ mRNA. Indeed, cycloheximide and several other protein synthesis inhibitors, all with different mechanisms of action, markedly upregulated mature TCRβ mRNA in the SL12.4 cell clone.[8,10] However, later experiments revealed that these translation inhibitors are probably not acting by removing an unstable repressor; instead, they act by blocking NMD, a translation-dependent process. This revelation came when it was discovered that the rearranged TCRβ gene in SL12.4 cells is nonproductively rearranged and therefore contains PTCs.[11] That NMD is responsible for downregulating TCRβ mRNA was made clear when PTC-bearing TCRβ genomic constructs were found to be expressed at much lower levels when transfected into SL12.4 cells than was an in-frame TCRβ genomic construct.[11] Cycloheximide strongly increased the steady-state level of mRNA from all the PTC-bearing constructs but not from the PTC-lacking control construct.

Having shown that an immortalized T-cell line downregulates PTC-bearing TCRβ transcripts, Wilkinson and colleagues next asked whether normal T cells can also do this. To address this, a large number of cDNAs corresponding to both TCRβ precursor mRNA and mature mRNA from fetal and adult thymocytes were sequenced.[11] About half of the precursor mRNAs were found to be out of frame, consistent with the known frequency of out-of-frame TCRβ genes in thymocytes. In contrast, a very small proportion of mature transcript cDNAs were out of frame, implying that thymocytes efficiently degrade out-of-frame mature TCRβ transcripts as a result of NMD.

The Second Signal for NMD

Once the phenomenon of NMD was firmly established, the next issue for the field was to understand the underlying mechanism. One initial question was how NMD discriminates between normal and premature termination codons. It turned out that a relatively simple mechanism is responsible for this discrimination: a second signal downstream of a stop codon signifies it as being premature. In mammalian cells, this second signal is an exon-exon junction downstream of the stop codon (Fig. 2). The initial hint that this is the case was the discovery that NMD is triggered by PTCs in internal exons but not by PTCs in the final exon. This was shown for several transcripts, including those encoding dihydrofolate reductase and Igμ.[6,12] More definitive evidence came from Lynne Maquat and colleagues, who showed that removing the introns downstream of a PTC in the triose phosphate isomerase (TPI) gene prevents its transcripts from being degraded by NMD.[13] Later, similar experiments showed that this two-signal rule applies to other mammalian transcripts, including those encoding TCRβ and β-globin.[14,15] We also performed a converse experiment in which we introduced introns downstream of the normal TCRβ stop codon. This triggered NMD, thereby providing further evidence for the two-signal rule.[14]

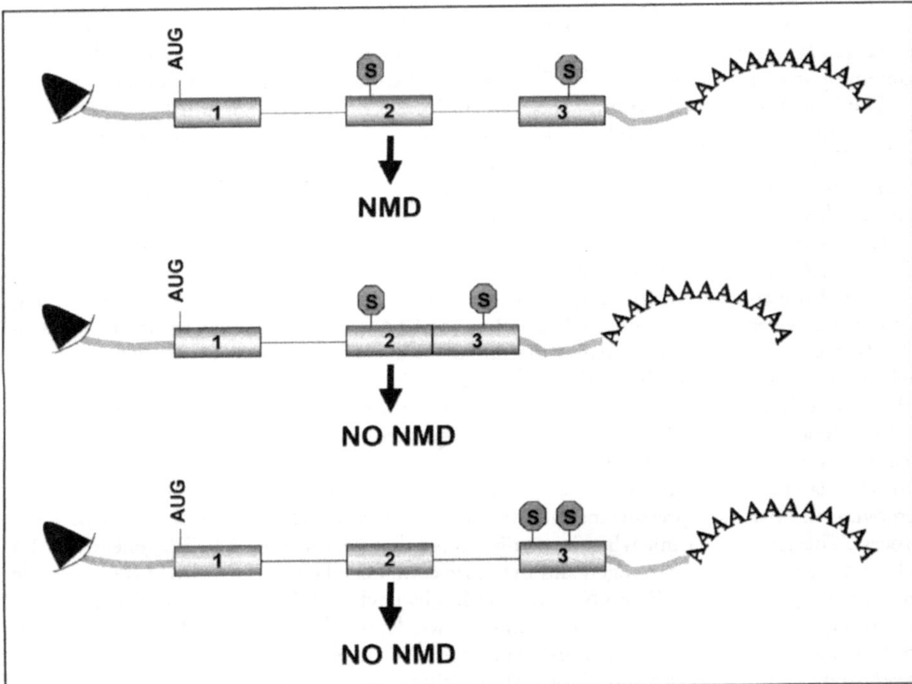

Figure 2. Mammalian NMD typically requires two signals within a transcript: (i) a stop codon (octagon) and (ii) at least one exon-exon junction downstream of the stop codon (top). Thus, the absence of an intron downstream of a stop codon abolishes NMD (middle) and introduction of an extra stop codon in the final exon does not trigger NMD (bottom). From left to right, the hat signifies the 5' cap structure, thick lines specify 5' and 3' untranslated regions, AUG represents the translation initiation codon, and the stretch of As denotes the 3' poly(A) tail.

For some time after the discovery of the two-signal rule, investigators were puzzled as to how introns could serve as the second signal for NMD, particularly since introns are spliced out in the nucleus and thus would not be present when the stop codons are recognized by the translation machinery in the cytoplasm. Experiments conducted by Lynne Maquat and colleagues using TPI transcripts led them to propose a model in which a specific intronic cis-acting element directs the formation of a "mark" left after RNA splicing that serves as the proximal second signal for NMD.[13] Later experiments conducted using TCRβ transcripts indicated that it was essential that the intron downstream of a PTC be spliceable to elicit NMD.[14] This suggested that the act of splicing generates the mark essential for NMD.[14]

This model was demonstrated to be correct when Herve Le Hir, Melissa Moore, and Lynne Maquat discovered that such a mark is indeed left near splice junctions after RNA splicing.[16] This mark is a protein complex—collectively called the exon junction complex (EJC)—that is deposited ~20-24 nucleotides upstream of exon-exon junctions in mature mRNA (see chapter by Maquat). Among the proteins in the EJC are ones that function in NMD (Fig. 3A). For example, both loss- and gain-of-function experiments have demonstrated that the EJC proteins UPF2, UPF3B, Y14, BTZ, and eIF4A3 have a role in NMD.[17-19] Other EJC components, such as RNPS1 and PYM, have been suggested to be involved in NMD based only on gain-of-function experiments.[20,21]

The discovery of the EJC in mammalian cells, along with investigations into the function and interactions of NMD factor orthologs in *Saccharomyces cerevisiae*, led to the development of a model for how NMD is elicited.[22,23] According to this model, when the translation apparatus recognizes a stop codon, it converts into a surveillance complex that scans or somehow interacts downstream in search of the EJC (Fig. 3B). If the surveillance complex makes contact with an EJC, this leads to decay of the mRNA. If no EJC is present downstream, the mRNA is protected from decay. This surveillance model explains why most normal mRNAs are not subject to NMD as the stop codon in most wild-type mammalian messages is in the final exon.[24] Conversely, PTC-bearing mRNAs typically harbor the PTC in an internal exon, meaning that one EJC will be downstream of the stop codon, thereby triggering NMD.

The surveillance model further posits that, unlike the surveillance complex, the ribosome does not elicit NMD when it interacts with the EJC; instead, the ribosome knocks the EJC off the mRNA. Thus, a ribosome traveling down a normal mRNA—one containing only an in-frame stop codon in the final exon—displaces all the EJCs. In contrast, in an aberrant PTC-containing mRNA, only the EJCs upstream of the PTC are displaced by the ribosome; the downstream EJCs are met by the surveillance complex, leading to mRNA decay.

A prediction of the surveillance model is that stop codons too close to the final exon-exon junction should not be able to elicit NMD, because the translation apparatus should not be able to read the stop codon until it has already displaced the last EJC. Indeed, this exception to the two-signal rule has been shown to be the case. Several groups observed that stop codons less than ~50 nucleotides upstream of the final exon-exon junction (abbreviated as -50) do not trigger the decay of most transcripts, including those encoding β-globin, TPI, mouse major urinary protein, and glutathione peroxidase-1[15,24,25] (Fig. 4A). This -50 boundary rule conforms with the surveillance model, as it is reasonable to suppose that a ribosome reading a stop codon at approximately the -50 position would make contact with the EJC, given the known position of the EJC at the -20 to -24 position and the length of mRNA (~30 nucleotides) bound by an actively translating ribosome.[26]

Unique Features of Ig and TCR NMD

Do Ig and TCR transcripts follow the general NMD rules established for other mammalian transcripts? In one important respect they do: the decay of TCRβ mRNA in response to a PTC requires a downstream intron[14] and correlative evidence suggests the same is true for Igμ mRNA.[6] However, in many other respects, these transcripts respond differently to PTCs than do other mammalian transcripts: (i) Igμ and TCRβ mRNAs do not obey the -50 boundary rule

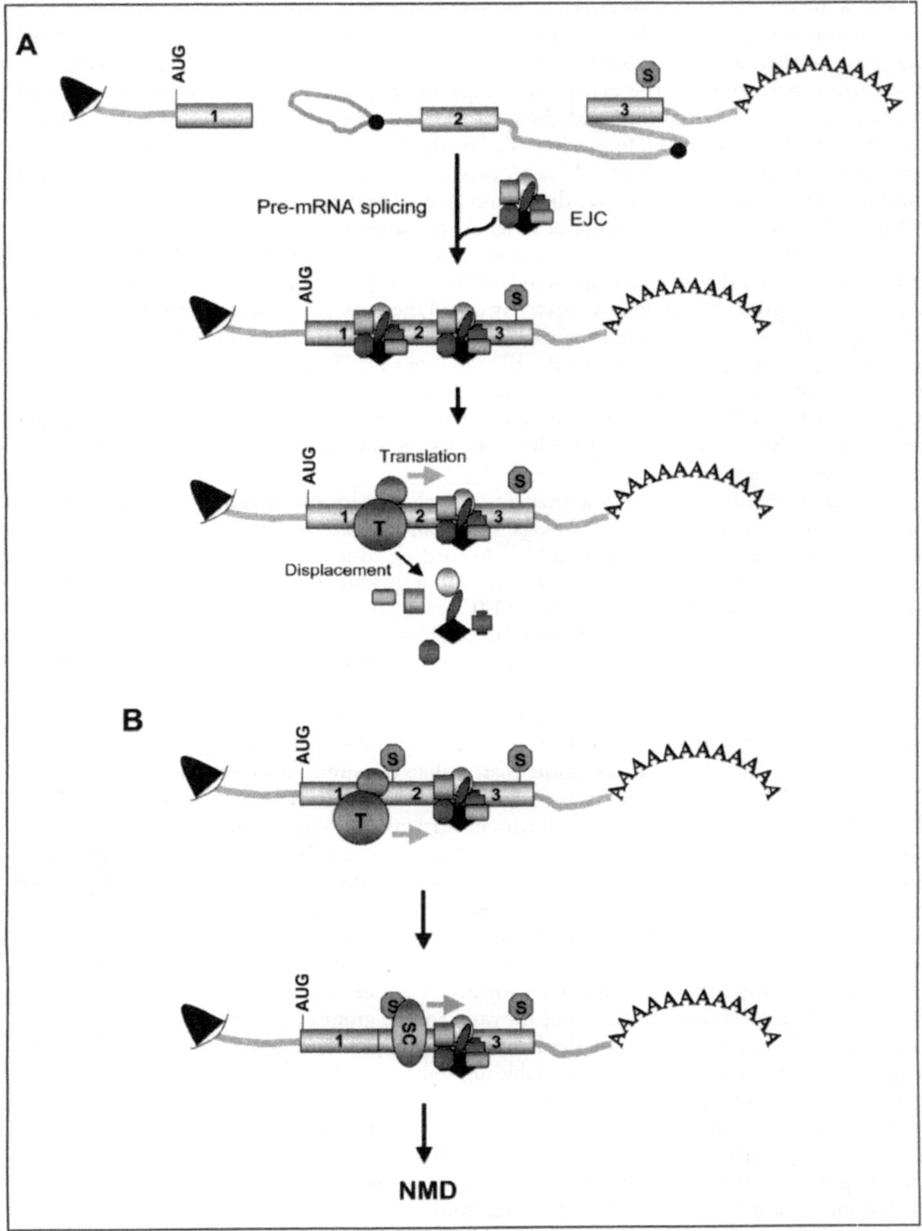

Figure 3. An exon junction complex is the second signal for NMD. A) An exon junction complex (EJC) is deposited ~20-24 nucleotides 5' of each exon-exon junction after precursor mRNA splicing. During the pioneer round of translation, the translation machinery (T) is thought to displace the EJCs. B) The surveillance model posits that when the translation machinery encounters a stop codon, a surveillance complex (SC) is formed that scans downstream of the stop codon (octagon). NMD is elicited if the SC encounters an EJC.

described above, implying that the classical EJC is not always responsible for the decay of those transcripts in response to PTCs; (ii) Igμ and TCRβ mRNAs are uniquely subject to polar regulation, in which the magnitude of mRNA downregulation differs for 5' and 3' PTCs; and (iii) Igμ and TCRβ mRNAs are more strongly downregulated by PTCs than are most mammalian transcripts (Fig. 4).

Disobedience of -50 Boundary Rule

TCRβ and Igμ transcripts have both been shown to be downregulated in response to PTCs at various positions, including those that reside less than 50 nucleotides upstream of the final exon-exon junction (Fig. 4B). In the case of TCRβ, nonsense, but not missense, mutations at the -48 and -36 positions in the penultimate exon (Cβ2.3) decrease transcript levels.[27] This is not a peculiarity of the Cβ2.3 exon, as the same is true for the rearranged VDJ exon if it is made the penultimate exon by deleting all but one of the downstream introns (Fig. 4C). In this context, nonsense, but not missense, mutations at the -40, -19, and -16 positions in the VDJ exon elicit NMD.[14,27] Igμ transcripts were recently shown to also disobey the -50 boundary rule, as nonsense mutations at either the -10 or -31 positions in the penultimate exon elicit NMD (Fig. 4D).[28]

A ribosome could only recognize stop codons at these positions if it had already displaced the EJC positioned at its normal position of -20 to -24. This implies that the classical EJC is not the second signal for TCRβ and Igμ transcripts, at least not when they are degraded in response to PTCs close to the 3' terminal exon-exon junction. It will be interesting to determine whether this novel second signal is the EJC at a more downstream position or, instead, a unique factor.

Polar Regulation

Most mammalian transcripts are downregulated to a similar extent regardless of PTC position (Fig. 4A). For example, PTCs in exons 1, 2, and 6 all decrease the level of TPI transcripts to ~30% of the level of control TPI transcripts.[29] Likewise, dihydrofolate reductase, β-globin, glutathione peroxidase-1, and BRCA1 transcripts exhibit a relatively uniform magnitude of NMD in response to PTCs at different position.[12,15,30-32] To our knowledge, the only mammalian transcripts that do not conform to this "uniform NMD magnitude" rule are those encoding TCRβ and Igμ. Both of these transcripts are downregulated much more strongly by PTCs at the 5' end than those at the 3' end of the penultimate exon[6,27] (Fig. 4B,D). The polarity is quite marked: the magnitude of mRNA downregulation varies by 5- to 10-fold. In the case of TCRβ, polarity is dictated by the distance to the final exon-exon junction, as polarity is conferred to a middle exon only when it is positioned in a penultimate context (Fig. 4C). This clearly demonstrates that the magnitude of NMD is not dictated by the position of a PTC per se; rather, it is dictated by the position of the PTC relative to the final exon-exon junction. The underlying molecular mechanism for polarity remains to be elucidated.

In contrast to the 3' half, the 5' half of Igμ transcripts exhibits "reverse polarity," such that 5' PTCs elicit weaker NMD than do 3' PTCs (Fig. 4D). For example, a PTC at codon position 3 reduces Igμ transcript levels to only about 30-50% of the level of control Igμ transcripts.[33] A graded increase in NMD then occurs as PTCs are introduced at progressively more 3' positions; by codon 108 in the VJD exon, NMD is maximal or near maximal (4% of control).[28] Reverse polarity in response to PTCs near the start codon has also been observed for TCRβ transcripts (Jun Wang and M.F. Wilkinson, unpublished observations). Reverse polarity may be the result of translation reinitiation after AUG-proximal stop codons. A polar response would be expected, as other studies have shown that the probability of reinitiation progressively decreases as the length of the upstream open reading frame increases.[34] Indeed, Igμ transcripts have been shown to have two AUG codons downstream of the normal start codon; these extra AUG codons reinitiate translation when translation from the first AUG is terminated by a PTC at codon position 3.[33] Translation reinitiation has also been observed in TPI transcripts in response to PTCs.[35]

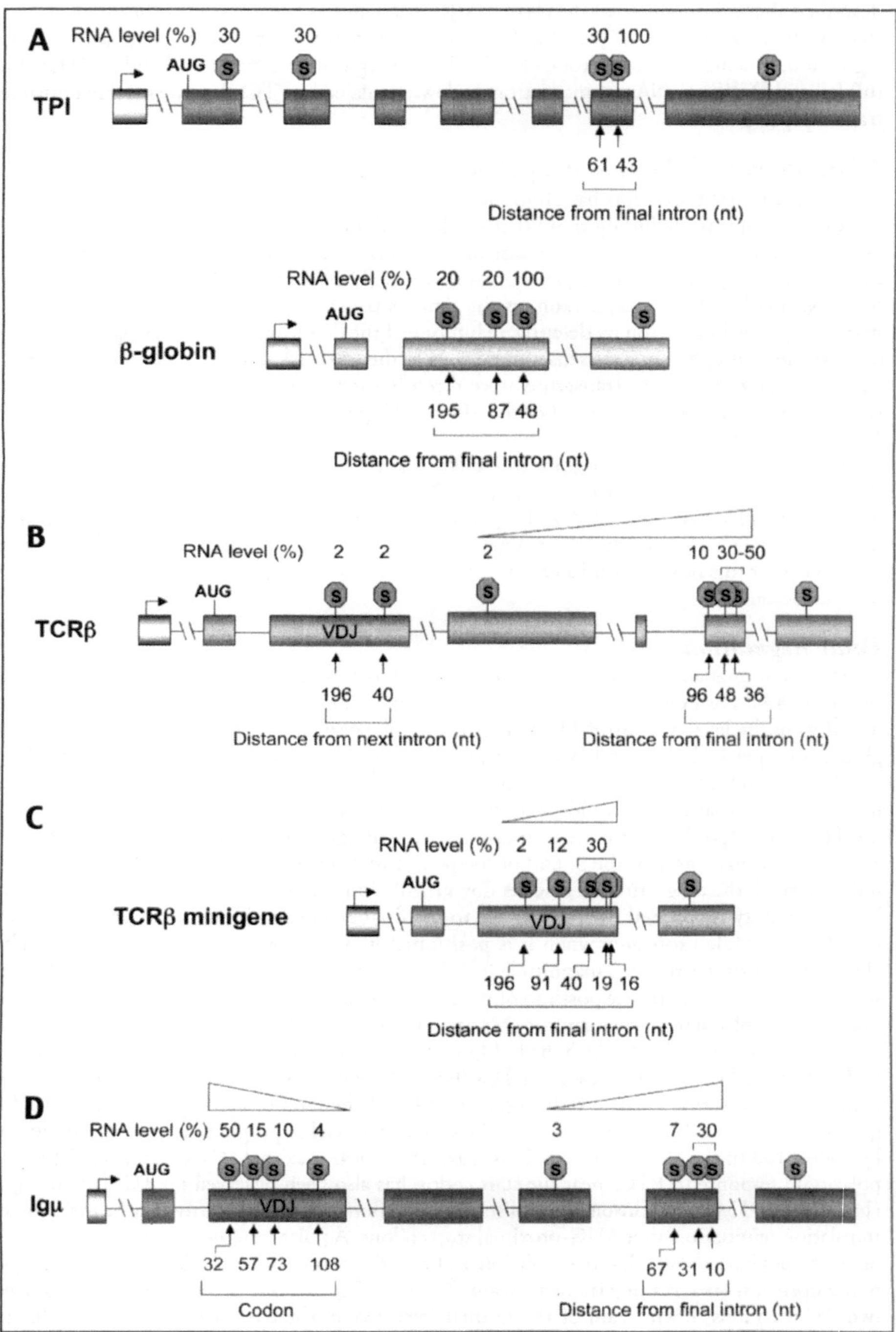

Figure 4. Unique features of TCR and Ig NMD. Figure legend continued on next page.

Figure 4, continued. A) Most mammalian transcripts are downregulated by PTCs to a similarly modest extent, regardless of PTC position (octagons designate in-frame stop codons; the normal stop codon is in the final exon). The only exception is PTCs ~50 nucleotides or less from the final exon-exon junction; such PTCs do not elicit NMD (the -50 boundary rule). Shown are TPI and β-globin transcripts harboring PTCs at various positions. The RNA levels indicated are relative to those of control transcripts that contain only the normal stop codon in the last exon. B) TCRβ transcripts do not abide by the -50 boundary rule and exhibit polar NMD (PTCs at the 5' end of the penultimate exon elicit stronger NMD than do PTCs at the 3' end). Also, TCRβ transcripts are more strongly downregulated in response to PTCs at most positions than are the majority of other transcripts. C) Polarity is dictated by distance to the final exon-exon junction, as polarity is observed in the VDJ exon only when it is made the penultimate exon (compare with panel B). D) Like TCRβ transcripts, Igμ transcripts are strongly downregulated by NMD, exhibit polar NMD, and do not obey the -50 boundary rule. Igμ transcripts also exhibit "reverse polarity" at their 5' end, such that 5' PTCs in this region elicit weaker NMD than do 3' PTCs.

Superdownregulation

The third unique feature of TCRβ and Igμ transcripts is that they are downregulated very strongly in response to PTCs at most positions (the only exceptions being at the extreme 5' and 3' ends, as described above). While the levels of most transcripts from nonrearranging genes are decreased by PTCs to 10-30%, the levels of TCRβ and Igμ transcripts are decreased to 2-5%.[6,7,14,28,33,36] In the case of TCRβ transcripts, this "superdownregulation" is not conferred by factors specific to T cells, nor is it a property of TCRβ promoters; instead, it is conferred by sequences present in TCRβ transcripts.[36] Mapping experiments have shown that deletion of the TCRβ VDJ exon and upstream intron sequences abolishes superdownregulation, demonstrating that this region contains one or more elements that dictate this regulation.[36]

Similar mapping experiments have been conducted on Igμ transcripts.[28] Deletion experiments identified the 5' half of the VDJ exon as housing sequences that mediate Igμ superdownregulation. Redundant elements may be present in this 177-nucleotide region, since deletions within this region do not abolish superdownregulation. This region does not appear to be sufficient for superdownregulation, because its introduction into a heterologous transcript does not confer robust mRNA downregulation in response to a PTC.

Recently, we obtained several lines of evidence that efficient RNA splicing is one molecular mechanism responsible for superdownregulation.[37] Decreasing the splicing efficiency of TCRβ transcripts by mutating their splice sites reduces TCRβ mRNA downregulation in response to PTCs. Conversely, improving the splicing efficiency of a poor NMD substrate greatly increases the magnitude of its downregulation in response to PTCs. The notion that more efficient splicing stimulates NMD allows many previous experiments to be interpreted in a new light. For example, we suggest that the previously identified NMD-promoting regulatory regions in Igμ and TCRβ transcripts may have this activity simply because they contain sequences that promote RNA splicing.

One clue to the mechanism behind this splicing efficiency-dependent effect was the discovery that efficient splicing also promotes translation. As evidence for this, splice-site mutations that impair (but do not prevent) TCRβ splicing only modestly decrease TCRβ mRNA levels but cause a dramatic decrease in TCRβ protein levels.[37] This result indicates that efficient splicing is essential for robust translation, providing a simple explanation for how efficient splicing might enhance NMD, as NMD is a translation-dependent process.[14,38-40] Interestingly, the magnitude of NMD does not appear to be dictated by the rate of translation per se. Working with Igμ transcripts in B cells, Marc Shulman and colleagues found that while the use of an inefficient alternative AUG leads to a dramatic > 20-fold decrease in the translation rate, it has no effect on the robust NMD typical of Igμ transcripts.[33]

Another clue towards understanding the underlying molecular mechanism was the discovery that the splicing efficiency of introns downstream, but not upstream, of a PTC dictates the magnitude of NMD.[37] This led us to propose a model in which an alternative EJC deposited after efficient splicing elicits more robust NMD than does the classical EJC.[37] We posit that this alternative EJC is deposited downstream of the classical EJC on TCR and Ig transcripts, thereby explaining why these transcripts do not obey the -50 boundary rule. The alternative EJC may also have special characteristics that dictate the polar NMD typical of Ig and TCR transcripts.

Developmental Regulation of NMD

In some instances, the magnitude of Ig and TCR mRNA downregulation appears to be developmentally regulated. A recent study was conducted to examine the expression of Igκ transcripts, which encode one of the two types of Ig light chains.[41] Immature B cells (pro-B cells) were found to weakly downregulate PTC-bearing Igκ transcripts (reduced to ~25% of in-frame transcripts), whereas splenic mature B cells very strongly downregulate PTC-bearing Igκ transcripts (59 of 59 Igκ cDNA clones in that study were in frame). It will be interesting to know whether B-cell maturation promotes NMD in general or NMD directed specifically at Igκ transcripts. Unlike the Igμ and TCRβ genes, Igκ genes have no introns in the constant region, so there are no introns downstream of the PTCs generated by nonproductive Igκ rearrangements. This strongly suggests that an EJC-independent mechanism is responsible for Igκ mRNA downregulation in response to nonsense codons.

Physiological Relevance and the Future

We do not yet know the physiological relevance of NMD for the immune system. One possibility is that NMD protects lymphocytes from the toxicity that would otherwise be elicited by truncated dominant-negative proteins expressed from nonproductively rearranged Ig and TCR genes. For proper function, the heavy and light chains of Ig and TCR must be paired. Because the truncated proteins expressed from nonproductively rearranged Ig and TCR genes would contain most of the variable region and, in most cases, part of the constant region, they could potentially pair with normal full-length Ig and TCR proteins and interfere with their processing or function. Numerous downstream consequences could then occur, including the generation of receptors unable to recognize antigens or receptors that do not initiate signal transduction. Conversely, the C-terminally truncated Ig or TCR chains may not be able to pair with their partner proteins and thereby cause disease. By analogy, heavy-chain disease is caused by mutant Igγ heavy chains that cannot pair with light chains.[42,43] Another possibility is that truncated Ig and TCR chains could be misfolded, causing them to be retained in the endoplasmic reticulum and induce a stress response. An unfolded protein-stress response has been well defined in a variety of organisms; when triggered, it causes translational inhibition, cell-cycle arrest, and, in some cases, apoptosis.[4]

Recent evidence suggests that NMD is required for normal T-cell development (Hal Dietz, personal communication). Transgenic mice expressing a dominant-negative form of the NMD factor UPF1 exhibit a blockade in thymocyte development such that they accumulate immature $CD4^-CD8^-$ thymocytes. This developmental block is not caused by the failure of NMD to downregulate the expression of truncated TCRβ proteins from nonproductively rearranged TCRβ genes, as transgenic mice engineered to express high levels of such truncated TCRβ proteins are not subject to this block. Instead, it appears to result from the ability of TCRβ mRNA itself to block TCRβ gene rearrangements and consequent thymocyte maturation. The evidence for this comes from the finding that transgenic mice expressing a full-length, stable TCRβ mRNA that is incapable of encoding TCRβ protein (because frameshifts were inserted downstream of the alternative AUGs, and all the PTCs were removed by site-directed mutagenesis) also exhibit a thymocyte maturation block.

In addition to aberrant TCR and Ig mRNAs, NMD may also regulate the level of mRNAs from some wild-type immune system genes. A recent microarray analysis conducted by Hal Dietz and colleagues revealed that transcripts for several cytokine-related proteins, including interleukin (IL)-6β, IL-27, IL-7 receptor, interferon-related developmental regulator-1 (IFRD1), and tumor necrosis factor receptor-2-associated protein, are probably targets of NMD[44] (see chapter by Shafari and Dietz). This is a vast underestimate of the number of immune system transcripts targeted by NMD, as only a small proportion of the human genome was analyzed. Furthermore, the microarray analysis was done in HeLa cells, not immune system cells. In the future, it will be important to determine why these transcripts are targeted for decay by NMD. One of them, the IFRD1 transcript, has a short open reading frame upstream of the IFRD1 reading frame. The stop codon in the upstream open reading frame may be responsible for eliciting NMD; by analogy, some yeast transcripts containing upstream open reading frames are subject to NMD.[45]

Interestingly, the HeLa-cell microarray analysis also identified immune system transcripts that appear to be upregulated by NMD. Both IL-11 and zinc-finger antiviral protein transcripts fit this category because their levels are downregulated in response to loss of the NMD factor UPF1. These transcripts may be indirect targets of NMD that are positively regulated by a factor that is the direct target of NMD.

It will be fascinating to identify all the direct and indirect NMD targets that impinge on the immune system, thereby allowing the elucidation of an NMD network. We speculate that slightly different NMD networks exist in different cell types of the immune system; for example, a B cell might have some unique NMD-regulated transcripts not present in a dendritic cell.

Little is known about the role of NMD in immune system defects. We speculate that there may be two classes of NMD-associated mutations that are detrimental to the human immune system: (i) nonsense and frameshift mutations in genes encoding proteins essential for the immune system, and (ii) mutations in genes encoding general NMD and EJC components. An example of the former class is a frameshift mutation identified in the interferon-γR1 gene of an infant that suffered from mycobacterial infections.[46] This frameshift generated downstream PTCs that were probably responsible for the decreased interferon-γR1 mRNA levels in B cells from this infant. It is possible that the truncated protein produced from the mutant allele had dominant-negative activity and that its extinction by NMD led to a less severe reduction in immune function than would otherwise occur. With regard to the latter class, no examples of mutations in NMD or EJC genes in humans have been reported. Such mutations, if they exist, are probably hypomorphic or heterozygous null alleles, as mice homozygous for a null mutation in the NMD gene *Upf1* die early during embryogenesis.[47] A possible exception would be genes encoding NMD factors that act on only a subset of PTC-bearing transcripts (as suggested by data indicating that the NMD of TCR and Ig transcripts is unique; see earlier). Homozygous null mutations in such genes may not be lethal.

Recently, it has come to light that NMD factors promote not only the decay of aberrant PTC-bearing transcripts but also the translation of normal transcripts.[48-50] Thus, individuals with a deficiency in an NMD factor may have both defective RNA surveillance and insufficient protein production from some transcripts. Such a decrease in the level of key immune system proteins could have profound effects on the ability of the immune system to function. There is also evidence from yeast that NMD factors promote normal translation termination.[48,51,52] If this is true in mammals, then deficiency of an NMD factor would increase translational read-through and result in more aberrant proteins with extended C-terminal ends. NMD factors have also been implicated in DNA damage surveillance and telomerase function,[53,54] (see chapters by Abraham and Oliveira, and Azzalin et al) both of which could also have a profound impact on the immune system.

While there are clearly many other interesting avenues for future investigation, we suggest that the following three issues are particularly important. First, why do Ig and TCR transcripts respond to PTCs in a robust, polar, and boundary-independent manner? Does this mean that unique NMD factors and/or an alternative EJC are responsible for their rapid decay? If so, are other mammalian transcripts subject to this alternative NMD pathway? Second, is the downregulation of PTC-bearing Ig and TCR transcripts by NMD essential for the normal function of lymphocytes? While it seems likely that high expression of truncated Ig and TCR proteins would be deleterious to lymphocytes, there is no direct evidence that this is the case. And if so, what specific abnormal events occur? Third, what immune system transcripts besides Ig and TCR mRNAs are targets of NMD? One particularly interesting class will be NMD targets transcribed from nonmutated genes. Clearly, we are just beginning to understand the relevance of NMD to the immune system.

References

1. Kronenberg M, Siu G, Hood LE et al. The molecular genetics of the T-cell antigen receptor and T-cell antigen recognition. Annu Rev Immunol 1986; 4:529-591.
2. Li S, Wilkinson MF. Nonsense surveillance in lymphocytes? Immunity 1998; 8(2):135-141.
3. Bruce SR, Wilkinson MF. Nonsense-mediated decay: A surveillance pathway that detects faulty TCR and BCR transcripts. Recent Res Devel Immunity Vol I. Trivandrum-695 023. Kerala, India: Research Signpost, 2003:1-12.
4. Ma Y, Hendershot LM. The unfolding tale of the unfolded protein response. Cell 2001; 107(7):827-830.
5. Baumann B, Potash MJ, Kohler G. Consequences of frameshift mutations at the immunoglobulin heavy chain locus of the mouse. EMBO J 1985; 4(2):351-359.
6. Connor A, Wiersma E, Shulman MJ. On the linkage between RNA processing and RNA translatability. J Biol Chem 1994; 269(40):25178-25184.
7. Jack HM, Berg J, Wabl M. Translation affects immunoglobulin mRNA stability. Eur J Immunol 1989; 19(5):843-847.
8. Qian L, Theodor L, Carter M et al. T cell receptor-beta mRNA splicing: Regulation of unusual splicing intermediates. Mol Cell Biol 1993; 13(3):1686-1696.
9. Wilkinson MF, MacLeod CL. Induction of T-cell receptor-alpha and -beta mRNA in SL12 cells can occur by transcriptional and post-transcriptional mechanisms. EMBO J 1988; 7(1):101-109.
10. Qian L, Vu MN, Carter MS et al. T cell receptor-beta mRNA splicing during thymic maturation in vivo and in an inducible T cell clone in vitro. J Immunol 1993; 151(12):6801-6814.
11. Carter MS, Doskow J, Morris P et al. A regulatory mechanism that detects premature nonsense codons in T-cell receptor transcripts in vivo is reversed by protein synthesis inhibitors in vitro. J Biol Chem 1995; 270(48):28995-29003.
12. Urlaub G, Mitchell PJ, Ciudad CJ et al. Nonsense mutations in the dihydrofolate reductase gene affect RNA processing. Mol Cell Biol 1989; 9(7):2868-2880.
13. Cheng J, Belgrader P, Zhou X et al. Introns are cis effectors of the nonsense-codon-mediated reduction in nuclear mRNA abundance. Mol Cell Biol 1994; 14(9):6317-6325.
14. Carter MS, Li S, Wilkinson MF. A splicing-dependent regulatory mechanism that detects translation signals. EMBO J 1996; 15(21):5965-5975.
15. Thermann R, Neu-Yilik G, Deters A et al. Binary specification of nonsense codons by splicing and cytoplasmic translation. EMBO J 1998; 17(12):3484-3494.
16. Le Hir H, Moore MJ, Maquat LE. Pre-mRNA splicing alters mRNP composition: Evidence for stable association of proteins at exon-exon junctions. Genes Dev 2000; 14(9):1098-1108.
17. Palacios IM, Gatfield D, St Johnston D et al. An eIF4AIII-containing complex required for mRNA localization and nonsense-mediated mRNA decay. Nature 2004; 427(6976):753-757.
18. Gehring NH, Neu-Yilik G, Schell T et al. Y14 and hUpf3b form an NMD-activating complex. Mol Cell 2003; 11(4):939-949.
19. Kim YK, Furic L, Desgroseillers L et al. Mammalian Staufen1 recruits Upf1 to specific mRNA 3'UTRs so as to elicit mRNA decay. Cell 2005; 120(2):195-208.
20. Bono F, Ebert J, Unterholzner L et al. Molecular insights into the interaction of PYM with the Mago-Y14 core of the exon junction complex. EMBO Rep 2004; 5(3):304-310.
21. Lykke-Andersen J, Shu MD, Steitz JA. Communication of the position of exon-exon junctions to the mRNA surveillance machinery by the protein RNPS1. Science 2001; 293(5536):1836-1839.
22. Singh G, Lykke-Andersen J. New insights into the formation of active nonsense-mediated decay complexes. Trends Biochem Sci 2003; 28(9):464-466.
23. Gonzalez CI, Bhattacharya A, Wang W et al. Nonsense-mediated mRNA decay in Saccharomyces cerevisiae. Gene 2001; 274(1-2):15-25.
24. Hawkins JD. A survey on intron and exon lengths. Nucleic Acids Res 1988; 16(21):9893-9908.
25. Nagy E, Maquat LE. A rule for termination-codon position within intron-containing genes: When nonsense affects RNA abundance. Trends Biochem Sci 1998; 23(6):198-199.
26. Steitz JA. Polypeptide chain initiation: Nucleotide sequences of the three ribosomal binding sites in bacteriophage R17 RNA. Nature 1969; 224(223):957-964.

27. Wang J, Gudikote JP, Olivas OR et al. Boundary-independent polar nonsense-mediated decay. EMBO Rep 2002; 3(3):274-279.
28. Buhler M, Paillusson A, Muhlemann O. Efficient downregulation of immunoglobulin mu mRNA with premature translation-termination codons requires the 5'-half of the VDJ exon. Nucleic Acids Res 2004; 32(11):3304-3315.
29. Daar IO, Maquat LE. Premature translation termination mediates triosephosphate isomerase mRNA degradation. Mol Cell Biol 1988; 8(2):802-813.
30. Kessler O, Chasin LA. Effects of nonsense mutations on nuclear and cytoplasmic adenine phosphoribosyltransferase RNA. Mol Cell Biol 1996; 16(8):4426-4435.
31. Perrin-Vidoz L, Sinilnikova OM, Stoppa-Lyonnet D et al. The nonsense-mediated mRNA decay pathway triggers degradation of most BRCA1 mRNAs bearing premature termination codons. Hum Mol Genet 2002; 11(23):2805-2814.
32. Sun X, Moriarty PM, Maquat LE. Nonsense-mediated decay of glutathione peroxidase 1 mRNA in the cytoplasm depends on intron position. EMBO J 2000; 19(17):4734-4744.
33. Buzina A, Shulman MJ. Infrequent translation of a nonsense codon is sufficient to decrease mRNA level. Mol Biol Cell 1999; 10(3):515-524.
34. Rajkowitsch L, Vilela C, Berthelot K et al. Reinitiation and recycling are distinct processes occurring downstream of translation termination in yeast. J Mol Biol 2004; 335(1):71-85.
35. Zhang J, Maquat LE. Evidence that translation reinitiation abrogates nonsense-mediated mRNA decay in mammalian cells. EMBO J 1997; 16(4):826-833.
36. Gudikote JP, Wilkinson MF. T-cell receptor sequences that elicit strong down-regulation of premature termination codon-bearing transcripts. EMBO J 2002; 21(1-2):125-134.
37. Gudikote JP, Imam JS, Garcia RF et al. RNA splicing promotes translation and RNA surveillance. Nat Struct Mol Biol 2005; 12(9):801-809.
38. Belgrader P, Cheng J, Maquat LE. Evidence to implicate translation by ribosomes in the mechanism by which nonsense codons reduce the nuclear level of human triosephosphate isomerase mRNA. Proc Natl Acad Sci USA 1993; 90(2):482-486.
39. Li S, Leonard D, Wilkinson MF. T cell receptor (TCR) mini-gene mRNA expression regulated by nonsense codons: A nuclear-associated translation-like mechanism. J Exp Med 1997; 185(6):985-992.
40. Wang J, Vock VM, Li S et al. A quality control pathway that down-regulates aberrant T-cell receptor (TCR) transcripts by a mechanism requiring UPF2 and translation. J Biol Chem 2002; 277(21):18489-18493.
41. Delpy L, Sirac C, Magnoux E et al. RNA surveillance down-regulates expression of nonfunctional kappa alleles and detects premature termination within the last kappa exon. Proc Natl Acad Sci USA 2004; 101(19):7375-7380.
42. Alexander A, Steinmetz M, Barritault D et al. Gamma Heavy chain disease in man: cDNA sequence supports partial gene deletion model. Proc Natl Acad Sci USA 1982; 79(10):3260-3264.
43. Wahner-Roedler DL, Witzig TE, Loehrer LL et al. Gamma-heavy chain disease: Review of 23 cases. Medicine (Baltimore) 2003; 82(4):236-250.
44. Mendell JT, Sharifi NA, Meyers JL et al. Nonsense surveillance regulates expression of diverse classes of mammalian transcripts and mutes genomic noise. Nat Genet 2004; 36(10):1073-1078.
45. He F, Li X, Spatrick P et al. Genome-wide analysis of mRNAs regulated by the nonsense-mediated and 5' to 3' mRNA decay pathways in yeast. Mol Cell 2003; 12(6):1439-1452.
46. Jouanguy E, Altare F, Lamhamedi S et al. Interferon-gamma-receptor deficiency in an infant with fatal bacille Calmette-Guerin infection. N Engl J Med 1996; 335(26):1956-1961.
47. Medghalchi SM, Frischmeyer PA, Mendell JT et al. Rent1, a trans-effector of nonsense-mediated mRNA decay, is essential for mammalian embryonic viability. Hum Mol Genet 2001; 10(2):99-105.
48. Wilkinson MF. A new function for nonsense-mediated mRNA-decay factors. Trends Genet 2005; 21(3):143-148.
49. Wiegand HL, Lu S, Cullen BR. Exon junction complexes mediate the enhancing effect of splicing on mRNA expression. Proc Natl Acad Sci USA 2003; 100(20):11327-11332.
50. Nott A, Le Hir H, Moore MJ. Splicing enhances translation in mammalian cells: An additional function of the exon junction complex. Genes Dev 2004; 18(2):210-222.
51. Wang W, Czaplinski K, Rao Y et al. The role of Upf proteins in modulating the translation read-through of nonsense-containing transcripts. EMBO J 2001; 20(4):880-890.
52. Keeling KM, Lanier J, Du M et al. Leaky termination at premature stop codons antagonizes nonsense-mediated mRNA decay in S. cerevisiae. RNA 2004; 10(4):691-703.
53. Brumbaugh KM, Otterness DM, Geisen C et al. The mRNA surveillance protein hSMG-1 functions in genotoxic stress response pathways in mammalian cells. Mol Cell 2004; 14(5):585-598.
54. Enomoto S, Glowczewski L, Lew-Smith J et al. Telomere cap components influence the rate of senescence in telomerase-deficient yeast cells. Mol Cell Biol 2004; 24(2):837-845.

Chapter 7

Role of SMG-1-Mediated Phosphorylation of Upf1 in NMD

Akio Yamashita, Isao Kashima and Shigeo Ohno*

Abstract

Nonsense-mediated mRNA decay (NMD) in higher eukaryotes requires the phosphorylation and dephosphorylation of Upf1, an RNA helicase. Recent studies suggest that sequential phosphorylation and dephosphorylation of Upf1 follows the recognition of premature termination codons within mRNAs and is required for NMD. Upf1 phosphorylation, which is mediated by the phosphoinositide 3-kinase-related protein kinase SMG-1, triggers remodeling of the Upf1-containing surveillance complex to form another complex. This second complex, which contains SMG-5, SMG-7 and protein phosphatase 2A, is required for Upf1 dephosphorylation. Upf1 phosphorylation requires a SMG-1-Upf1 complex that interacts directly with the exon junction complex component Upf2, supporting the conclusion that Upf1 phosphorylation occurs on mRNA during the mRNA surveillance process.

Introduction

NMD is a surveillance mechanism that detects and degrades mRNAs containing premature termination codons (PTCs). Seven *smg* gene products have been identified as trans-acting factors that are required for NMD in *Caenorhabditis elegans* (Table 1; see chapter by Anderson), and each is conserved in mammals (Table 1; see chapters by Maquat, and Singh and Lykke-Andersen). One of these mammalian factors is Upf1 (SMG-2 in *C. elegans*), which undergoes a cycle of phosphorylation and dephosphorylation that is required for NMD. The other mammalian *smg* gene products can be divided into two groups based on function (Table 1). One group consists of SMG-1, Upf2 (SMG-3 in *C. elegans*) and Upf3 (SMG-4 in *C. elegans*), which are needed to phosphorylate Upf1. The other group consists of SMG-5, SMG-6, and SMG-7, which are required to dephosphorylate Upf1. Upf1 forms a surveillance complex with other Upf/SMG proteins. Other lines of evidence suggest that an exon junction complex (EJC) helps distinguish PTCs from normal translation termination codons in vertebrate cells. However, the biochemical nature of the surveillance complex, as well as the biological significance of the interactions between the surveillance complex and the EJC, are not well understood. Recent studies that aimed to clarify the roles of Upf1 phosphorylation and dephosphorylation during NMD and the interactions among proteins required for NMD not only establish the importance of both Upf1 phosphorylation and dephosphorylation but also reveal the mechanism by which the surveillance complex recognizes a PTC within mRNA.

*Corresponding Author: Shigeo Ohno—Department of Molecular Biology, Yokohama City University School of Medicine, Yokohama 236-0004, Japan.
Email: ohnos@med.yokohama-cu.ac.jp

Table 1. Evolutionary relationships of Upf/SMG proteins

Yeast	Nematodes	Flies	Mammals	Upf1 Complex	Role in Upf1 Phosphorylation
(Ton1/2p)	SMG-1	SMG-1	SMG-1	IC/LC	Kinase
Upflp	SMG-2	Upf1	Upf1	–	–
Upf2p	SMG-3	Upf2	Upf2	LC	Phosphorylation
			Upf3b	Unknown	Phosphorylation
Upf3p	SMG-4	Upf3	Upf3aL	LC	Unknown
			Upf3aS	WC	Unknown
(Nmd4P)	SMG-5	SMG-5	SMG-5	WC	Dephosphorylation
(Yor166cp)	SMG-6	SMG-6	SMG-6	Unknown	Dephosphorylation
(Est1p)	SMG-7	–	SMG-7	WC	Dephosphorylation
(Ebslp)					

IC, initial complex; LC, large complex; WC, wee complex. See text for details.

Structure and Biochemical Characterization of SMG-1

SMG-1 is the fifth member of a family of phosphoinositide 3-kinase-related protein kinases (PIKKs) that includes ATM (ataxia telangiectasia mutated), ATR (ATM and Rad3 related), mTOR (target of rapamycin) and DNA-PK (DNA-dependent protein kinase)[1] (see chapter by Abraham et al). Human and *C. elegans* SMG-1 share the highest overall similarity among PIKKs. SMG-1 shares PIKK and helical repeat regions (containing FAT domains) with all the PIKK family members (Fig. 1). SMG-1 activity shows preference for Mn^{2+} over Mg^{2+}, is inhibited in vitro by wortmannin (IC_{50} of 60-105 nM) and caffeine (IC_{50} of 0.3 mM), but is not inhibited in vitro by staurosporin nor rapamycin.[2-4] Wortmannin inhibits SMG-1 with an IC_{50} value of 1-2 μM in intact cells.[3,4] ATM, ATR and DNA-PK are S/T-Q-directed kinases,[1] and they show strong preference for phosphorylating serine (S)/threonine (T) followed by a glutamine (Q). Similarly to these PIKKs, SMG-1 preferentially phosphorylates S/T-Q, and it can phosphorylate p53 at serine 15.[3,4] SMG-1 and mTOR contain a TS (TOR, SMG-1 homology) domain, which consists of a binding site for the FKBP12-rapamycin complex in mTOR. However, there is no evidence suggesting that SMG-1 interacts with the FKBP12-rapamycin complex. The FKBP12-rapamycin-binding (FRB) domain essential for mTOR activation has been reported to bind phosphatidic acid (PA).[5] However, again, there is no evidence that SMG-1 interacts with PA.

In addition to the sequences conserved among all PIKK family members, there are SMG-1-specific sequences that may be involved in SMG-1-specific protein-protein interactions. Consistent with this notion, there are proteins that bind specifically to the N-terminal half of SMG-1 (Kashima et al, in preparation). Another unique structural feature of SMG-1 relative to other PIKKs is a large insert between the PIKK domain and the FAT-C domain. Although we do not know the role of this insert, its presence within the SMG-1 proteins of all species examined to date implies functional significance.

Human SMG-1 has at least two isoforms, p430 and p400, which have different N-terminal sequences (see also chapter by Abraham and Oliveira). An antibody against the N-terminal 106 amino acids of the 3657 amino acid-protein encoded by cDNA recognizes only the p430 isoform.[3] The p400 isoform lacks antibody recognition. The biological significance of this N-terminal sequence variation remains unclear.

SMG-1-Mediated Phosphorylation of Upf1

While genetic screens of *C. elegans* identified the seven *smg* genes required for NMD, genetic screens of *Saccharomyces cerevisiae* identified only the three *upf* genes required for NMD (Table 1; see also chapter by Baker and Parker). Therefore only the Upf1/SMG-2, Upf2/SMG-3 and Upf3/SMG-4 core components of the NMD surveillance complex are conserved in both eukaryotes. Knockdown of each *Drosophila melanogaster*[6] or human Upf/SMG protein suppresses NMD.[4,6-8] Additionally, components of the vertebrate exon junction complex (EJC) in addition to the Upf proteins have also been shown to be required for NMD (see chapter by Maquat). *C. elegans* SMG-2 is a phosphoprotein, and genetic studies have classified the six other *smg* proteins as regulators of SMG-2 phosphorylation.[9] SMG-1, SMG-3 and SMG-4 are required for SMG-2 phosphorylation, whereas SMG-5, SMG-6 and SMG-7 are required for SMG-2 dephosphorylation.[9]

Similar to *C. elegans* SMG-2, mammalian Upf1 is a phosphoprotein,[3,10] and SMG-1 kinase activity is essential for Upf1/SMG-2 phosphorylation and NMD in vivo.[3,11] SMG-1 phosphorylates Upf1/SMG-2 in vitro, suggesting that SMG-1 is a direct kinase of Upf1/SMG-2.[2,3,11] SMG-1 phosphorylates N-terminal and C-terminal regions of Upf1 in vitro.[3] Four serine residues, LS^{1073}QP, LS^{1078}QD, LS^{1096}QD and LS^{1116}QY, in the metazoan-specific C-terminal SQ-rich region of human Upf1 are efficiently phosphorylated by SMG-1 in vitro.[3] At least two of these residues, S1078 and S1096, are also phosphorylated in vivo in an SMG-1-dependent manner.[3] The sequence motifs surrounding S1078 and S1096 are conserved among metazoan Upf1s, supporting the importance of phosphorylation at these sites (Fig. 2). Furthermore, expression of phospho-resistant Upf1, in which the four serine residues targets of SMG-1-mediated phosphorylation were replaced with alanine, suppresses NMD in mammalian cells (Kashima et al, in preparation). All these results strongly suggest that NMD involves Upf1 phosphorylation at the four sites.

Only a very small fraction of Upf1 is phosphorylated in mammalian cells.[3,10,12] Importantly, overexpression of SMG-1 induces accumulation of highly phosphorylated Upf1 and accelerates NMD in mammalian cells.[3] Thus, SMG-1-mediated phosphorylation of Upf1 seems to be a critical rate-limiting step during NMD.

Mechanism of SMG-1-Mediated Phosphorylation of Upf1

SMG-3/Upf2 is essential for the SMG-1-mediated phosphorylation of Upf1 in both *C. elegans* and mammals[9] (Kashima et al, in preparation). Interestingly, SMG-1 associates with SMG-2 in a SMG-3-independent manner in *C. elegans*,[11] and human SMG-1 associates with Upf1 in a Upf2-independent manner (Kashima et al, in preparation). However, even though SMG-1 forms a complex with SMG-2, SMG-1 fails to phosphorylate SMG-2 in a *smg-3 C. elegans*.[9,11] A very similar situation applies to mammalian-cell counterparts (Kashima et al, in preparation). Additionally, mutated Upf1 that cannot associate with Upf2 is scarcely phosphorylated by SMG-1, although its association with SMG-1 remains intact in mammalian cells. These observations led us to hypothesize a model for Upf1 phosphorylation during NMD (Fig. 3). In the model, prior to phosphorylation by SMG-1, Upf1 associates with SMG-1 to form an initial SMG-1-Upf1 complex (IC). Subsequently, Upf2 associates with the IC and triggers Upf1 phosphorylation.

Upf1 can associate with eukaryotic translation release factor (eRF) 1 and 3 in yeast[13] and mammalian cells (Y.K. Kim and L.E. Maquat, unpublished data; G. Singh and J. Lykke-Andersen, unpublished data; I. Kashima and S. Ohno, unpublished data), and Upf2 is most likely an EJC component (see chapters by Maquat and Singh and Lykke-Andersen). Taken together, the finding that SMG-1-mediated phosphorylation of SMG-2/Upf1 requires SMG-3/Upf2 in *C. elegans*[9] and mammals (Kashima et al, in preparation) suggests that SMG-1-mediated Upf1 phosphorylation occurs on EJCs after PTC recognition (Fig. 3). This notion is supported by the following observations. First, the phosphorylated form of Upf1 is

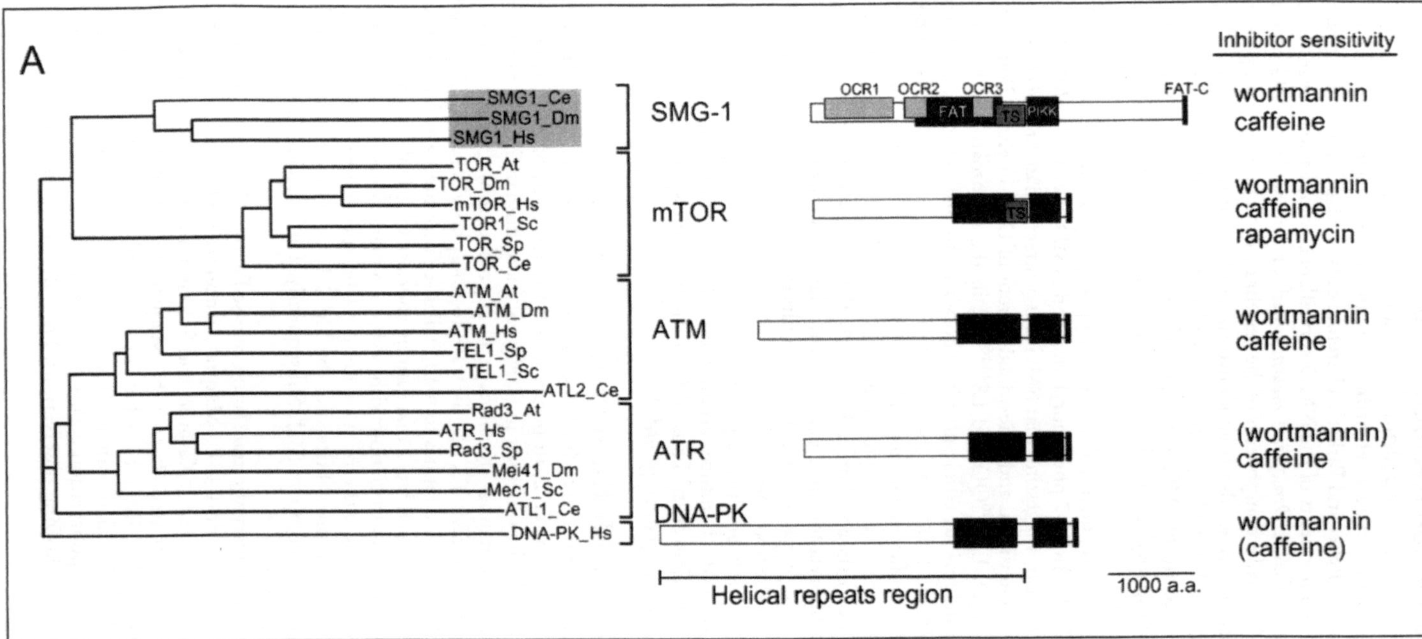

Figure 1. A) Molecular phylogenic tree of PIKK (PI3K-related protein kinase) domains of PIKK family members (left). Schematic representation of mammalian PIKK family members (right). The putative PIKK, FAT-C, and FAT domains are shown as black boxes. These domains are conserved among PIKK family members. The TS (TOR-SMG-1 conserved) domains are shown as dark gray boxes. SMG-1 OCR (one-specific conserved region) domains are shown as light gray boxes. Inhibitor sensitivity is shown to the right. Abbreviations: Ce, *Caenorhabditis elegans*; Dm, *Drosophila melanogaster*, Hs, *Homo sapiens*; At, *Arabodopsis thalia*; Sc, *Saccharomyces cerevisiae*. Figure continued on next page.

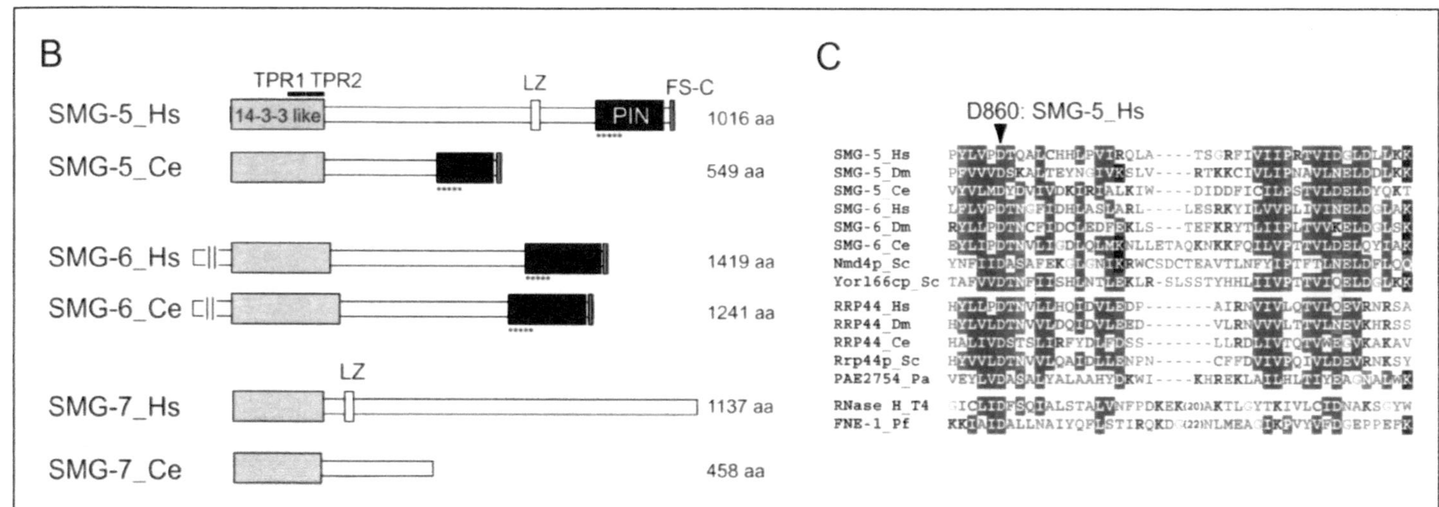

Figure 1, continued. B) Schematic structures of human and *C. elegans* SMG-5, SMG-6 and SMG-7. 14-3-3-like domains are shown as light gray boxes. TPR motifs of the 14-3-3-like domains are shown as black bars. PIN domains are shown as black boxes. FS-C (SMG-5 [five]-SMG-6 [six] C-terminus) domains are shown as dark gray boxes. Putative LZ (leucine zipper) motifs are shown as white boxes. Asterisks indicate alignment regions shown in (C). C) Sequence comparison of the N-terminal half of the human SMG-5 PIN domain with other PIN domain-containing proteins. The arrowhead shows the asparatic acid that is essential for SMG-5 function and is conserved in PIN domain-containing proteins and nucleases. Abbreviations: Pa, *Pyrobaculum aerophilum*, T4, T4 bacteriophage; Pf, *Pyrococcus furiosus*. RRP44, 3'-5' exonuclease (exosome component). PAE2754, Mg^{2+}-dependent exonuclease. FNE1, 5'-3' flap endonuclease.

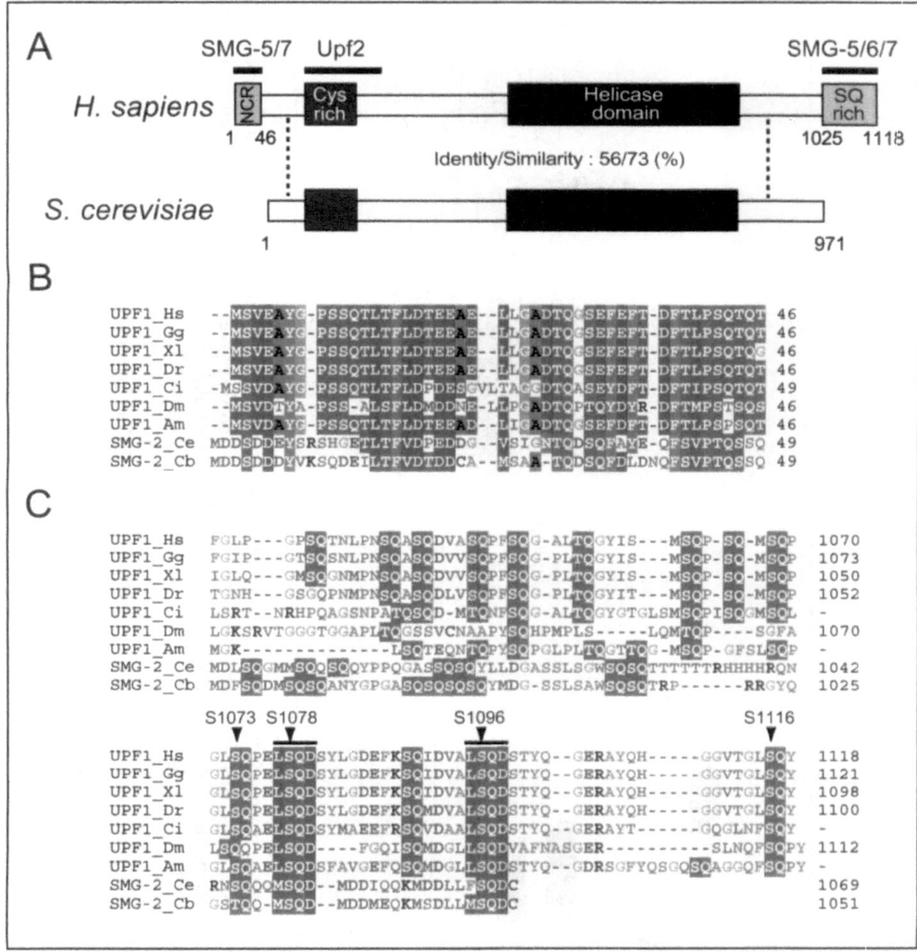

Figure 2. A) Schematic structures of human and yeast Upf1. Metazoan-specific NCR (N-terminal conserved region) and SQ-rich region of Upf1 are shown as gray boxes. Helicase domains are shown as black boxes. Cysteine-rich regions are shown as dark gray boxes. Putative Upf/SMG protein-interacting regions are indicated as black bars. B) Alignment of the NCR of Upf1. Identical and similar amino acid residues are shaded with color. Abbreviations: Hs, *Homo sapience*; Gg, *Gallus gallus*; Xl, *Xenopus laevis*; Dr, *Danio rerio*; Ci, *Ciona intestinalis*; Dm, *Drosophila melanogaster*; Am, *Apis mellifera*; Ce, *Caenorhabditis elegans*; Cb, *Caenorhabditis briggsae*. C) Alignment of SQ-rich region of Upf1. SQ/TQ motifs are shaded with color. Conserved amino acids surrounding S1078 and S1096 are shaded with color. Arrow-heads show serine residues that can be phosphorylated by SMG-1.

primarily present in the polysomal fraction of mammalian cells.[10] Second, SMG-1 immunoprecipitates with EJC components in a Upf2-dependent manner (Kashima et al, in preparation). Third, SMG-1 immunoprecipitates with poly(A) binding protein (PABP) and cap binding protein (CBP) 20 only when cell extracts are not treated with RNase (Kashima et al, in preparation).

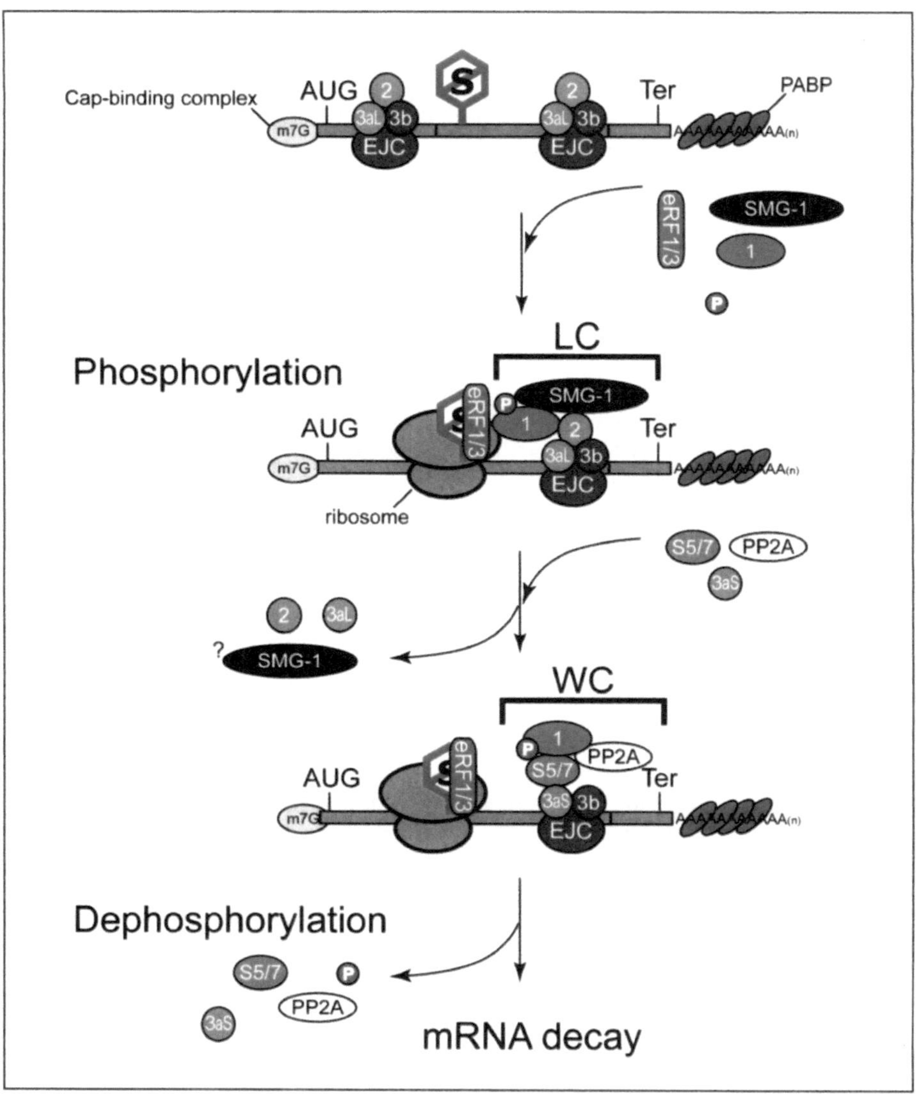

Figure 3. Hypothetical model showing the roles of Upf1 phosphorylation and dephosphorylation during mRNA surveillance in metazoans. A fraction of nonphosphorylated Upf1 associates with SMG-1 to form an initial SMG-1-Upf1 complex (IC). The ribosome that recognizes a termination codon recruits IC to mRNA. If IC can associate with the Upf2 component of the cytoplasmic EJC to form the large complex (LC), then SMG-1 phosphorylates Upf1. Upf1 phosphorylation triggers remodeling of the LC to form the wee complex (WC) and causes dephosphorylation of Upf1 by PP2A. Dephosphorylation is followed by degradation of the corresponding mRNA. See text for details. NMD components are shown as color-coded circles. Abbreviations: P, Phosphate group; U1, Upf1; U2, Upf2; 3b, Upf3b; 3aL, Upf3a long; 3aS, Upf3a short; S5/7, SMG-5-SMG-7 complex; eRF1/3, eRF1-eRF3 complex. AUG, start codon; Ter, termination codon; S, stop codon that triggers NMD.

Consequences of SMG-1-Mediated Phosphorylation of Upf1

To date, the ATPase activities of nonphosphorylated and phosphorylated Upf1 are not detectably different (I. Kashima and T. Ohnishi, unpublished data). On the other hand, there is evidence that phosphorylation induces remodeling of the Upf1-containing surveillance complex. SMG-5 and SMG-7 interact and associate with phosphorylated (P)-Upf1 to form the wee complex (WC)[12,14] (Fig. 3). WC contains the catalytic subunit of protein phosphatase 2A (PP2A) and Upf3aS.[12] As noted above, SMG-5 and SMG-7 are essential for Upf1/SMG-2 dephosphorylation[12,14] (Kashima et al, in preparation).

Although sequence analysis does not predict SMG-5 and SMG-7 to be either phosphatases or kinases, they appear to be targeting subunits of PP2A. Okadaic acid, a potent inhibitor of PP2A, increases accumulation of P-Upf1 and the presence of WC in vivo.[3,12] Thus, PP2A is most likely a P-Upf1 phosphatase. The N-terminal sequence of SMG-7, including the TPR (tetratricopeptide repeat) motif, shows structural similarity to 14-3-3[15] and associates with the SQ-rich sequence within the Upf1 C-terminus in a phospho-specific manner.[15] Consistently, a phosphorylation-resistant variant of Upf1 in which four serine residues that are normally targeted for phosphorylation by SMG-1 are replaced by alanine shows reduced affinity for the SMG-5-SMG-7 complex (Kashima et al, in preparation). The 14-3-3-like domains are conserved in SMG-5, SMG-6 and SMG-7, suggesting that they can also bind the phosphorylated Upf1 peptide sequence (Fig. 1C). Additionally, SMG-5 and SMG-6 share PIN (PilT amino-terminal) and FS-C (SMG-5 [Five]-SMG-6 [Six] C-terminus) domains within their C-termini (Fig. 1C). A PIN domain is found in many proteins having nuclease activity[16] (Fig. 1C). SMG-5 harboring an asparatic acid (D)-860 to alanine (A) substitution with the PIN domain or a deletion that removes the FS-C domain maintain the ability to selectively bind P-Upf1, Upf3aS, SMG-7 and PP2A but fail to support P-Upf1 dephosphorylation and, when overexpressed, block NMD. Thus, both PIN and FS-C domain are essential for SMG-5 function in P-Upf1 dephosphorylation and NMD. D860 within the PIN domain is strictly conserved within SMG-5 and SMG-6 as well as PIN domains from other proteins and is thought to be essential for catalytic activity[16] (Fig. 1C). Therefore, it is possible that SMG-5 and SMG-6 have nucleotide binding and/or nuclease activity, which might be essential to trigger dephosphorylation of P-Upf1 and NMD. SMG-6 has been shown to associate with Upf1, Upf2, Upf3b and PP2A, although it is unclear whether it also associates with P-Upf1, SMG-5 and SMG-7.[17] To date, we do not know the role of SMG-6 on the mRNA surveillance complex.

The SMG-1-independent interaction between SMG-2 and SMG-5 in *C. elegans*, and the association of recombinant, nonphosphorylated amino acids 1-462 of human Upf1 with human SMG-5 in vitro suggest that SMG-5, and probably SMG-7, can also associate with Upf1/SMG-2 in a phosphorylation-independent manner.[12,14] Upf1 harboring a truncated metazoan-specific, N-terminal region (NCR, N-terminal conserved region, see Fig. 2B) has a reduced affinity for the SMG-5-SMG-7 complex and is more highly phosphorylated than wild-type Upf1 in vivo.[12] The above observations not only support the importance of the SMG-5-SMG-7 complex for P-Upf1 dephosphorylation, but also suggest that the NCR is a primary binding domain of the SMG-5-SMG-7 complex. Initially, the SMG-5-SMG-7 complex associates with the NCR, possibly in a phosphorylation-independent manner, and then binds the phosphorylated SQ-motif within the C-terminal region of Upf1. How can the SMG-5-SMG-7 complex support dephosphorylation of Upf1 by PP2A? One possibility is that the SMG-5-SMG-7 complex supports the action of PP2A in WC. Another possibility is that, similar to the situation for IC, WC needs additional factor(s) to trigger Upf1 dephosphorylation, and the SMG-5-SMG-7 complex recruits the activator(s) of PP2A. At present, we cannot determine which, if either, of these possibilities is applicable.

Upf2 makes a large, ~1000-kDa protein complex (LC) that contains Upf1 and Upf3aL[18] but not SMG-5, SMG-7, Upf3aS[12,18] (Fig. 3). Intriguingly, Upf1 harboring an NCR deletion associates with both Upf2 and Upf3aL more strongly than does wild type Upf1, suggesting that the association of the SMG-5-SMG-7 complex with P-Upf1 results in release of

Up2, Upf3aL and possibly SMG-1. WC complex formation then induces the dephosphorylation of P-Upf1. We do not know much about SMG-5-SMG-7 complex function except in P-Upf1 dephosphorylation. Yellow fluorescent protein-tagged SMG-7 recruits Cyan fluorescent protein-tagged Upf1 to cytoplasmic foci (known as P-bodies or GW-bodies) (see chapters by Baker and Parker, and Singh and Lykke-Anderson),[19] suggesting that SMG-7 might target mRNA to P-bodies. The specific association of P-Upf1 and SMG-7 suggests that Upf1 dephosphorylation might occur in P-bodies. Consistent with this possibility, artificially tethering SMG-7 to mRNA results in mRNA decay independently of an upstream termination codon.[19] The role of P-Upf1 dephosphorylation might be to trigger the mRNA degradation machinery.

Regulation of SMG-1-Mediated Upf1 Phosphorylation

The amount of P-Upf1 in mammalian cells is reduced by serum starvation.[10] Additionally, NMD is inhibited by the amino acid starvation.[20] Serum and amino acid starvation suppresses translation.[21,22] Therefore, inhibition of Upf1 phosphorylation and NMD by starvation may be simply because of reduced translation. Serum and amino acid starvation suppress the mTOR signaling cascade, which positively regulates translation.[1] On the other hand, rapamycin, which is a specific inhibitor of mTOR, does not inhibit NMD or Upf1 phosphorylation in mammalian cells.[3] Overproduction of 4E-binding protein (BP)1, which is regulated by mTOR, cannot suppress NMD, since it inhibits eukaryotic translation initiation factor (eIF)4E-dependent but not CBP80/CBP20-dependent translation[23] (see chapter by Maquat). Taken together, these observations suggest that SMG-1 might respond to extracellular stimulation and be inactivated by serum and/or amino acid starvation. Indeed, SMG-1 is a stress-activated kinase that is stimulated by ionizing radiation (IR) and UV-B.[4] Genotoxic stress induces SMG-1- and ATM-dependent Upf1 phosphorylation in mammalian cells[4] It will be interesting to define the role of Upf1 phosphorylation after genotoxic stress. Unlike in *C. elegans*, SMG-1-deficiency in adult *D. melanogaster* does not abolish NMD.[24] In this situation, ATM might regulate Upf1 phosphorylation, although no significant role of ATM in NMD has been identified in mammalian cells. SMG-1 can also contribute to the regulation of p53 activation (see chapter by Abraham and Oliveira).

Suppression of SMG-1-Mediated Upf1 Phosphorylation Inhibits NMD

The finding that NMD requires Upf1 phosphorylation and dephosphorylation raises the promise of experimental strategies that specifically suppress NMD in mammalian cells. Inhibitors of SMG-1, the overproduction of kinase-inactive SMG-1 or SMG-1 knockdown can inhibit both Upf1 phosphorylation and NMD in mammalian cells[3,4,25] (Usuki et al, in preparation). These inhibitors also prevent the decay of endogenous nonsense-containing p53 mRNA in two cancer cells[3] and nonsense-containing collagen VI $\alpha 2$ mRNA in Ullrich's disease fibroblasts.[25] The inhibition of SMG-1 results in accumulation of C-terminus truncated proteins in vivo.[3,25] Those results were confirmed by SMG-1 knockdown[4] (Usuki et al, in preparation). Notably, in case of Ullrich's disease fibroblasts, up-regulation of a truncated collagen VI $\alpha 2$ subunit results in collagen VI assembly and partially functional extracellular matrix formation[25] (Usuki et al, in preparation). NMD protects cells by preventing aberrant truncated proteins that manifest gain-of-function and dominant-negative effects that can result in genetic diseases (see chapter by Holbrook et al). In some cases, such as Ullrich's fibroblasts, truncated protein retains normal functions at least partially, and the selective inhibition of NMD may provide a novel strategy to rescue the disease phenotype in this and other PTC-related diseases where the truncated protein shows no dominant-negative effect. However, pharmacological approaches using wortmannin or caffeine, which is another inhibitor of SMG-1, cannot be administered to patients because of toxicity and nonspecific side effects. Identification of specific inhibitors of SMG-1 needs to be established for useful clinical applications.

Anyway, suppression or inhibition of Upf1 phosphorylation has provided a rather selective suppression of NMD under a variety of experimental conditions. This has greatly not only clarified the physiological role of NMD in mammalian cells but also identified target genes regulated by NMD in a variety of physiological and pathological conditions.

Perspectives

As described above, accumulating evidence suggests that Upf1 phosphorylation by SMG-1 occurs upon PTC recognition during the mRNA surveillance process. Recently, phosphorylation of Upf1 protein (p) was also described in yeast.[26] As in metazoans, phosphorylated Upf1p in yeast is less abundant than unphosphorylated Upf1p. Yeast Upf1p does not contain the C-terminal SQ-rich and NCR regions of Upf1/SMG-2 that mediate SMG-5-SMG-7 complex binding (Fig. 2). One intriguing possibility is that Tor1p and Tor2p are Upf1p kinases. Tor1p and Tor2p are most closely related to the SMG-1 PIKK family member (Fig. 1A), but they do not preferentially target the S/T-Q motif.[1] While no obvious orthologs to SMG-1, SMG-5, SMG-6 and SMG-7 are present in the yeast genome, there are four proteins, Nmd4p, Yor166cp, Est1p and EBS1p, that resemble SMG-5, SMG-6 and SMG-7 that could function like metazoan SMG-5, SMG-6 and SMG-7. For instance, Nmd4p and Yor166cp contain a PIN domain related to that of SMG-5 and SMG-6[16] (Fig. 1C), and Nmd4p interacts with Upf1p.[27] Est1p, which has been identified as a putative homolog to SMG-6 (see chapter by Azzalin et al), and Ebs1p, both of which are essential for telomerase function, have a 14-3-3 like domain, although they lack several phosphoserine binding residues.[15] It remains to be clarified whether the cycle of Upf1 phosphorylation and dephosphorylation is critical for NMD in yeast as it is in metazoans.

Acknowledgements

We thank L. Maquat and J. Lykke-Andersen for communicating unpublished results at the time of writing. This work was supported in part by grants to S. Ohno from the Japan Society for the Promotion of Science, from the Ministry of Education, Culture, Sports, Science and Technology of Japan, and from the Collaboration of Regional Entities for the Advancement of Technological Excellence program of Japan Science and Technology Agency. A. Yamashita received support from the Yokohama Foundation for Advancement of Medical Science. I. Kashima is a Research Fellow of the Japan Society for the Promotion of Science.

References

1. Abraham RT. PI 3-kinase related kinases: 'Big' players in stress-induced signaling pathways. DNA Repair (Amst) 2004; 3(89):883-887.
2. Denning G, Jamieson L, Maquat LE et al. Cloning of a novel phosphatidylinositol kinase-related kinase: Characterization of the human SMG-1 RNA surveillance protein. J Biol Chem 2001; 276(25):22709-22714.
3. Yamashita A, Ohnishi T, Kashima I et al. Human SMG-1, a novel phosphatidylinositol 3-kinase-related protein kinase, associates with components of the mRNA surveillance complex and is involved in the regulation of nonsense-mediated mRNA decay. Genes Dev 2001; 15(17):2215-2228.
4. Brumbaugh KM, Otterness DM, Geisen C et al. The mRNA surveillance protein hSMG-1 functions in genotoxic stress response pathways in mammalian cells. Mol Cell 2004; 14(5):585-598.
5. Fang Y, Vilella-Bach M, Bachmann R et al. Phosphatidic acid-mediated mitogenic activation of mTOR signaling. Science 2001; 294(5548):1942-1945.
6. Gatfield D, Unterholzner L, Ciccarelli FD et al. Nonsense-mediated mRNA decay in Drosophila: At the intersection of the yeast and mammalian pathways. EMBO J 2003; 22(15):3960-3970.
7. Mendell JT, ap Rhys CM, Dietz HC. Separable roles for rent1/hUpf1 in altered splicing and decay of nonsense transcripts. Science 2002; 298(5592):419-422.
8. Kim YK, Furic L, Desgroseillers L et al. Mammalian Staufen1 recruits Upf1 to specific mRNA 3'UTRs so as to elicit mRNA decay. Cell 2005; 120(2):195-208.

9. Page MF, Carr B, Anders KR et al. SMG-2 is a phosphorylated protein required for mRNA surveillance in Caenorhabditis elegans and related to Upf1p of yeast. Mol Cell Biol 1999; 19(9):5943-5951.
10. Pal M, Ishigaki Y, Nagy E et al. Evidence that phosphorylation of human Upfl protein varies with intracellular location and is mediated by a wortmannin-sensitive and rapamycin-sensitive PI 3-kinase-related kinase signaling pathway. RNA 2001; 7(1):5-15.
11. Grimson A, O'Connor S, Newman CL et al. SMG-1 is a phosphatidylinositol kinase-related protein kinase required for nonsense-mediated mRNA Decay in Caenorhabditis elegans. Mol Cell Biol 2004; 24(17):7483-7490.
12. Ohnishi T, Yamashita A, Kashima I et al. Phosphorylation of hUPF1 induces formation of mRNA surveillance complexes containing hSMG-5 and hSMG-7. Mol Cell 2003; 12(5):1187-1200.
13. Czaplinski K, Ruiz-Echevarria MJ, Paushkin SV et al. The surveillance complex interacts with the translation release factors to enhance termination and degrade aberrant mRNAs. Genes Dev 1998; 12(11):1665-1677.
14. Anders KR, Grimson A, Anderson P. SMG-5, required for C. elegans nonsense-mediated mRNA decay, associates with SMG-2 and protein phosphatase 2A. EMBO J 2003; 22(3):641-550.
15. Fukuhara N, Ebert J, Unterholzner L et al. SMG7 is a 14-3-3-like adaptor in the nonsense-mediated mRNA decay pathway. Mol Cell 2005; 17(4):537-547.
16. Clissold PM, Ponting CP. PIN domains in nonsense-mediated mRNA decay and RNAi. Curr Biol 2000; 10(24):R888-890.
17. Chiu SY, Serin G, Ohara O et al. Characterization of human Smg5/7a: A protein with similarities to Caenorhabditis elegans SMG5 and SMG7 that functions in the dephosphorylation of Upf1. RNA 2003; 9(1):77-87.
18. Schell T, Kocher T, Wilm M et al. Complexes between the nonsense-mediated mRNA decay pathway factor human upfl (up-frameshift protein 1) and essential nonsense-mediated mRNA decay factors in HeLa cells. Biochem J 2003; 373(Pt 3):775-783.
19. Unterholzner L, Izaurralde E. SMG7 acts as a molecular link between mRNA surveillance and mRNA decay. Mol Cell 2004; 16(4):587-596.
20. Mendell JT, Sharifi NA, Meyers JL et al. Nonsense surveillance regulates expression of diverse classes of mammalian transcripts and mutes genomic noise. Nat Genet 2004; 36(10):1073-1078.
21. Pain VM. Translational control during amino acid starvation. Biochimie 1994; 76(8):718-728.
22. Duncan R, Hershey JW. Regulation of initiation factors during translational repression caused by serum depletion. Abundance, synthesis, and turnover rates. J Biol Chem 1985; 260(9):5486-5492.
23. Chiu SY, Lejeune F, Ranganathan AC et al. The pioneer translation initiation complex is functionally distinct from but structurally overlaps with the steady-state translation initiation complex. Genes Dev 2004; 18(7):745-754.
24. Chen Z, Smith KR, Batterham P et al. Smg1 nonsense mutations do not abolish nonsense-mediated mRNA decay in Drosophila melanogaster. Genetics 2005; In press.
25. Usuki F, Yamashita A, Higuchi I et al. Inhibition of nonsense-mediated mRNA decay rescues the phenotype in Ullrich's disease. Ann Neurol 2004; 55(5):740-744.
26. de Pinto B, Lippolis R, Castaldo R et al. Overexpression of Upf1p compensates for mitochondrial splicing deficiency independently of its role in mRNA surveillance. Mol Microbiol 2004; 51(4):1129-1142.
27. He F, Jacobson A. Identification of a novel component of the nonsense-mediated mRNA decay pathway by use of an interacting protein screen. Genes Dev 1995; 9(4):437-454.

CHAPTER 8

Physiologic Substrates and Functions for Mammalian NMD

Neda A. Sharifi and Harry C. Dietz*

Abstract

Nonsense-mediated mRNA decay (NMD) is ubiquitous among eukaryotes. While this degree of conservation certainly attests to strong evolutionary pressure for maintenance of NMD, the mechanistic basis for this maintenance has remained both speculative and controversial. The fact that NMD was initially discovered, and has been most comprehensively studied, in the context of acquired premature termination codons (PTCs) that disrupt protein coding potential and exert overt phenotypic consequences (including disease), has certainly flavored (if not biased) interpretation of its basic physiologic purpose(s). The prevailing view has been that NMD protects the organism from deleterious truncated peptides with dominant-negative or gain-of-function potential that would be expressed from nonsense alleles if the corresponding transcripts were stable. Given that the fitness of individual organisms, as opposed to the population at large, is of little consequence to evolution, this restricted function of NMD cannot plausibly explain its conservation. This chapter will outline the emerging view that NMD plays a broad role in the regulation of gene expression, and, as a direct or indirect consequence, in the maintenance of critical developmental and homeostatic functions including orchestration of the response to environmental stress. The biologic and medical implications of this insight will also be discussed.

Lessons from Lower Eukaryotes

NMD is not required for the survival of *Saccharomyces cerivisiae*, *Schizosaccharomyces pombe*, or *Caenorhabditis elegans*. NMD-deficient yeast strains show a minor respiratory defect when grown with a suboptimal carbon source, while NMD-deficient worms (*smg⁻* strains) show minor morphologic abnormalities of genitalia and reduced brood sizes.[1-4] Studies using *C. elegans* have been particularly informative regarding the biological importance of at least one function of NMD. Heterozygous nonsense mutations in a number of genes have been found to be recessive in NMD-competent strains but dominant in NMD-deficient strains. The best-characterized example involves the UNC-54 gene that encodes myosin heavy chain in worm body wall muscles.[5] Worms heterozygous for nonsense mutations in *unc-54* show significant degradation of nonsense transcripts and are phenotypically normal. Targeted disruption of NMD results in substantial accumulation of mutant *unc-54* mRNA, expression of dominant-negative truncated myosin fragments, and an "uncoordinated" (abnormal movement and reduced mobility in the posterior portion of the body) phenotype.[3,6] Rare (but informative) examples of physiologic substrates for NMD in nematodes have been described. For

*Corresponding Author: Harry C. Dietz—McKusick-Nathans Institute of Genetic Medicine, Howard Hughes Medical Institute, Johns Hopkins University School of Medicine, Baltimore, Maryland 21205, U.S.A. Email: hdietz@jhmi.edu

Nonsense-Mediated mRNA Decay, edited by Lynne E. Maquat. ©2006 Eurekah.com.

example, *smg*⁻ strains show accumulation of aberrantly spliced SRp20 and SRp30b mRNAs that result due to the inclusion of alternative exons that contain PTCs relative to the extended translational open reading frame (ORF). Curiously, other mRNAs containing alternatively spliced exons encoding PTCs did not appear to be substrates for NMD, suggesting either a context- or gene-specified influence.[7] Multiple naturally occurring unproductively spliced ribosomal protein mRNAs were also shown to be degraded by NMD. These transcripts were unified by the incomplete removal of intronic sequences due to utilization of alternative splice sites. Intriguingly, overexpression of one of these proteins (RPL-12) resulted in increased production of the corresponding unproductive mRNA that was subsequently decayed by NMD, allowing autoregulation of protein abundance and function.[8] NMD also clears pseudogene-derived RNAs that exert no selectable function and have therefore acquired PTCs through nematode evolution.[9] Nevertheless, these examples do not extend the repertoire of mechanisms by which NMD influences gene expression beyond elimination of "trash" transcripts with no demonstrated or inferred function.

Studies using yeast have been more comprehensive, and hence more revealing. To a variable extent, splicing is inherently inefficient, resulting in a low but significant level of incompletely spliced pre-mRNAs that encode PTCs and are predictable substrates for NMD. Indeed, the yeast CYH2, MER2 and RP51B pre-mRNAs all have intron-encoded PTCs and show a tremendous increase in steady-state abundance when NMD is inactivated. In this light, aberrant mRNAs resulting from other natural sources of error or inefficiency, including faulty transcription initiation or RNA processing, would also be subject to nonsense surveillance.[10]

Transcripts containing a short upstream translational open reading frame (uORF) in their 5' untranslated region (UTR) would also be expected to be targets for NMD, because the stop codon of the uORF can be recognized as a premature stop codon.[11] The important distinction here is that these mRNAs retain full ability to encode their intended protein and function. Competition for initiation of scanning to the AUG of the uORF or to the AUG of the downstream "intended" ORF would dictate both the steady-state abundance and translational yield of a pool of nascent uORF-encoding mRNAs. It was also inherently obvious, and later demonstrable, that this process is subject to regulation and provides a dynamic mechanism to fine tune the expression of this class of mRNAs. For example, the CPA1 gene in yeast contains a uORF that encodes a 25 amino acid peptide. Under nutrient-replete conditions, arginine induces ribosomal stalling at the termination codon of the uORF and prevents scanning to and translation of the downstream ORF encoding carbamoylphosphate synthetase (CPSase A), culminating in NMD of the mRNA. In the absence of arginine, there is efficient translation of the downstream CPSase A ORF and hence avoidance of NMD. Deletion of any of the UPF genes that encode trans-factors essential for yeast NMD prevents arginine repression of CPSase synthesis.[12] For GCN4 mRNA, which encodes multiple uORFs, the abundance of eukaryotic translation initiation factor (eIF)2 influences alternative ribosomal initiation at uORFs that either do or do not promote reinitiation at the downstream Gcn4p ORF and suppression of NMD.[13] Furthermore, both the GCN4 and YAP1 uORF-containing mRNAs harbor a stabilizing element that binds the RNA binding protein Pub1p, preventing NMD.[14,15] Thus, through uORFs, NMD provides a mechanism to alter gene expression in response to variation in either transcriptional programs or environmental stimuli.

Recent chip-based transcriptional profiling strategies have attempted to elucidate physiologic substrates for NMD on a genome-wide scale. Although discussed in greater detail in other chapters (see chapter by He and Jacobson), a brief review here is warranted. The first such study found that about 8% of all yeast mRNAs showed significant changes in steady-state abundance after deletion of any or all of the UPF genes, with the vast majority showing an increase. Curiously, the majority of these effects occurred at the level of transcript production rather than stability, inferring that the direct effect was limited to a relatively small subset of mRNAs that secondarily promote increased transcriptional yield of other targets.[16] In a refinement of this strategy, Jacobson and colleagues found that NMD-deficient yeast strains showed increased expression of about 10% of all transcripts assayed, that the majority of randomly selected NMD-regulated transcripts show a corresponding increase in stability, and that about one-third of all NMD-regulated mRNAs had an architecture that allowed inference regarding the mechanism of initiation of NMD.[17] As predicted from prior work, these included those with known nonsense mutations in their coding sequence, retained intronic sequences, uORFs or out-of-frame 5' UTR AUGs that lead to translation termination a short distance into the downstream "intended" ORF, and leaky scanning that promotes translation initiation at an out-of-frame AUG in the intended ORF. Additional classes of mRNAs included those that utilize +1 frameshifting in their translation, bicistronic mRNAs, mRNAs encoded by expressed pseudogenes, and transposable elements or their long terminal repeat sequences. Once again, there were numerous and intuitive possibilities for physiologic regulation of NMD. As might be predicted, the NMD-regulated mRNAs had relatively low baseline abundance. RNAs in two functional MIPS (Munich Information Center for Protein Sequences) categories, cell rescue and defense and transport facilitation, were enriched among those that were regulated by NMD. Interestingly, telomere maintenance had been previously shown to be defective in NMD-deficient strains.[18] Two separate microarray studies showed that multiple mRNAs that contribute to telomere maintenance are increased in abundance after NMD inhibition; some showed increased stability (e.g., EST1), while others did not (e.g., EST3 and STN1), suggesting both direct and indirect NMD regulation, respectively[17,19] (see chapter by Azzalin et al).

Mammalian NMD

Given the similarities between the mechanism and outcome of NMD in higher and lower eukaryotes that is described extensively in other chapters, should we expect the repertoire of physiologic targets and purposes to be any different? The answer is both no and yes. No, in that the basic rule that any mRNA that satisfies the structural constraints for initiation of NMD will likely do so, irrespective of origin. Yes, in that these structural constraints differ between organisms. Furthermore, the increased complexity of genome organization and processing events underlying diversification of the output of mammalian genomes offers many more possibilities for the generation of "trash" RNAs and the regulated production of alternative mRNAs. Finally, given the evolutionary timeframe within which NMD has been an integral part of eukaryotic cellular function, it seems likely that NMD has been integrated into higher developmental and homeostatic functions in mammals.

Lessons from Animal Models: A Glimpse into This Increased Complexity Derived from Gene Targeting Experiments in Mice

The yeast UPF1 gene has defined homologues that derive orthologous proteins in worms (SMG-2), mice and humans (UPF1, also known as RENT1).[20-22] As previously mentioned, UPF1 or SMG-2 is not required for viability in yeast and worms, respectively. Both expression of a dominant-negative form of Upfl and gene silencing using RNA interference (i) have demonstrated that this gene is essential for mammalian NMD.[23,24] Mice heterozygous for a disrupted *Upfl* allele are born at the expected Mendelian frequency, are phenotypically normal, and show full efficiency of NMD, suggesting that the concentration of Upfl is not limiting in mammalian NMD.[25] The surprising finding was that homozygotes were never observed in heterozygote matings. A developmental survey revealed that wild-type, heterozygous, and homozygous targeted embryos were found at the expected 1:2:1 ratio in the preimplantation stage (3.5 days post coitum, dpc). Homozygosity for the disrupted allele was under-represented at 6.5 dpc, and was never observed at or after 7.5 dpc. Homozygous embryos showed loss of NMD in early preimplantation stages, diffuse regression due to apoptosis by 5.5 dpc, and complete absence from the decidua by 7.0 dpc. In that loss of viability correlated temporally with the complex process of implantation, it remained possible that Upfl function is required for establishment of a nutrient supply as opposed to basic cellular metabolism. Contrary to this hypothesis, null embryos placed in culture at 3.5 dpc performed normally until 5.5 dpc, but showed complete regression of the inner cell mass by 7.5 dpc as the pool of oocyte-derived Upfl mRNA was waning. These data suggest that Upfl function is required for mammalian cellular viability. The inference was that mammalian cells are dependent upon NMD. Since this time, however, it has become evident that Upfl has functions that are distinct from NMD including nonsense-mediated altered splicing (NAS) and Staufen1-mediated mRNA decay (SMD)[25,26] (see chapters by Zhang and Krainer, and Kim and Maquat). The former relates to the ability of certain PTCs to influence splicing events, while the latter relates to the induced decay after Staufen1-mediated recruitment of Upfl to the 3' UTR of certain transcripts (e.g., Arfl). Importantly, neither of these processes requires the function of other factors known to be essential for NMD (e.g., Rent2/Upf2). Thus, it remains formally possible that the dependency of mammalian cells upon Upfl function relates to abrogation of NAS, SMD or another pathway, as opposed to NMD. The observation that mammalian cells appear more sensitive to RNAi-mediated silencing of Upfl, as opposed to Upf2 (our unpublished data), may add credence to this hypothesis; alternatively, it may simply relate to the exceptional efficiency of Upfl RNAi-mediated gene silencing.

The finding that NMD factors are essential for mouse survival was somewhat disappointing in that it precluded use of NMD-deficient animals to further investigate the role of NMD in the modification of human genetic disease. The question remained whether a relative (as opposed to complete) attenuation of NMD efficiency might be both tolerable and sufficient for this purpose. To address this issue, a transgenic mouse line was created that constitutively expresses a known dominant-negative form of Upfl (R844C) on an otherwise normal genetic background (our unpublished data). While both viable and fertile, these mice showed a substantial (~3-fold) stabilization of nonsense mRNAs in all tissues tested. With the exception of a variable and transient stunting of growth, there were no outward phenotypic abnormalities. There was, however, a major hint regarding particularly susceptible tissues, which were presumably those that have the highest physiologic burden of nonsense transcripts. One physiologic process unique to higher organisms where NMD is likely to be of paramount importance is maturation of the immune system (see chapter by Gudikote and Wilkinson). Immunoglobulin variable chain (Ig) and T cell receptor (TCR) genes undergo programmed rearrangement that involves the joining of V, (D) and J segments in different combinations. Additional diversity is generated from the addition of a variable number of nontemplated nucleotides at the junction of the variable gene segments.[27,28] The random nature of these events contributes to the immunologic diversity needed for recognition of a vast repertoire of antigen specificities; it also dictates that two of three rearrangements will be nonproductive, i.e., will be out-of-frame and generate a downstream PTC. While it is known that these nonsense transcripts are subject to extreme downregulation by NMD, perhaps making use of defined cis-elements that dictate this efficiency, the purpose of this event, and the consequence of its loss, remains speculative.[29] In the mice harboring the *Upfl*-dominant negative allele, we observed a crisis in thymic development, characterized by the synchronous dropout of clonal cell populations that coincides temporally with the onset of TCRβ allele rearrangement (our unpublished data). In theory, this could simply reflect the deleterious consequence of expression of truncated TCRβ peptides from stabilized nonsense transcripts. However, we found that simple transgenic expression of stable TCRβ mRNA that could not derive TCRβ protein (due to frameshifting) on a normal (NMD-sufficient) background had identical phenotypic consequences. There are two important lessons. First, this provides clear evidence that NMD has been functionally incorporated into mammalian-specific developmental programs. Second, it suggests that stabilized nonsense mRNAs per se, as opposed to the proteins they derive, can exert phenotypic effects.

A Genome-Wide Approach

A major distinguishing feature of mammalian NMD is that it is predominantly dependent on splicing.[30] In mammalian cells, a termination codon is interpreted as premature if there is a

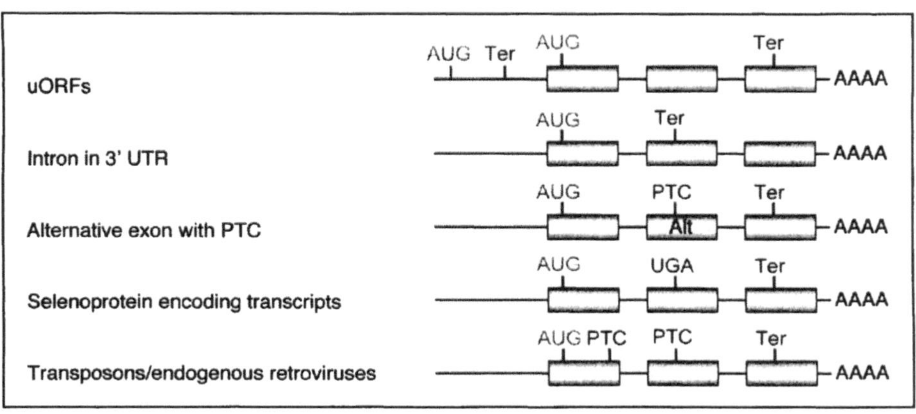

Figure 1. Identifiable NMD-inducing features. Putative NMD-inducing features include upstream open reading frames (uORFs), introns in the 3' untranslated region (UTR), alternative splicing that introduces premature termination codons (PTCs) either directly or by shifting the translational readion frame, transcripts encoding selenocysteine-containing proteins because they contain at least one UGA codon that supports the incorporation of selenocysteine, albeit with less than 100% efficiency, and nonfunctional transcripts such as endogenous retroviruses and transposons. These transcripts are unified by the presence of a spliced intron downstream of a termination codon, a context known sufficient to initiate NMD. AUG specifies an initiation codon, Ter denotes a termination codon, and AAAA signifies the poly(A) tail.

downstream exon-exon junction more than a critical distance (~50 bases for most tested transcripts, as few as 8 bases for TCRβ RNA) downstream.[29,30] As in yeast and flies NMD can occur independent of splicing,[31] it seemed likely that unique classes of endogenous NMD substrates exist in mammals. To address this issue, chip-based analyses were used to determine the relative expression levels of about 4,000 transcripts before and after RNAi-mediated silencing of Upf1 expression in HeLa cells. About 5% of all mRNAs assayed showed significant and consistent upregulation of expression in the NMD-deficient state. About half of the NMD-regulated transcripts had identifiable features that satisfied the structural constraints for initiation of mammalian NMD (Fig. 1). As in yeast, these included those with uORFs (67%) or transcripts derived from noncoding evolutionary remnants such as transposons or endogenous retroviruses (2%). Our lack of observation of other classes identified in yeast (pseudogenes, polycistronic transcripts and transcripts with inefficient pre-mRNA splicing, leaky scanning, or programmed frameshifting) likely reflects the limited composition of current mammalian microarrays and incomplete annotation of the human genome.[32]

Other classes that were observed in mammals included transcripts with an intron 50 or more bases downstream of the bona fide termination codon in the 3' UTR (9%), those with an alternatively spliced exon that harbors or induces a PTC (20%) and those encoding a selenoprotein where the UGA codon can be alternatively interpreted as a stop codon or as encoding selenocysteine (2%). Many factors suggest that the majority of these transcripts are direct targets for NMD. First, over half of the NMD-regulated transcripts were unified by an architecture that is predicted to initiate NMD, i.e., a termination codon upstream of a splicing-generated exon-exon junction. Second, a random sample of 14 Upf1-responsive transcripts, with and without this architecture, all showed a prolonged half-life in Upf1-depleted cells. Finally, five randomly selected transcripts (again with and without an obvious NMD-initiating architecture) showed concordant upregulation after RNAi-mediated silencing of Upf2, a second essential mediator of mammalian NMD.[32]

NMD and Splicing

The finding of a large number of NMD substrates with alternative splicing that generates a PTC was both predictable and informative. Using a pure bioinformatics approach, Brenner and colleagues have shown that as many as one-third of observed or reliably inferred transcripts derived from alternative splicing of human genes result in PTCs, either by direct introduction of a PTC or by frameshifting the extended ORF to create a PTC at least 50 nucleotides upstream of the last exon-exon junction (see chapter by Soergel et al). Interestingly, there was apparent enrichment of NMD-inducing alternative splicing among genes encoding ribosomal proteins, providing an apparent parallel to the regulation of similar (and even orthologous) genes in *C. elegans*.[33]

An obvious question is "Why bother?" In essence, if the only purpose of nonproductive alternative splicing and subsequent NMD is to reduce the level of expression of a given gene product, why not achieve this through transcriptional regulation alone? There are many explanations to consider, all predicated on the assumption that there is some greater meaning to the elimination of alternatively spliced nonsense mRNAs other than cleaning up the irrelevant trash. One benefit would be if NMD provided a more rapid mechanism to reduce the expression level of a functional transcript. This would not seem the case if the "productively" spliced mRNA were the only one with functional importance. At face value, in this scenario the abundance of the functional mRNA and gene product remains largely dependent on the transcriptional regulation of expression of splicing factors, offering little intuitive benefit. One exception, which has not been previously discussed in this context, is exemplified when environmental variations, such as fluctuations in temperature, have the ability to induce rapid changes in splicing profiles that may occur independent of compensatory transcriptional changes.[34,35] Still, NMD is relegated as the clean-up mechanism. It seems that diversification, rather than speed, is the more critical factor. The coupling of NMD with alternative splicing allows the cell to use the complex machinery for alternative splicing to achieve a diverse and regulated output from a single gene. Our preexisting view that alternative splicing is biased toward the production of functional transcripts and proteins is, in itself, essentially biased by the clean-up capacity of NMD. An emerging view, as exemplified by the work of Brenner and colleagues, is that alternative splicing may be inherently messy. In this view, NMD confers tolerance by precluding a protein output and phenotypic consequences from these misadventures. Interestingly, Krainer and colleagues recently reported that selected serine/arginine-rich (SR-proteins) that bind to exonic splicing enhancers (ESEs) and control alternative splicing also have the capacity to boost the efficiency of NMD (see Chapter by Zhang and Krainer). Initial observations suggest that this may be a global effect, indicating that any cell expressing high levels of these proteins (and presumably showing increased alternative splicing) would have a corresponding increase in NMD efficiency.[36] Another observation is that SR-proteins can specifically enhance the translational efficiency of ESE-dependent RNAs,[37-39] resulting in increased nonsense surveillance, but this does not reconcile all of the observations made by Krainer. Finally, SR-proteins may specifically recruit NMD-promoting factors to transcripts undergoing alternative splicing. Although this has not been formally demonstrated, it would be analogous to the apparent situation for Ig and TCRβ mRNAs, i.e., a mechanism to ensure the efficient decay of predictable and high-burden substrates.

This situation is altered radically if the truncated proteins encoded by alternatively spliced nonsense transcripts have intended functions that require regulation. How might these proteins be expressed? First, many nonsense transcripts are only partially degraded, leading to variable quantities of residual mRNAs. Whether or not these mRNAs are translationally competent remains speculative. For example, nascent mRNAs may have bypassed NMD by segregation to a translationally-incompetent pool. Alternatively, Parker and colleagues have shown that yeast transcripts that have undergone nonsense surveillance but are rescued from decay by isolated inactivation of a terminal effector of degradation (e.g., decapping and exonuclease functions) do not derive significant amounts of protein.[40] Second, it remains unclear whether some or many apparent nonsense alternatively-spliced mRNAs also undergo alternative transcriptional initiation, 3'-end processing, or translational initiation (by leaky scanning, programmed frameshifting, or internal ribosome entry), culminating in restoration of an ORF and avoidance of NMD. Certainly, the most interesting scenario would involve regulated variation in the efficiency of nonsense surveillance itself. While most surveys have suggested that NMD is constitutively active, there has been no comprehensive spatial, developmental, or environmental assessment. Informative, albeit preliminary, findings include the observation by Bateman and colleagues that a specific nonsense mRNA was degraded in chondrocytes but not in lymphoblasts or bone,[41] and our observation that the cell type-specific decay of TCOF1 nonsense transcripts varied in an allele-specified manner (i.e., NMD occurred for all TCOF1 nonsense transcripts in fibroblasts, but for only some in lymphoblasts; our unpublished data).

The coupling of NMD to the clearance of alternatively spliced transcripts and those derived from transposons, endogenous retroviruses and pseudogenes has clear evolutionary implications. In essence, it seems likely that NMD confers tolerance to mechanisms that drive evolution. Advantageous events will be selected and maintained, while nonproductive events will be rendered silent and inconsequential. One might ask why apparently nonproductive events have not been permanently squelched over evolution by the baseline rate of random mutation in critical cis-elements (e.g., alternative splice sites). While this undoubtedly occurs, one answer may be that NMD buffers and hence maintains variation that may only prove beneficial under extremes of environmental conditions. This theme is significantly expanded below.

NMD and Response to Stress

Among the physiologic NMD substrates identified in mammalian cells, those encoding proteins involved in amino acid transport, amino acid synthesis or amino acid activation were significantly enriched ($P < 10^{-4}$).[32] There was also upregulation of the mRNAs encoding two transcription factors that coordinate the cellular response to amino acid starvation (ATF3 and ATF4[42,43]). Given that amino acid starvation inhibits translation,[44] and that translation is critical for nonsense surveillance, it stood to reason that regulation of these transcripts by NMD couples their expression levels to translational efficiency. Thus, under conditions of amino acid starvation, inhibition of translation and NMD initiates the specific expression of mRNAs that promote restoration of amino acid homeostasis. Moreover, there is adequate precedent that starvation-induced translational inhibition can be associated with increased translational yield for selected transcripts, including ATF3 and ATF4.[42,45] Further support for this model came with the observation that simple and transient amino acid starvation causes significant attenuation of NMD, as assessed by the increased abundance of nonsense transcripts derived from a β-globin minigene and the increased abundance of representative NMD-regulated transcripts both with and without functions related to amino acid homeostasis.[32] As expected, some transcripts related to amino acid homeostasis that were not upregulated after inhibition of NMD showed increased abundance in response to starvation, while others did not. These data attest to the complexity of the response to nutrient deprivation.[32]

Further insight came from a more comprehensive curation of yeast transcripts that showed increased abundance upon NMD inhibition.[10] This showed clear evidence that this mechanism to preserve amino acid homeostasis in response to environmental cues is evolutionarily conserved.[32] Upregulated transcripts included SSY1, encoding a sensor of external amino acid concentration that also coordinates transcription of amino acid permease genes in response to nutrient deprivation;[46] LST7, encoding a positive regulator of transport of amino acid permeases from the Golgi to the cell surface;[47] and APG3, encoding a low-affinity permease that plays a prominent role in amino acid transport induced by nitrogen starvation.[48] Other upregulated transcripts encode factors involved in the transport and catabolism of secondary nitrogen sources,[49] including the only known allantoate permease, two putative allantoate permeases, five of six factors known to mediate allantoin degradation, a urea transport and degradation enzyme, and a sensor and transporter of ammonia.[32] NMD inhibition was also associated with increased expression of crucial mediators of autophagy (bulk vacuolar degradation of

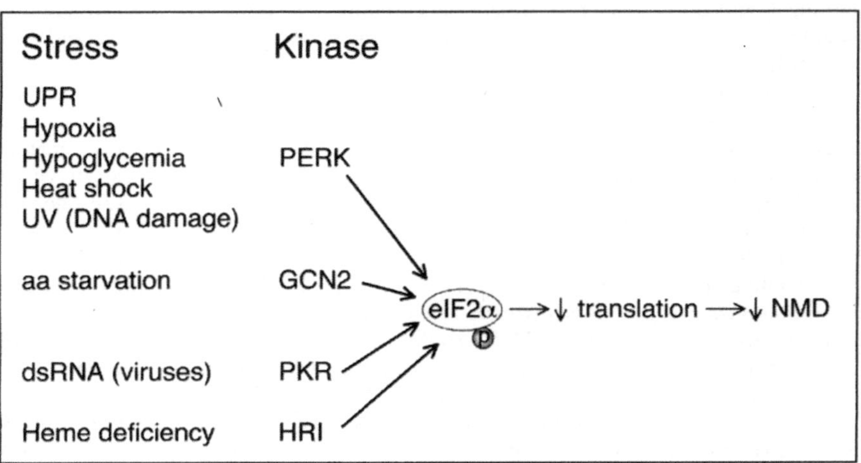

Figure 2. Inhibition of NMD in response to various stress signals. Many stress signals result in phosphorylation of eIF2α through specific kinases. eIF2α phosphorylation inhibits NMD through global translational repression. In this manner, upregulation of the nonsense transcriptome can contribute to the response to a wide variety of stress signals.

cytosolic proteins to recycle amino acids in response to starvation) and sporulation (the ability of a cell to enter a dormant state to wait out better times). Thus, NMD inhibition contributes to a complex and productive response to starvation.[32] Multiple NMD-regulated transcripts that are central to this response have been shown experimentally to lack responsiveness to Gcn4p, a master transcriptional activator of amino acid biosynthetic genes in multiple pathways,[50] suggestive of complementarity between transcriptional and post-transcriptional responses to amino acid deprivation.[32]

Cells have evolved adaptive mechanisms to cope with different stress conditions. Phosphorylation of translation initiation factor 2 (eIF2) on serine 51 of its α subunit allows integration of various stress signals with translation regulation[51] (Fig. 2). This response impacts the cell by promoting global translation inhibition and transcriptional induction of "stress response genes".[52] The coordinated expression of stress response genes affects cell survival, cell-cycle progression and differentiation.[53-55] Distinct mammalian kinases have been shown to couple specific stress signals to eIF2α phosphorylation. PERK is activated by hypoxia,[56] hypoglycemia,[57,58] DNA damage[59] and accumulation of unfolded proteins in the endoplasmic reticulum (ER) lumen.[60] GCN2 is activated by uncharged tRNAs during amino acid starvation,[61] PKR is activated by dsRNA during viral infection[62] and HRI is activated during heme depletion or exposure to toxins such as arsenite.[63,64]

Accumulation of unfolded proteins in the lumen of the ER induces an orchestrated adaptive program known as the unfolded protein response (UPR).[65] Recent microarray analysis of physiologic substrates of mammalian NMD revealed that many transcripts induced by the UPR pathway are also induced in NMD-deficient cells.[32] These include ATF3,[66] ATF4[67] and HSJ1.[68] One would predict that induction of UPR would result in inhibition of NMD due to global translational repression (Fig. 3). Indeed, we have now demonstrated this experimentally (our unpublished data). ATF3 and ATF4 are master transcriptional factors that are also induced by amino acid starvation,[42,45] hypoxia[69] and oxidative stress.[67,70] They induce the transcription of a wide variety of proteins that constitute a productive cellular response.[71] HSJ1, a cochaperone protein, can target misfolded proteins to the proteasome through its ubiquitin interaction motifs.[72] These examples underscore an elegant and coordinated network of pathways to promote organismal survival. In essence, a drop in translational efficiency represents

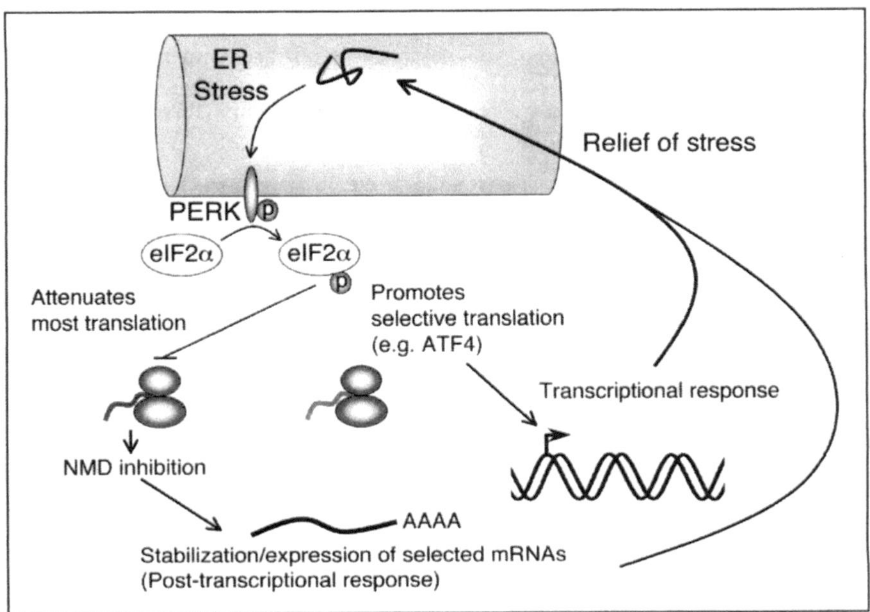

Figure 3. Accumulation of unfolded proteins in the lumen of the ER induces transcriptional and post-transcriptional responses. PERK is a specific eIF2α kinase that is activated in response to UPR. Phosphorylation of eIF2α results in: (1) Selective translational induction of stress response genes; and (2) Global translational attenuation and inhibition of NMD. Thus, inhibition of NMD may be an integral component of the UPR, contributing to restoration of ER homeostasis through stabilization and expression of selected mRNAs. Strategies by which selected transcripts both maintain translation and avoid NMD are discussed in the text.

both a marker of reduced fidelity of the environment and a mechanism to mount a productive response to perceived stress.

This model presents a number of inherent conundrums. How can the cell achieve selective translation of stress-relieving factors in the face of the global translational inhibition induced by phosphorylation of eIF2α? If translation is allowed, how does the mRNA, a natural NMD substrate, avoid degradation? A number of strategies are possible. For example, both ATF4 and GCN4 mRNAs contain multiple uORFs that are alternatively utilized in a regulated fashion. Phosphorylation of eIF2α promotes ribosomal reinitiation at the productive start codon downstream of NMD-inducing uORFs.[13,45,73] Hepatitis C virus (HCV) infection induces the UPR and initiates cap-independent internal ribosome entry site (IRES)-mediated translation of HCV and GRP78 RNA, the latter of which encodes a heat shock ER chaperone.[74] Likewise, heat shock, a potent inducer of the UPR, also enhances IRES-dependent translation of HCV, GRP78, and encephalomyovarditis virus RNA.[75] IRES-dependent translation of uORF containing cationic amino acid transporter (Cat-1) mRNA is induced by eIF2α phosphorylation by GCN2 kinase or PERK in response to amino acid starvation or glucose deprivation, respectively.[76] Many uORF-burdened mRNAs involved in cell differentiation also contain a downstream IRES that is selectively utilized in the face of inhibition of global protein synthesis, an event that correlates with increased phosphorylation of eIF2α.[77] Given that IRES-mediated translation is resistant to the translational inhibition imposed by eIF2α phosphorylation, this presents an elegant and complete solution to our conundrums; translation is maintained and initiates downstream of NMD-inducing uORFs. The relevance of this mechanism to most of the demonstrated physiologic NMD substrates remains to be elucidated.

Recent characterization of human (h) SMG1, an essential effector of mammalian NMD, provided further insight that there is cross-talk between NMD and stress-induced signaling pathways[78-80] (see chapter by Abraham and Oliveira). hSmg1 is a member of the phosphoinositide 3-kinase (PI3-kinase)-related kinase (PIKK) family.[79] Interestingly, several members of this family play a central role in mediating cellular responses to genotoxic stress such as ionizing radiation. Exposure to ionizing radiation causes hSmg1 and ATM (another PIKK family member) to phosphorylate p53, the tumor suppressor known as "the guardian of the genome".[80] p53, in turn, mediates cell-cycle arrest at several check points in order to maintain genome stability and overall cellular viability.[81] In addition to p53, hSmg1 and ATM phosphorylate Upf1 upon exposure to ionizing radiation.[80] Phosphorylation of Upf1 is required for NMD,[20,78] possibly affecting subcellular localization of Upf1[23,82,83] or influencing the assembly/disassembly of the NMD machinery[20] (see chapter by Yamashita et al). Integration of hSmg1 into the NMD pathway provides evidence that Upf1 contributes to the maintenance of genome and transcriptome stability in the face of environmental insults.

Conclusions and Perspectives

In summary, we have highlighted the importance of NMD as a critical post-transcriptional control mechanism that orchestrates a wide variety of cellular functions and responses. This exercise is also informative regarding the tendency of evolution to co-opt functions established in lower organisms during the synthesis of complex developmental and homeostatic schemes that are uniquely relevant to higher eukaryotes. As discussed in other chapters (see Holbrook et al, Keeling et al), there is a price to pay from a medical perspective. In selected examples, NMD clears mutant transcripts that would otherwise encode proteins with sufficient residual function to rescue disease phenotypes.[84-86] The overall viability of mice with relative (but not complete) loss of NMD function suggests that there maybe a therapeutic window within which to manipulate NMD for therapeutic gain.

Acknowledgements

H.C. Dietz is an investigator in the Howard Hughes Medical Institute, which supported this work.

References

1. Leeds P, Peltz SW, Jacobson A et al. The product of the yeast UPF1 gene is required for rapid turnover of mRNAs containing a premature translational termination codon. Genes Dev 1991; 5(12A):2303-2314.
2. Altamura N, Groudinsky O, Dujardin G et al. NAM7 nuclear gene encodes a novel member of a family of helicases with a Zn-ligand motif and is involved in mitochondrial functions in Saccharomyces cerevisiae. J Mol Biol 1992; 224(3):575-587.
3. Pulak R, Anderson P. mRNA surveillance by the Caenorhabditis elegans smg genes. Genes Dev 1993; 7(10):1885-1897.
4. Hodgkin J, Papp A, Pulak R et al. A new kind of informational suppression in the nematode Caenorhabditis elegans. Genetics 1989; 123(2):301-313.
5. Brenner S. The genetics of Caenorhabditis elegans. Genetics 1974; 77(1):71-94.
6. Cali BM, Anderson P. mRNA surveillance mitigates genetic dominance in Caenorhabditis elegans. Mol Gen Genet 1998; 260(2-3):176-184.
7. Morrison M, Harris KS, Roth MB. smg mutants affect the expression of alternatively spliced SR protein mRNAs in Caenorhabditis elegans. Proc Natl Acad Sci USA 1997; 94(18):9782-9785.
8. Mitrovich QM, Anderson P. Unproductively spliced ribosomal protein mRNAs are natural targets of mRNA surveillance in C. elegans. Genes Dev 2000; 14(17):2173-2184.
9. Mitrovich QM, Anderson P. mRNA Surveillance of Expressed Pseudogenes in C. elegans. Curr Biol 2005; 15(10):963-967.
10. He F, Peltz SW, Donahue JL et al. Stabilization and ribosome association of unspliced pre-mRNAs in a yeast upf1- mutant. Proc Natl Acad Sci USA 1993; 90(15):7034-7038.
11. Linz B, Koloteva N, Vasilescu S et al. Disruption of ribosomal scanning on the 5'-untranslated region, and not restriction of translational initiation per se, modulates the stability of nonaberrant mRNAs in the yeast Saccharomyces cerevisiae. J Biol Chem 1997; 272(14):9131-9140.

12. Messenguy F, Vierendeels F, Pierard A et al. Role of RNA surveillance proteins Upf1/CpaR, Upf2 and Upf3 in the translational regulation of yeast CPA1 gene. Curr Genet 2002; 41(4):224-231.
13. Gaba A, Wang Z, Krishnamoorthy T et al. Physical evidence for distinct mechanisms of translational control by upstream open reading frames. EMBO J 2001; 20(22):6453-6463.
14. Ruiz-Echevarria MJ, Peltz SW. The RNA binding protein Pub1 modulates the stability of transcripts containing upstream open reading frames. Cell 2000; 101(7):741-751.
15. Ruiz-Echevarria MJ, Peltz SW. Utilizing the GCN4 leader region to investigate the role of the sequence determinants in nonsense-mediated mRNA decay. EMBO J 1996; 15(11):2810-2819.
16. Lelivelt MJ, Culbertson MR. Yeast Upf proteins required for RNA surveillance affect global expression of the yeast transcriptome. Mol Cell Biol 1999; 19(10):6710-6719.
17. He F, Li X, Spatrick P et al. Genome-wide analysis of mRNAs regulated by the nonsense-mediated and 5' to 3' mRNA decay pathways in yeast. Mol Cell 2003; 12(6):1439-1452.
18. Lew JE, Enomoto S, Berman J. Telomere length regulation and telomeric chromatin require the nonsense-mediated mRNA decay pathway. Mol Cell Biol 1998; 18(10):6121-6130.
19. Dahlseid JN, Lew-Smith J, Lelivelt MJ et al. mRNAs encoding telomerase components and regulators are controlled by UPF genes in Saccharomyces cerevisiae. Eukaryot Cell 2003; 2(1):134-142.
20. Page MF, Carr B, Anders KR et al. SMG-2 is a phosphorylated protein required for mRNA surveillance in Caenorhabditis elegans and related to Upf1p of yeast. Mol Cell Biol 1999; 19(9):5943-5951.
21. Perlick HA, Medghalchi SM, Spencer FA et al. Mammalian orthologues of a yeast regulator of nonsense transcript stability. Proc Natl Acad Sci USA 1996; 93(20):10928-10932.
22. Applequist SE, Selg M, Raman C et al. Cloning and characterization of HUPF1, a human homolog of the Saccharomyces cerevisiae nonsense mRNA-reducing UPF1 protein. Nucleic Acids Res 1997; 25(4):814—821.
23. Mendell JT, ap Rhys CM, Dietz HC. Separable roles for rent1/hUpf1 in altered splicing and decay of nonsense transcripts. Science 2002; 298(5592):419-422.
24. Sun X, Perlick HA, Dietz HC et al. A mutated human homologue to yeast Upf1 protein has a dominant-negative effect on the decay of nonsense-containing mRNAs in mammalian cells. Proc Natl Acad Sci USA 1998; 95(17):10009-10014.
25. Medghalchi SM, Frischmeyer PA, Mendell JT et al. Rent1, a trans-effector of nonsense-mediated mRNA decay, is essential for mammalian embryonic viability. Hum Mol Genet 2001; 10(2):99-105.
26. Kim YK, Furic L, Desgroseillers L et al. Mammalian Staufen1 recruits Upf1 to specific mRNA 3'UTRs so as to elicit mRNA decay. Cell 2005; 120(2):195-208.
27. Hood L, Kronenberg M, Hunkapiller T. T cell antigen receptors and the immunoglobulin supergene family. Cell 1985; 40(2):225-229.
28. Kronenberg M, Siu G, Hood LE et al. The molecular genetics of the T-cell antigen receptor and T-cell antigen recognition. Annu Rev Immunol 1986; 4:529-591.
29. Carter MS, Li S, Wilkinson MF. A splicing-dependent regulatory mechanism that detects translation signals. EMBO J 1996; 15(21):5965-5975.
30. Zhang J, Sun X, Qian Y et al. Intron function in the nonsense-mediated decay of beta-globin mRNA: Indications that pre-mRNA splicing in the nucleus can influence mRNA translation in the cytoplasm. RNA 1998; 4(7):801-815.
31. Gatfield D, Unterholzner L, Ciccarelli FD et al. Nonsense-mediated mRNA decay in Drosophila: At the intersection of the yeast and mammalian pathways. EMBO J 2003; 22(15):3960-3970.
32. Mendell JT, Sharifi NA, Meyers JL et al. Nonsense surveillance regulates expression of diverse classes of mammalian transcripts and mutes genomic noise. Nat Genet 2004; 36(10):1073-1078.
33. Lewis BP, Green RE, Brenner SE. Evidence for the widespread coupling of alternative splicing and nonsense-mediated mRNA decay in humans. Proc Natl Acad Sci USA 2003; 100(1):189-192.
34. Majercak J, Sidote D, Hardin PE et al. How a circadian clock adapts to seasonal decreases in temperature and day length. Neuron 1999; 24(1):219-230.
35. Ars E, Serra E, de la Luna S et al. Cold shock induces the insertion of a cryptic exon in the neurofibromatosis type 1 (NF1) mRNA. Nucleic Acids Res 2000; 28(6):1307-1312.
36. Zhang Z, Krainer AR. Involvement of SR proteins in mRNA surveillance. Mol Cell 2004; 16(4):597-607.
37. Sanford JR, Gray NK, Beckmann K et al. A novel role for shuttling SR proteins in mRNA translation. Genes Dev 2004; 18(7):755-768.
38. Lim LP, Sharp PA. Alternative splicing of the fibronectin EIIIB exon depends on specific TGCATG repeats. Mol Cell Biol 1998; 18(7):3900-3906.
39. Schell T, Kulozik AE, Hentze MW. Integration of splicing, transport and translation to achieve mRNA quality control by the nonsense-mediated decay pathway. Genome Biol 2002; 3(3):Reviews1006.1-1006.6.

40. Muhlrad D, Parker R. Recognition of yeast mRNAs as "nonsense containing" leads to both inhibition of mRNA translation and mRNA degradation: Implications for the control of mRNA decapping. Mol Biol Cell 1999; 10(11):3971-3978.
41. Bateman JF, Freddi S, Nattrass G et al. Tissue-specific RNA surveillance? Nonsense-mediated mRNA decay causes collagen X haploinsufficiency in Schmid metaphyseal chondrodysplasia cartilage. Hum Mol Genet 2003; 12(3):217-225.
42. Pan Y, Chen H, Siu F et al. Amino acid deprivation and endoplasmic reticulum stress induce expression of multiple activating transcription factor-3 mRNA species that, when overexpressed in HepG2 cells, modulate transcription by the human asparagine synthetase promoter. J Biol Chem 2003; 278(40):38402-38412.
43. Siu F, Bain PJ, LeBlanc-Chaffin R et al. ATF4 is a mediator of the nutrient-sensing response pathway that activates the human asparagine synthetase gene. J Biol Chem 2002; 277(27):24120-24127.
44. Pain VM. Translational control during amino acid starvation. Biochimie 1994; 76(8):718-728.
45. Averous J, Bruhat A, Jousse C et al. Induction of CHOP expression by amino acid limitation requires both ATF4 expression and ATF2 phosphorylation. J Biol Chem 2004; 279(7):5288-5297.
46. Forsberg H, Ljungdahl PO. Sensors of extracellular nutrients in Saccharomyces cerevisiae. Curr Genet 2001; 40(2):91-109.
47. Roberg KJ, Bickel S, Rowley N et al. Control of amino acid permease sorting in the late secretory pathway of Saccharomyces cerevisiae by SEC13, LST4, LST7 and LST8. Genetics 1997; 147(4):1569-1584.
48. Schreve JL, Garrett JM. Yeast Agp2p and Agp3p function as amino acid permeases in poor nutrient conditions. Biochem Biophys Res Commun 2004; 313(3):745-751.
49. Hardwick JS, Kuruvilla FG, Tong JK et al. Rapamycin-modulated transcription defines the subset of nutrient-sensitive signaling pathways directly controlled by the Tor proteins. Proc Natl Acad Sci USA 1999; 96(26):14866-14870.
50. Natarajan K, Meyer MR, Jackson BM et al. Transcriptional profiling shows that Gcn4p is a master regulator of gene expression during amino acid starvation in yeast. Mol Cell Biol 2001; 21(13):4347-4368.
51. Shi Y, Vattem KM, Sood R et al. Identification and characterization of pancreatic eukaryotic initiation factor 2 alpha-subunit kinase, PEK, involved in translational control. Mol Cell Biol 1998; 18(12):7499-7509.
52. Prostko CR, Brostrom MA, Brostrom CO. Reversible phosphorylation of eukaryotic initiation factor 2 alpha in response to endoplasmic reticular signaling. Mol Cell Biochem 1993; 127-128:255-265.
53. Pearce AK, Humphrey TC. Integrating stress-response and cell-cycle checkpoint pathways. Trends Cell Biol 2001; 11(10):426-433.
54. Holcik M, Yeh C, Korneluk RG et al. Translational upregulation of X-linked inhibitor of apoptosis (XIAP) increases resistance to radiation induced cell death. Oncogene 2000; 19(36):4174-4177.
55. Harding HP, Zhang Y, Bertolotti A et al. Perk is essential for translational regulation and cell survival during the unfolded protein response. Mol Cell 2000; 5(5):897-904.
56. Koumenis C, Naczki C, Koritzinsky M et al. Regulation of protein synthesis by hypoxia via activation of the endoplasmic reticulum kinase PERK and phosphorylation of the translation initiation factor eIF2alpha. Mol Cell Biol 2002; 22(21):7405-7416.
57. Harding HP, Zeng H, Zhang Y et al. Diabetes mellitus and exocrine pancreatic dysfunction in perk-/- mice reveals a role for translational control in secretory cell survival. Mol Cell 2001; 7(6):1153-1163.
58. Scheuner D, Song B, McEwen E et al. Translational control is required for the unfolded protein response and in vivo glucose homeostasis. Mol Cell 2001; 7(6):1165-1176.
59. Wu S, Hu Y, Wang JL et al. Ultraviolet light inhibits translation through activation of the unfolded protein response kinase PERK in the lumen of the endoplasmic reticulum. J Biol Chem 2002; 277(20):18077-18083.
60. Harding HP, Zhang Y, Ron D. Protein translation and folding are coupled by an endoplasmic-reticulum-resident kinase. Nature 1999; 397(6716):271-274.
61. Dever TE, Feng L, Wek RC et al. Phosphorylation of initiation factor 2 alpha by protein kinase GCN2 mediates gene-specific translational control of GCN4 in yeast. Cell 1992; 68(3):585-596.
62. Kostura M, Mathews MB. Purification and activation of the double-stranded RNA-dependent eIF-2 kinase DAI. Mol Cell Biol 1989; 9(4):1576-1586.
63. Chen JJ, Throop MS, Gehrke L et al. Cloning of the cDNA of the heme-regulated eukaryotic initiation factor 2 alpha (eIF-2 alpha) kinase of rabbit reticulocytes: Homology to yeast GCN2 protein kinase and human double-stranded-RNA-dependent eIF-2 alpha kinase. Proc Natl Acad Sci USA 1991; 88(17):7729-7733.

64. McEwen E, Kedersha N, Song B et al. Heme-regulated inhibitor kinase-mediated phosphorylation of eukaryotic translation initiation factor 2 inhibits translation, induces stress granule formation, and mediates survival upon arsenite exposure. J Biol Chem 2005; 280(17):16925-16933.
65. Kozutsumi Y, Segal M, Normington K et al. The presence of malfolded proteins in the endoplasmic reticulum signals the induction of glucose-regulated proteins. Nature 1988; 332(6163):462-464.
66. Jiang HY, Wek SA, McGrath BC et al. Activating transcription factor 3 is integral to the eukaryotic initiation factor 2 kinase stress response. Mol Cell Biol 2004; 24(3):1365-1377.
67. Harding HP, Novoa I, Zhang Y et al. Regulated translation initiation controls stress-induced gene expression in mammalian cells. Mol Cell 2000; 6(5):1099-1108.
68. Schnaider T, Soti C, Cheetham ME et al. Interaction of the human DnaJ homologue, HSJ1b with the 90 kDa heat shock protein, Hsp90. Life Sci 2000; 67(12):1455-1465.
69. Blais JD, Filipenko V, Bi M et al. Activating transcription factor 4 is translationally regulated by hypoxic stress. Mol Cell Biol 2004; 24(17):7469-7482.
70. Hai T, Wolfgang CD, Marsee DK et al. ATF3 and stress responses. Gene Expr 1999; 7(4-6):321-335.
71. Harding HP, Zhang Y, Zeng H et al. An integrated stress response regulates amino acid metabolism and resistance to oxidative stress. Mol Cell 2003; 11(3):619-633.
72. Chapple JP, Cheetham ME. The chaperone environment at the cytoplasmic face of the endoplasmic reticulum can modulate rhodopsin processing and inclusion formation. J Biol Chem 2003; 278(21):19087-19094.
73. Hinnebusch AG. Gene-specific translational control of the yeast GCN4 gene by phosphorylation of eukaryotic initiation factor 2. Mol Microbiol 1993; 10(2):215-223.
74. Tardif KD, Mori K, Siddiqui A. Hepatitis C virus subgenomic replicons induce endoplasmic reticulum stress activating an intracellular signaling pathway. J Virol 2002; 76(15):7453-7459.
75. Kim YK, Jang SK. Continuous heat shock enhances translational initiation directed by internal ribosomal entry site. Biochem Biophys Res Commun 2002; 297(2):224-231.
76. Fernandez J, Bode B, Koromilas A et al. Translation mediated by the internal ribosome entry site of the cat-1 mRNA is regulated by glucose availability in a PERK kinase-dependent manner. J Biol Chem 2002; 277(14):11780-11787.
77. Gerlitz G, Jagus R, Elroy-Stein O. Phosphorylation of initiation factor-2 alpha is required for activation of internal translation initiation during cell differentiation. Eur J Biochem 2002; 269(11):2810-2819.
78. Yamashita A, Ohnishi T, Kashima I et al. Human SMG-1, a novel phosphatidylinositol 3-kinase-related protein kinase, associates with components of the mRNA surveillance complex and is involved in the regulation of nonsense-mediated mRNA decay. Genes Dev 2001; 15(17):2215-2228.
79. Denning G, Jamieson L, Maquat LE et al. Cloning of a novel phosphatidylinositol kinase-related kinase: Characterization of the human SMG-1 RNA surveillance protein. J Biol Chem 2001; 276(25):22709-22714.
80. Brumbaugh KM, Otterness DM, Geisen C et al. The mRNA surveillance protein hSMG-1 functions in genotoxic stress response pathways in mammalian cells. Mol Cell 2004; 14(5):585-598.
81. Lane DP. Cancer. p53, guardian of the genome. Nature 1992; 358(6381):15-16.
82. Pal M, Ishigaki Y, Nagy E et al. Evidence that phosphorylation of human Upfl protein varies with intracellular location and is mediated by a wortmannin-sensitive and rapamycin-sensitive PI 3-kinase-related kinase signaling pathway. RNA 2001; 7(1):5-15.
83. Unterholzner L, Izaurralde E. SMG7 acts as a molecular link between mRNA surveillance and mRNA decay. Mol Cell 2004; 16(4):587-596.
84. Frischmeyer PA, Dietz HC. Nonsense-mediated mRNA decay in health and disease. Hum Mol Genet 1999; 8(10):1893-1900.
85. Mendell JT, Dietz HC. When the message goes awry: Disease-producing mutations that influence mRNA content and performance. Cell 2001; 107(4):411-414.
86. Hentze MW, Kulozik AE. A perfect message: RNA surveillance and nonsense-mediated decay. Cell 1999; 96(3):307-310.

Chapter 9

NMD and Human Disease

Jill A. Holbrook, Gabriele Neu-Yilik, Matthias W. Hentze and Andreas E. Kulozik*

Abstract

We discuss in this chapter how nonsense-mediated mRNA decay (NMD) affects the expression of human genetic diseases resulting from premature termination codons (PTCs) by considering how NMD alters disease phenotype. NMD may exert a beneficial, neutral, or harmful effect, depending on the location of the PTC in the transcript and the properties of the truncated protein. In the case of many PTCs, the resulting truncated protein might be nonfunctional and could be degraded without harmful effects. In these instances, NMD probably does not significantly influence phenotype. In other cases, NMD can prevent expression of potentially dominant negative proteins. Therefore, NMD can sometimes exert a protective effect that benefits heterozygous carriers of PTCs. NMD can also contribute to a disease phenotype when it inhibits expression of partially functional proteins. Therapies that could affect gene expression in such cases are under development and may, in the future, provide avenues for effective clinical treatment of diseases that involve NMD.

"NMD-Neutral" PTCs

In general, whether NMD alters the phenotype of a disease depends on both the affected gene and the location of the disease-causing PTC. PTCs upstream of the so-called "NMD boundary," which is located 50 to 55 bases 5' of the last exon-exon junction, generally trigger NMD of the affected transcript. In contrast, PTCs 3' of this boundary generally escape detection by the NMD machinery, leading to transcript survival. Perhaps surprisingly, it is likely that NMD often has no discernible effect on disease phenotype, since in the absence of NMD many PTCs would probably lead to expression of inactive peptides rather than dominant-negative proteins. Although it might be expected that many PTCs would fall into this "NMD-neutral" category, experimental data are lacking to demonstrate that particular PTCs lead to an identical phenotype in both the presence and absence of NMD. Such "neutral" PTCs will therefore not be discussed further.

NMD-Mediated Protection of Heterozygotes: The Example of β-Thalassemia

NMD may be expected to be beneficial when it inhibits expression of truncated proteins that could exert deleterious effects. The prototypical genetic condition illustrating the protective effects of NMD is β-thalassemia, which is a disorder of hemoglobin production. Normal hemoglobin, which is necessary for oxygen transport, is a tetramer composed of two α-globin and two β-globin subunits. The common recessive form of β-thalassemia occurs in homozygotes

*Corresponding Author: Andreas E. Kulozik—Department of Pediatric Oncology, Hematology and Immunology, University of Heidelberg, D-69120 Heidelberg, Germany, and Molecular Medicine Partnership Unit, D-69120 Heidelberg, Germany.
Email: andreas.kulozik@med.uni-heidelberg.de

Nonsense-Mediated mRNA Decay, edited by Lynne E. Maquat. ©2006 Eurekah.com.

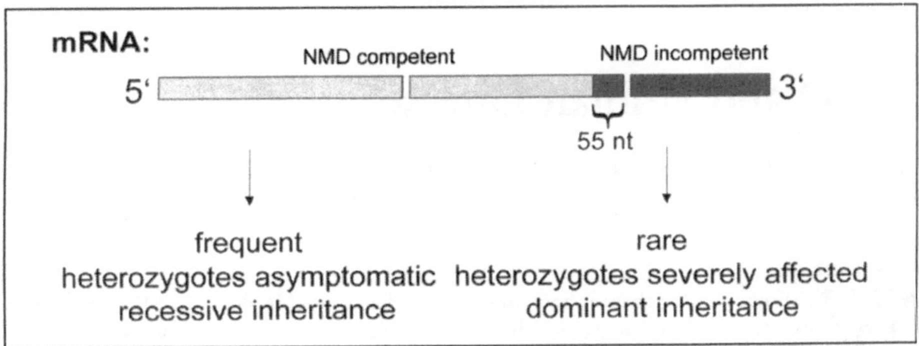

Figure 1. Position-dependent effects of nonsense mutations of NMD correlate with inheritance pattern and clinical severity of disease. Human β-globin mRNAs that contain PTCs within their 5' portion are generally targeted by NMD, protecting heterozygotes from the disease β-thalassemia. In contrast, mRNAs that contain PTCs within their 3' portion are generally not targeted for NMD and, thus, are translated into truncated proteins that result in symptomatic, dominantly inherited β-thalassemia. Other conditions listed in Table 1 have similar characteristics. Reprinted form Holbrook et al, 2004.[1]

who possess NMD-competent PTCs in both copies of the β-globin gene. The resulting defective β-globin mRNA is degraded by NMD. Free α-globin, which is toxic, is then present in excess and is degraded proteolytically.[2] Therefore, the quantity of tetrameric hemoglobin is insufficient, causing severe anemia in affected persons. In comparison, heterozygous carriers of a single NMD-competent PTC generally produce enough β-globin from the normal allele to maintain sufficient amounts of tetrameric hemoglobin, and they are clinically healthy. Rare forms of dominant β-thalassemia, in contrast, are caused by NMD-incompetent PTCs within the last exon of the β-globin gene. These PTCs give rise to a large amount of truncated β-globin that cannot be sufficiently degraded and precipitates in toxic inclusion bodies.[3] The remarkable difference between healthy heterozygotes possessing NMD-competent PTCs and anemic heterozygotes possessing NMD-incompetent PTCs indicates that NMD protects many healthy heterozygotes from manifesting clinical disease[4] (Fig. 1).

Although β-thalassemia is the only genetic condition so far in which the protective effect of NMD has been thoroughly investigated by experiment, a similar protective role of NMD for heterozygotes with NMD-competent PTCs can be reasonably postulated from similar genotype-phenotype relationships in a number of other diseases (Table 1). These include:

- Dominantly and recessively inherited susceptibility to mycobacterial infections caused by mutations in the *IFNGR1* gene.[5,6] The recessive form of this condition is often fatal, with patients succumbing to disseminated mycobacterial infections at a young age. This form is caused by PTCs that are probably NMD-competent, since no IFNGR protein was found in a patient with recessive disease. Heterozygous carriers of these PTCs are healthy. The dominant form of the disease arises from PTCs that are predicted to be NMD-incompetent. This form results in production of a truncated IFNGR receptor, as may be expected for a transcript that survives NMD. Heterozygotes manifest increased susceptibility to mycobacterial infection, although such infections are generally not as severe as in the recessive case.
- Brachydactyly type B (see ref. 7), which is a dominant condition involving malformed hands and feet, and Robinow syndrome, which is recessive and characterized by more severe skeletal malformation. Both diseases are caused by mutations in the *ROR2* gene. Brachydactyly type B is caused by PTCs that are expected to be NMD-incompetent. Robinow syndrome is characterized by PTCs that should be NMD-competent, and heterozygous carriers are unaffected.

Table 1. Genetic conditions in which NMD can modulate phenotype

Gene	PTC Location	Effect of PTC	Reference(s)
β-globin (*HBB*)	5' to NMD boundary	Recessively inherited β-thalassemia major; heterozygotes healthy	2,3
	3' to NMD boundary	Dominantly inherited β-thalassemia intermedia	2,3
Interferon gamma receptor1 (*IFNGR1*)	5' to NMD boundary	Recessively inherited susceptibility to mycobacterial infection; heterozygotes healthy	5
	3' to NMD boundary	Dominantly inherited susceptibility to mycobacterial infection	6
Receptor tyrosine kinase-like orphan receptor 2 (*ROR2*)	5' to NMD boundary	Recessively inherited Robinow syndrome (orodental abnormalities, hypoplastic genitalia, multiple rib/vertebral anomalies); heterozygotes healthy	7
	3' to NMD boundary	Dominantly inherited brachydactyly type B (shortening of digits and metacarpals)	7
Cone-rod homeobox (*CRX*)	5' to NMD boundary	No homozygotes to date; PTC found in unaffected heterozygote	10
	3' to NMD boundary	Dominantly inherited retinal disease	10
von Willebrand factor (VWF)	5' to NMD boundary	Recessively inherited type 3 von Willebrand disease; heterozygotes healthy	von Willebrand database www.shef.ac.uk/vwf
	3' to NMD boundary	Dominantly inherited type 2A disease	8
Coagulation factor X (*F10*)	5' to NMD boundary	Recessively inherited bleeding tendency; heterozygotes healthy	9
	3' to NMD boundary	Dominantly inherited bleeding tendency	9
Rhodopsin (*RHO*)	5' to NMD boundary	Recessively inherited blindness; heterozygotes have abnormalities on retinogram but no clinical disease	11
	3' to NMD boundary	Dominantly inherited blindness	12
SRY-Box 10 (*SOX10*)	5' to NMD boundary	Haploinsufficiency leading to congenital neurosensory deafness and colonic agangliosis	13
	3' to NMD boundary	Dominantly inherited neural developmental defect including neurosensory deafness, colonic agangliosis, peripheral neuropathy and central dysmyelinating leukodystrophy	13

Reprinted form Holbrook et al, 2004.[1]

- Dominant and recessive von Willebrand disease,[8] which are disorders of blood clotting. PTCs that should be NMD competent lead to severe disease, but only in a pattern of recessive inheritance. In contrast, a mutation that removes the physiological stop codon and results in a new stop codon downstream creates a mutated transcript that should not be an NMD substrate and results in dominant disease.
- Dominant and recessive factor X deficiency,[9] which is another disorder of blood clotting. Factor X deficiency is usually transmitted in an autosomal recessive pattern so that heterozygous carriers are healthy. Among the causative mutations, some generate NMD-competent PTCs. A splice site mutation that results in a PTC that should be NMD-incompetent, however, leads to moderate disease in heterozygotes.
- Retinal degeneration.[10-12] PTCs in the 3′ end of the *CRX* gene, downstream of the NMD boundary, lead to disease in heterozygotes. In contrast, a PTC upstream of the NMD boundary was found in a healthy heterozygote. Mutations in the rhodopsin gene follow a similar pattern.

A somewhat different example that also demonstrates the protective effect of NMD is provided by the *SOX10* gene. In this case, both NMD-competent and NMD-incompetent PTCs give rise to disease. The NMD-competent PTCs appear to cause simple haploinsufficiency, and they result in hereditary neurosensory deafness and intestinal obstruction. However, the NMD-incompetent PTCs results in the production of a dominantly acting truncated protein, which causes a more severe syndrome that includes central and peripheral demyelination.[13]

Taken in sum, these genetic conditions provide considerable evidence that NMD protects heterozygous carriers of PTCs from expressing dominantly acting deleterious truncated proteins. It is also noteworthy that in many of these conditions, the most severe disease is evident in patients harboring two autosomal recessive mutations. These patients are usually typified by complete or near-complete protein deficiency. Patients harboring dominant mutations, in contrast, usually retain some protein function, which tends to ameliorate the disease phenotype.

NMD in Acquired Genetic Conditions[1]

Mutations in tumor-suppressor genes are common steps in the development and progression of cancer. As with inherited genetic conditions, NMD appears to provide protection against expression of mutated, truncated tumor-suppressor peptides. For example, NMD has been shown to degrade PTC-containing transcripts arising from the *BRCA1* gene.[14] PTC-containing mRNAs from the *TP53* and Wilms tumor (*WT1*) loci[15-19] are also reduced in abundance compared to wildtype or missense mutated transcripts, presumably due to the action of NMD.

Evidence that NMD protects against dominant truncated forms of these tumor-suppressor proteins derives from experiments in which intronless cDNA constructs encoding unspliced mRNAs that are NMD-incompetent are expressed in cell lines or animals. The resulting C-terminally truncated proteins exert dominant detrimental effects, such as increased chemoresistance, decreased apoptosis, increased tumorigenicity,[20-22] interference with transcription-activating ability, and mislocalization of the corresponding cellular tumor suppressor protein[23,24] (Table 2). These studies indicate that if abnormal transcripts containing PTCs were not degraded by NMD, clinically recessive tumor-suppressor mutations could instead result in dominant disease due to the synthesis of truncated, dominantly acting oncoproteins. NMD may thus protect heterozygous carriers of PTC-mutated tumor-suppressor genes from developing cancer, at least for as long as the other tumor-suppressor allele remains intact.

NMD may also affect expression of certain tumor-suppressor genes by modulating the quantity of splice variants produced by these genes, although this hypothesis remains somewhat speculative. For example, the *TP53* gene produces a splice variant that contains a PTC at low levels in normal tissues.[25] In contrast, a much larger amount of this splice variant—accounting for approximately half of all *TP53* transcripts—was found in a leukemic cell line.[26] Although a causal relationship has not yet been established, large amounts of the splice variant could potentially contribute to leukemogenesis, which would be prevented under normal conditions by NMD-induced degradation of the variant.

Table 2. Effect of NMD on expression of tumor-suppressor genes

Gene	Experimental Data	Reference(s)
BRCA1	NMD demonstrated for many transcripts containing PTCs.	14
	Expression of NMD-insensitive C-terminal truncation mutations leads to increased chemoresistance, decreased apoptosis, increased tumorigenicity and decreased survival time.	20,21
TP53	PTCs associated with decreased or absent mRNA and protein.	15-18
	Expression of NMD-insensitive C-terminal truncation mutations leads to tumorigenesis.	22
WT1	PTCs associated with absent mRNA.	19
	NMD-insensitive PTC-containing mRNAs encode truncated protein that inhibits ability of wild-type protein to activate transcription and alters subnuclear localization of wild-type protein.	23,24

Reprinted form Holbrook et al, 2004.[1]

Transcripts from the *WT1* gene also undergo alternative splicing to produce two major splice forms. The more abundant splice form (+KTS) encodes three additional amino acids at the 3' end of the penultimate exon, while the less abundant splice form (-KTS) lacks these residues. Interestingly, a mutation found in acute myelogenous leukaemia,[27] Wilms tumor[28] and Frasier syndrome (male pseudohermaphroditism and progressive glomerulopathy)[29] is also located within the penultimate exon of the *WT1* gene. This mutation introduces an NMD-competent PTC, but only for the +KTS form. In contrast, for the -KTS form, the PTC is within the terminal 50 bases of the penultimate exon. Therefore, -KTS transcripts probably escape NMD and are translated to produce a truncated protein (Fig. 2). NMD could thus limit expression of the primary transcript. This is important because, in general, Frasier syndrome is

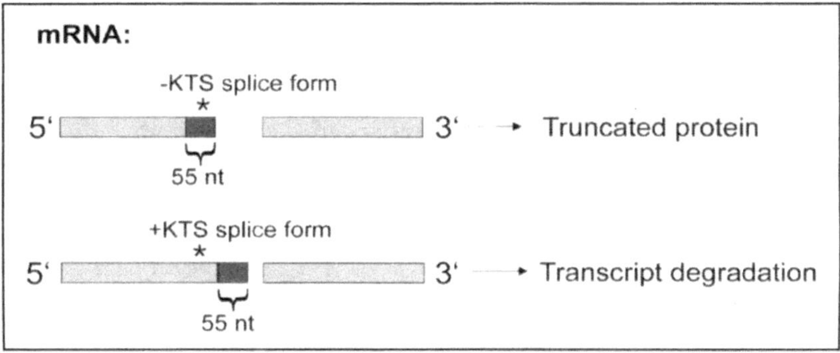

Figure 2. Different fates of PTC-containing splice variants derived from the *WT1* gene. The penultimate exon of the +KTS splice variant (lower) encodes an extra three amino acids, whereas the -KTS variant (upper) lacks these amino acids. Therefore, NMD should result in degradation of the +KTS variant, since the PTC (shown as an asterisk) is situated more than 55-nucleotides (nt) upstream of the end of the penultimate exon (see chapter by Maquat). In contrast, NMD should not occur in the case of the -KTS variant, where the PTC is situated less than 55-nt upstream of the end of the penultimate exon, and a truncated protein should therefore be produced. Only the last two exons of the *WT1* gene are represented, and the shaded region represents the part of the penultimate exon within which a resident PTC will not trigger NMD.

attributable to mutations that selectively reduce the abundance of the +KTS isoform.[30,31] Therefore, Frasier syndrome in individuals carrying this particular PTC mutation could be caused by NMD-induced degradation of the +KTS isoform. In addition, the truncated form of the minor splice variant might act in a dominant fashion to promote tumorigenesis.

Medical Therapies for PTC-Related Disease

In contrast to its role in preventing dominant disease, NMD can also eliminate mRNAs that would otherwise result in the production of partly or fully functional truncated protein, thereby contributing to the protein deficiency that is the hallmark of many recessive genetic conditions. In such instances, interventions to prevent degradation of transcripts containing PTCs may be therapeutically useful. Drug therapy with these aims is the subject of Chapter 10 in this book and therefore will be discussed only briefly here.

At this time, the therapeutic approach that is closest to clinical applicability is the use of aminoglycoside antibiotics that allow read-through of nonsense codons. These drugs bind to the decoding center of the ribosome[32] and decrease the accuracy requirements for codon-anticodon pairing, thereby resulting in incorporation of an amino acid into the polypeptide chain instead of chain termination. Thus, full-length, albeit missense-mutated, proteins are synthesized. Aminoglycosides have been used to treat Duchenne muscular dystrophy, Hurler syndrome, X-linked nephrogenic diabetes insipidus, ataxia-telangiectsia, and cystinosis, resulting in some functional improvement in cell lines[33-39] and animal models.[36,40] Some trials of aminoglycoside therapy also have been carried out in humans with PTC-associated diseases. Promising results have been obtained in a controlled clinical trial for cystic fibrosis, in which full-length CFTR protein was detected in nasal epithelial cells of two treated individuals.[41] In contrast, clinical studies of individuals with muscular dystrophy have not shown functional improvement.[42,43] Overall, although early results indicate that aminoglycoside treatment may have potential applicability, a therapeutic benefit has yet to be demonstrated. The effect of prolonged treatment with aminoglycosides is also a concern. First, there is the problem of toxicity. Second, there is the issue of whether the general, long-term suppression of PTCs, which is accompanied by the suppression of physiologic termination codons and the potential for pseudogene transcript translation, will result in the build-up of abnormal proteins that could trigger other cellular problems.

Other approaches to modulating PTC-induced transcript degradation are also under investigation. One potential approach is to use antisense oligoribonucleotides to redirect splicing, thereby avoiding the production of PTCs in the first place. Initially described as a method for correcting the in vitro aberrant splicing of a disease-associated beta globin gene,[44] this strategy employs antisense 2'-O-methylribonucleotides (2OMeAO) that hybridize to splice sites or branch point junctions of aberrantly spliced pre-mRNA, thereby restoring normal splicing in a significant fraction of molecules. This approach was modified for use with a disease-associated dystrophin transcript.[45] In this case, the targeted PTC was located within an exon coding for a dispensable protein region. Antisense 2OMeAO that targeted splice sites flanking the PTC promoted in-frame skipping of the affected exon, effectively removing the PTC. Treatment of *mdx* mice with these antisense oligonucleotides resulted in low-level expression of shortened but functional dystrophin. In a further step toward the clinic, the efficiency of oligonucleotide delivery to tissues has been enhanced by the use of vehicles such as block copolymer.[46] Before trials of 2OMeAO are feasible in humans, however, a systemic delivery method needs to be developed. As with all forms of gene therapy, the issues of transfection efficiency, potential immune responses, and side effects must be addressed. Unfortunately, this sort of treatment would be feasible only for mutations in which manipulation of splicing maintains in-frame translation, and—if exon skipping is the result—that does not remove essential protein regions or result in protein mis-folding. Therefore, this treatment will probably be limited to specific cases rather than provide a general therapy for PTC-associated diseases.

A further potential approach—although currently very far from realization—involves modulating NMD per se, rather than modulating recognition of PTCs. Down-regulating the central NMD protein Upf1 using RNA interference has been shown to inhibit NMD in cultured cells, and it might constitute a starting point for therapeutic developments. Additionally, some evidence exists to suggest that different individuals with identical genetic mutations may exhibit distinct phenotypic severities due to differences in the efficiency of NMD.[47] Thus, identifying factors that regulate the efficiency of NMD could permit development of therapies that fine-tune NMD, potentially allowing more targeted interventions in patients with PTC-associated diseases.

Conclusions

NMD plays an important role in modulating the manifestation of hereditary and acquired genetic diseases, and a deepening understanding of NMD will augment the establishment of genotype-phenotype relationships in a number of conditions. The contribution of NMD to genetic diseases may be beneficial, neutral or harmful, depending on the specific PTC, its location, and the type of mutation that generates the PTC. Therefore, the potential role of NMD in disease must be appreciated when the functional effect of a mutation is considered. Therapies that target PTC-containing transcripts are under development, and continued research in this direction should help lead to viable strategies to treat PTC-associated diseases.

References

1. Holbrook JA, Neu-Yilik G, Hentze MW et al. Nonsense-mediated decay approaches the clinic. Nat Genet 2004; 36(8):801-808.
2. Hall GW, Thein S. Nonsense codon mutations in the terminal exon of the beta-globin gene are not associated with a reduction in beta-mRNA accumulation: A mechanism for the phenotype of dominant beta-thalassemia. Blood 1994; 83(8):2031-2037.
3. Thein SL, Hesketh C, Taylor P et al. Molecular basis for dominantly inherited inclusion body beta-thalassemia. Proc Natl Acad Sci USA 1990; 87(10):3924-3928.
4. Kugler W, Enssle J, Hentze MW et al. Nuclear degradation of nonsense mutated beta-globin mRNA: A post-transcriptional mechanism to protect heterozygotes from severe clinical manifestations of beta-thalassemia? Nucleic Acids Res 1995; 23(3):413-418.
5. Jouanguy E, Altare F, Lamhamedi S et al. Interferon-gamma-receptor deficiency in an infant with fatal bacille Calmette-Guerin infection. N Engl J Med 1996; 335(26):1956-1961.
6. Jouanguy E, Lamhamedi-Cherradi S, Lammas D et al. A human IFNGR1 small deletion hotspot associated with dominant susceptibility to mycobacterial infection. Nat Genet 1999; 21(4):370-378.
7. Schwabe GC, Tinschert S, Buschow C et al. Distinct mutations in the receptor tyrosine kinase gene ROR2 cause brachydactyly type B. Am J Hum Genet 2000; 67(4):822-831.
8. Schneppenheim R, Budde U, Obser T et al. Expression and characterization of von Willebrand factor dimerization defects in different types of von Willebrand disease. Blood 2001; 97(7):2059-2066.
9. Millar DS, Elliston L, Deex P et al. Molecular analysis of the genotype-phenotype relationship in factor X deficiency. Hum Genet 2000; 106(2):249-257.
10. Rivolta C, Berson EL, Dryja TP. Dominant Leber congenital amaurosis, cone-rod degeneration, and retinitis pigmentosa caused by mutant versions of the transcription factor CRX. Hum Mutat 2001; 18(6):488-498.
11. Rosenfeld PJ, Cowley GS, McGee TL et al. A null mutation in the rhodopsin gene causes rod photoreceptor dysfunction and autosomal recessive retinitis pigmentosa. Nat Genet 1992; 1(3):209-213.
12. Sung CH, Davenport CM, Hennessey JC et al. Rhodopsin mutations in autosomal dominant retinitis pigmentosa. Proc Natl Acad Sci USA 1991; 88(15):6481-6485.
13. Inoue K, Khajavi M, Ohyama T et al. Molecular mechanism for distinct neurological phenotypes conveyed by allelic truncating mutations. Nat Genet 2004.
14. Perrin-Vidoz L, Sinilnikova OM, Stoppa-Lyonnet D et al. The nonsense-mediated mRNA decay pathway triggers degradation of most BRCA1 mRNAs bearing premature termination codons. Hum Mol Genet 2002; 11(23):2805-2814.
15. Kawasaki T, Tomita Y, Watanabe R et al. mRNA and protein expression of p53 mutations in human bladder cancer cell lines. Cancer Lett 1994; 82(1):113-121.

16. Williams C, Norberg T, Ahmadian A et al. Assessment of sequence-based p53 gene analysis in human breast cancer: Messenger RNA in comparison with genomic DNA targets. Clin Chem 1998; 44(3):455-462.
17. Magnusson KP, Sandstrom M, Stahlberg M et al. p53 splice acceptor site mutation and increased HsRAD51 protein expression in Bloom's syndrome GM1492 fibroblasts. Gene 2000; 246(1-2):247-254.
18. Usuda J, Inomata M, Fukumoto H et al. Restoration of p53 gene function in 12-O-tetradecanoylphorbor 13-acetate-resistant human leukemia K562/TPA cells. Int J Oncol 2003; 22(1):81-86.
19. King-Underwood L, Pritchard-Jones K. Wilms' tumor (WT1) gene mutations occur mainly in acute myeloid leukemia and may confer drug resistance. Blood 1998; 91(8):2961-2968.
20. Fan S, Yuan R, Ma YX et al. Mutant BRCA1 genes antagonize phenotype of wild-type BRCA1. Oncogene 2001; 20(57):8215-8235.
21. Sylvain V, Lafarge S, Bignon YJ. Dominant-negative activity of a Brca1 truncation mutant: Effects on proliferation, tumorigenicity in vivo, and chemosensitivity in a mouse ovarian cancer cell line. Int J Oncol 2002; 20(4):845-853.
22. Cardinali M, Kratochvil FJ, Ensley JF et al. Functional characterization in vivo of mutant p53 molecules derived from squamous cell carcinomas of the head and neck. Mol Carcinog 1997; 18(2):78-88.
23. Reddy JC, Morris JC, Wang J et al. WT1-mediated transcriptional activation is inhibited by dominant negative mutant proteins. J Biol Chem 1995; 270(18):10878-10884.
24. Englert C, Vidal M, Maheswaran S et al. Truncated WT1 mutants alter the subnuclear localization of the wild-type protein. Proc Natl Acad Sci USA 1995; 92(26):11960-11964.
25. Flaman JM, Waridel F, Estreicher A et al. The human tumour suppressor gene p53 is alternatively spliced in normal cells. Oncogene 1996; 12(4):813-818.
26. Chow VT, Quek HH, Tock EP. Alternative splicing of the p53 tumor suppressor gene in the Molt-4 T-lymphoblastic leukemia cell line. Cancer Lett 1993; 73(2-3):141-148.
27. King-Underwood L, Renshaw J, Pritchard-Jones K. Mutations in the Wilms' tumor gene WT1 in leukemias. Blood 1996; 87(6):2171-2179.
28. Little MH, Prosser J, Condie A et al. Zinc finger point mutations within the WT1 gene in Wilms tumor patients. Proc Natl Acad Sci USA 1992; 89(11):4791-4795.
29. Kohsaka T, Tagawa M, Takekoshi Y et al. Exon 9 mutations in the WT1 gene, without influencing KTS splice isoforms, are also responsible for Frasier syndrome. Hum Mutat 1999; 14(6):466-470.
30. Barbaux S, Niaudet P, Gubler MC et al. Donor splice-site mutations in WT1 are responsible for Frasier syndrome. Nat Genet 1997; 17(4):467-470.
31. Klamt B, Koziell A, Poulat F et al. Frasier syndrome is caused by defective alternative splicing of WT1 leading to an altered ratio of WT1 +/-KTS splice isoforms. Hum Mol Genet 1998; 7(4):709-714.
32. Eustice DC, Wilhelm JM. Fidelity of the eukaryotic codon-anticodon interaction: Interference by aminoglycoside antibiotics. Biochemistry 1984; 23(7):1462-1467.
33. Zsembery A, Jessner W, Sitter G et al. Correction of CFTR malfunction and stimulation of Ca-activated Cl channels restore HCO3- Secretion in cystic fibrosis bile ductular cells. Hepatology 2002; 35(1):95-104.
34. Bedwell DM, Kaenjak A, Benos DJ et al. Suppression of a CFTR premature stop mutation in a bronchial epithelial cell line. Nat Med 1997; 3(11):1280-1284.
35. Keeling KM, Brooks DA, Hopwood JJ et al. Gentamicin-mediated suppression of Hurler syndrome stop mutations restores a low level of alpha-L-iduronidase activity and reduces lysosomal glycosaminoglycan accumulation. Hum Mol Genet 2001; 10(3):291-299.
36. Barton-Davis ER, Cordier L, Shoturma DI et al. Aminoglycoside antibiotics restore dystrophin function to skeletal muscles of mdx mice. J Clin Invest 1999; 104(4):375-381.
37. Sangkuhl K, Schulz A, Rompler H et al. Aminoglycoside-mediated rescue of a disease-causing nonsense mutation in the V2 vasopressin receptor gene in vitro and in vivo. Hum Mol Genet 2004.
38. Lai CH, Chun HH, Nahas SA et al. Correction of ATM gene function by aminoglycoside-induced read-through of premature termination codons. Proc Natl Acad Sci USA 2004; 101(44):15676-15681, (Epub 12004 Oct 15621).
39. Helip-Wooley A, Park MA, Lemons RM et al. Expression of CTNS alleles: Subcellular localization and aminoglycoside correction in vitro. Mol Genet Metab 2002; 75(2):128-133.
40. Du M, Jones JR, Lanier J et al. Aminoglycoside suppression of a premature stop mutation in a Cftr-/- mouse carrying a human CFTR-G542X transgene. J Mol Med 2002; 80(9):595-604.
41. Wilschanski M, Yahav Y, Yaacov Y et al. Gentamicin-induced correction of CFTR function in patients with cystic fibrosis and CFTR stop mutations. N Engl J Med 2003; 349(15):1433-1441.
42. Wagner KR, Hamed S, Hadley DW et al. Gentamicin treatment of Duchenne and Becker muscular dystrophy due to nonsense mutations. Ann Neurol 2001; 49(6):706-711.

43. Politano L, Nigro G, Nigro V et al. Gentamicin administration in Duchenne patients with premature stop codon. Preliminary results. Acta Myol 2003; 22(1):15-21.
44. Dominski Z, Kole R. Restoration of correct splicing in thalassemic pre-mRNA by antisense oligonucleotides. Proc Natl Acad Sci USA 1993; 90(18):8673-8677.
45. Mann CJ, Honeyman K, Cheng AJ et al. Antisense-induced exon skipping and synthesis of dystrophin in the mdx mouse. Proc Natl Acad Sci USA 2001; 98(1):42-47.
46. Lu QL, Mann CJ, Lou F et al. Functional amounts of dystrophin produced by skipping the mutated exon in the mdx dystrophic mouse. Nat Med 2003; 9(8):1009-1014.
47. Kerr TP, Sewry CA, Robb SA et al. Long mutant dystrophins and variable phenotypes: Evasion of nonsense-mediated decay? Hum Genet 2001; 109(4):402-407.

Chapter 10

Therapies of Nonsense-Associated Diseases

Kim M. Keeling, Ming Du and David M. Bedwell*

Abstract

A large number of diseases are caused by premature stop mutations that often lead to a complete loss of protein function and a severe reduction in mRNA levels due to nonsense-mediated mRNA decay (NMD). Two main approaches to develop treatments for diseases caused by premature stop mutations have been investigated. The first is to reduce the efficiency of translation termination through the use of pharmacological agents or by the expression of suppressor tRNAs. The second approach is to replace the premature stop mutation with wild-type sequence using gene repair techniques. Although each of these approaches have been demonstrated using in vitro studies or mouse models, currently only strategies using pharmacological agents to suppress stop mutations have reached preliminary clinical trials. The future of suppression therapy will require finding ways to increase the efficacy of current compounds to suppress premature stop mutations without side effects, or to design or discover safe new compounds that suppress premature stop mutations with increased efficiency. In addition, combined therapies that simultaneously suppress a premature stop mutation and inhibit NMD of the nonsense-containing mRNA may be the most effective way to increase the efficiency of suppression therapy.

Introduction

A large number of diseases including cystic fibrosis, Duchenne muscular dystrophy, β-thalassemia, and many types of cancers are caused by the presence of premature stop mutations in mRNAs. Premature stop mutations can arise as a result of mutations within germline or somatic DNA, inaccurate or inefficient pre-mRNA splicing, or improper RNA editing. According to the Human Gene Mutation Database, 12% of all mutations reported are single point mutations that result in a premature stop codon.[1] If mutations that alter the translational reading frame such as deletions, insertions, and splicing mutations are also considered, premature stop mutations may be responsible for as many as one-third of all inherited genetic disorders or cancers.[2] Furthermore, the disease phenotypes caused by premature stop mutations are frequently more severe than those that result from missense mutations, since premature stop mutations often result in a complete loss of protein function.

Suppression of Premature Stop Mutations

One approach to treat diseases that result from in-frame premature stop mutations is to reduce the efficiency of translation termination so production of some full-length, functional protein is restored. Translation termination in eukaryotic cells occurs when one of the three stop codons, UAA (ochre), UAG (amber), or UGA (opal), enters the ribosomal A site. Stop codon recognition is not carried out by codon-anticodon interactions since no tRNA anticodons are

*Corresponding Author: David M. Bedwell—Department of Microbiology, BBRB 432 / Box 8, 1530 3rd Avenue, South, The University of Alabama at Birmingham, Birmingham, Alabama 35294-2170, U.S.A. Email: dbedwell@uab.edu

Nonsense-Mediated mRNA Decay, edited by Lynne E. Maquat. ©2006 Eurekah.com.

complementary to any of the stop codons. Rather, stop codon recognition is mediated by a protein known as eukaryotic release factor 1 (eRF1). eRF1 recognizes each of the three stop codons in the ribosomal A site,[3,4] and RNA cross-linking studies suggest that this interaction is direct.[5] Upon recognition of a stop codon, eRF1 transmits a signal to the ribosomal peptidyl transferase center that leads to release of the nascent polypeptide from the peptidyl-tRNA located in the ribosomal P site.[6] Another eukaryotic release factor, eRF3, is a GTPase that binds eRF1[7] and facilitates the efficiency and accuracy of stop codon recognition.[8]

Under normal conditions, translation termination is a very efficient process with an estimated error rate of approximately 0.1%.[9-12] However, eRF1 and near-cognate aminoacyl-tRNAs (aminoacyl-tRNAs with an anticodon complementary to two of the three nucleotides of the stop codon) normally compete for A site binding. Under certain conditions, the rate at which near-cognate aminoacyl-tRNAs successfully compete with eRF1 at a stop codon can be increased, resulting in incorporation of an amino acid carried by a near-cognate aminoacyl-tRNA into the nascent polypeptide.[13] This process is termed "termination suppression" or "readthrough". In the case of a premature stop mutation, readthrough normally results in the continued elongation of the polypeptide chain in the correct reading frame and the production of full-length protein.

Obvious perturbations to the efficiency of translation termination include mutations in the translational machinery such as ribosomal proteins,[14-19] ribosomal RNAs (rRNAs),[20-23] termination factors,[24-31] and aminoacyl-tRNAs.[32-35] Interestingly, the identity of the stop codon itself also affects termination efficiency. Generally, termination is most efficient at UAA stop codons, followed by the UAG and UGA stop codons. In addition, the sequence context both upstream and downstream of the stop codon also influences termination efficiency.[12,36-42] In particular, the first nucleotide downstream of the stop codon plays an important role in determining the efficiency of translation termination, and sequence analysis of natural stop codons has revealed a strong bias at that position in many species (including humans). This observation led to the proposal that eRF1 may normally recognize a tetranucleotide termination signal.[42] RNA cross-linking studies have confirmed that eRF1 contacts the first nucleotide following the stop codon.[43]

Aminoglycoside-Mediated Nonsense Suppression

The efficiency of translation termination can also be reduced through the action of a large class of structurally related antibiotics called aminoglycosides. Aminoglycosides bind to a region of the small subunit rRNA known as the decoding site[44] that normally monitors proper codon-anticodon interactions. Several nucleotides in the decoding site act to probe the conformation of the codon-anticodon helix to ensure that tRNA selection is correct. When aminoglycosides bind to the decoding site, they induce a conformational change that reduces the ability of rRNA to discriminate between cognate and near-cognate aminoacyl-tRNAs.[45-50] This reduction in the accuracy of codon recognition increases the probability that translational misreading will occur, including the readthrough of stop codons.

Remarkably, the rRNA sequences that make up the prokaryotic and eukaryotic decoding sites are very similar (Fig. 1). One of the main differences lies within the major groove of the decoding site where aminoglycosides bind. A key residue for aminoglycoside binding to the prokaryotic decoding site, the A1408 nucleotide, is a G nucleotide in the corresponding position of the eukaryotic decoding site.[51] Introduction of the A1408G mutation in the bacterial decoding site has been shown to reduce the affinity for aminoglycoside binding significantly.[52] This suggests that this key nucleotide difference at least partially accounts for the specificity of aminoglycosides to inhibit the bacterial ribosome, resulting in their utility as antibiotics in humans.

However, several studies have shown that some aminoglycosides can also stimulate low levels of misreading that leads to termination suppression in eukaryotic systems.[53-63] A recent yeast study revealed that aminoglycosides induced little or no misreading at sense codons,

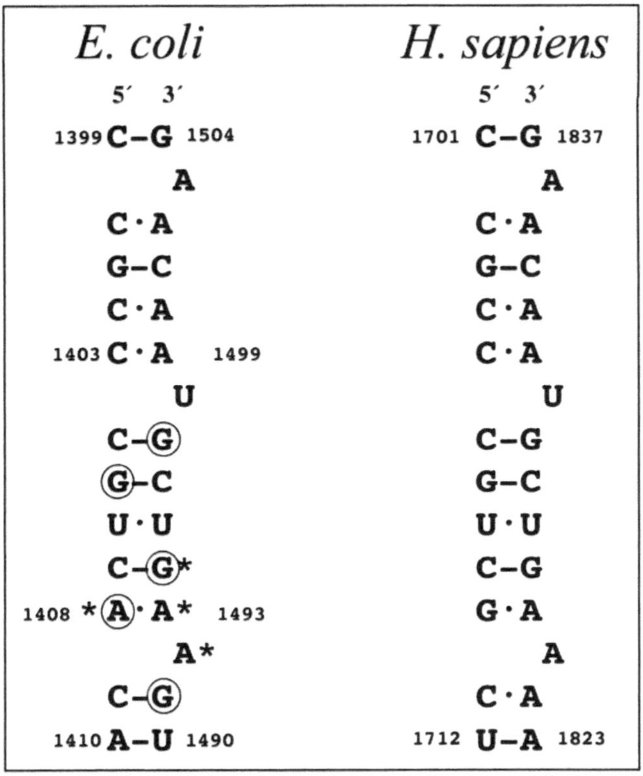

E. coli	*H. sapiens*
5′ 3′	5′ 3′
1399 C–G 1504	1701 C–G 1837
A	A
C·A	C·A
G–C	G–C
C·A	C·A
1403 C·A 1499	C·A
U	U
C–G	C–G
G–C	G–C
U·U	U·U
C–G*	C–G
1408 *A·A* 1493	G·A
A*	A
C–G	C·A
1410 A–U 1490	1712 U–A 1823

Figure 1. Comparison of the decoding sites in *E. coli* 16S rRNA and human 18S rRNA. Residues in the *E. coli* decoding site that are protected from dimethyl sulfoxide (DMS) modification by paromomycin are circled.[45] Residues that are protected from DMS modification by mRNA in a model decoding structure and also protected by an mRNA-dependent interaction of tRNA with the A site of ribosomes are marked by asterisks.[159]

while the suppression of nonsense mutations was generally robust.[64] This result suggests that stop codons are generally much more susceptible to aminoglycoside-induced misreading than sense codons in eukaryotes. This selectivity could be due to inherent differences in the fidelity of the elongation and termination processes.

The observations that aminoglycosides suppress premature stop codons in eukaryotic systems have led to a number of investigations to determine whether aminoglycosides could provide sufficient readthrough of premature stop mutations to suppress the phenotypes associated with human diseases. In most cases this question remains to be answered. However, it has been shown in mammalian cells that aminoglycosides can induce the suppression of nonsense mutations that cause many diseases, resulting in the restoration of low levels of functional protein. Aminoglycosides have been shown to induce readthrough of nonsense mutations that cause cystic fibrosis,[65-67] Duchenne muscular dystrophy,[68,69] Hurler syndrome,[70,71] infantile neuronal ceroid lipofuscinosis,[72] cystinosis,[73] X-linked nephrogenic diabetes insipidus,[74] recessive spinal muscular atrophy,[75] and polycystic kidney disease.[76] They have also been shown to suppress nonsense mutations in the *p53*[60] and *ATM*[77] tumor suppressor genes.

Particularly promising results have been obtained using mouse models for both Duchenne muscular dystrophy (DMD) and cystic fibrosis (CF). A mouse model for Duchenne muscular

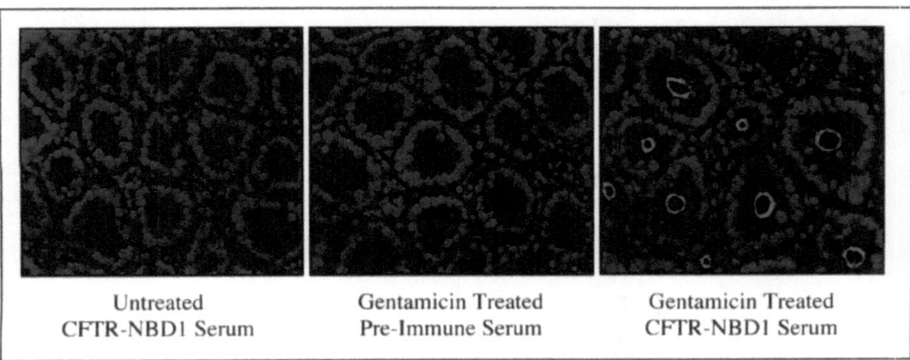

Figure 2. Immunofluorescence staining of submucosal glands in the duodenum shows the appearance of cystic fibrosis transmembrane conductance regulator (CFTR) protein following gentamicin treatment. Samples from homozygous *Cftr-/- hCFTR*-G542X transgenic mice (harvested from untreated or gentamicin-treated animals) were stained with a nuclear stain (blue) and incubated with either preimmune serum or CFTR-NBD1 serum. After incubation with a secondary antibody conjugated to a fluorescent dye (green), samples were visualized by fluorescence microscopy. CFTR protein was observed at the apical surface of epithelial cells only in tissues incubated with CFTR-specific serum that had been harvested from gentamicin-treated animals.

dystrophy known as the *mdx* mouse carries a naturally occurring UAA nonsense mutation in the dystrophin gene. It was shown that the administration of gentamicin via subcutaneous injections resulted in the partial restoration of dystrophin protein in muscle tissue of the *mdx* mouse.[69] In addition, muscle contraction assays demonstrated that gentamicin treatment restored enough functional dystrophin protein to significantly reduce muscle contractile injury that is the hallmark of DMD. In another study, a transgenic CF mouse model that carried a UGA nonsense mutation in the cystic fibrosis transmembrane conductance regulator (*CFTR*) gene was used to examine whether subcutaneous injections of gentamicin could restore CFTR protein expression. Treated mice showed a partial restoration of CFTR protein expression by immunofluoresence (Fig. 2) and a partial restoration of cAMP-activated chloride channel activity.[78] These findings indicate that the administration of gentamicin stimulated readthrough of the *CFTR* premature stop mutation, resulting in the production of functional CFTR protein.

Several small clinical trials have also administered aminoglycosides to CF or DMD patients who carry premature stop mutations to determine whether any restoration of protein function occurred. To date, three such trials with CF patients have been reported. In two of the trials, the administration of gentamicin via nasal droplets restored some CFTR activity in the nasal epithelia of CF patients with a *CFTR* nonsense mutation.[79,80] In a third study, a partial restoration of CFTR function was detected in the nasal epithelium of CF patients that carried a *CFTR* nonsense mutation when gentamicin was administered intravenously.[81] Decidedly more mixed results were obtained in two clinical trials in which DMD patients with nonsense mutations were administered intravenous gentamicin. In one trial with four DMD patients, no increase in dystrophin levels or physical improvement could be ascertained after aminoglycoside treatment.[82,83] However, the results of another clinical trial were more promising, since three of the four treated patients showed a partial restoration of dystrophin protein expression.[83]

The chemical structure of aminoglycosides determines their ability to suppress nonsense mutations (Fig. 3). Only a subset of these compounds is effective at suppressing nonsense mutations in eukaryotes, and only three aminoglycosides from this group (gentamicin, amikacin, and tobramycin) are approved for internal human use. In addition, the susceptibility of premature stop codons to aminoglycoside-mediated suppression depends on the identity of the stop codon and its surrounding mRNA sequence context.[59,60] This codon and context dependence

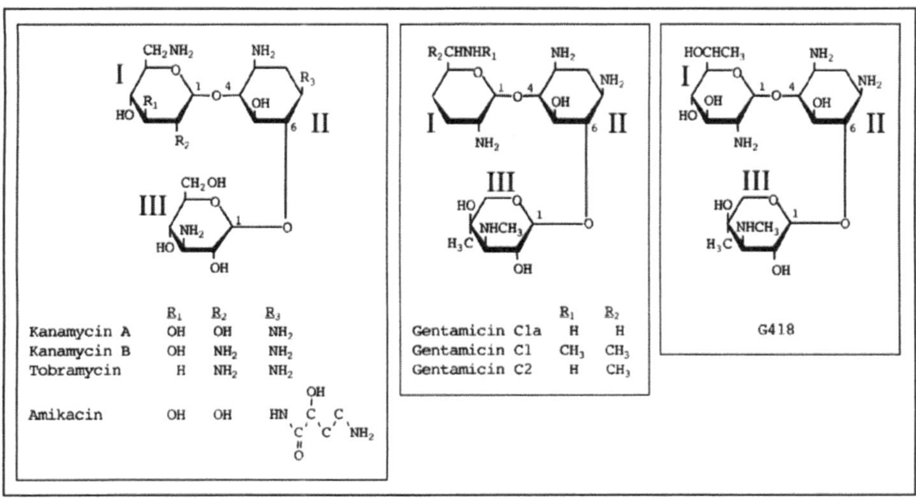

Figure 3. Structures of aminoglycosides commonly used in a clinical setting. The 2-deoxystreptamine ring is labeled ring II in each structure.

of aminoglycoside suppression is consistent with the results of a DMD clinical trial in which three DMD patients that carried UGA premature stop mutations showed a partial restoration of dystrophin expression after gentamicin treatment. However, no dystrophin could be detected in a patient with a UAA stop mutation after receiving the same gentamicin regimen.[83] These results suggest that aminoglycoside suppression of a UGA stop codon may result in significantly more readthrough in patients than aminoglycoside suppression of a UAA codon, as previously concluded from in vitro studies.[59,68,84] Thus, it may be necessary to alter the efficiency or stop codon specificity of nonsense suppression if a greater level of readthrough is to be obtained for diseases caused by certain stop mutations.

A major hurdle to the long-term use of aminoglycosides as a therapy to treat diseases caused by nonsense mutations is their toxicity, which can lead to kidney damage and hearing loss. However, the majority of these side effects do not appear to be due to their ability to induce translational misreading, but rather to other consequences related to their charged nature. Aminoglycosides are taken into cells via megalin, a multi-ligand, endocytic receptor that is particularly abundant in the proximal tubules of the kidney and the hair cells of the inner ear.[85] Upon entering kidney cells, the positively charged nature of aminoglycoside molecules promotes their binding to acidic phospholipids in the lysosomal membrane,[85,86] which alters the activity of a number of enzymes. In addition, aminoglycosides have been shown to promote the generation of free radical species that leads to tissue damage. Approaches that may reduce aminoglycoside toxicity include: altering the route and duration of their administration;[87,88] coadministering compounds such as antioxidants to circumvent free radical damage;[89-92] and coadministering polyanions such as poly-L-aspartate[93,94] or daptomycin[95,96] to sequester aminoglycosides away from the lysosomal membrane.

Another potential drawback associated with the suppression of premature stop mutations using aminoglycosides is the potential suppression of native stop codons. If this occurred on a global basis, it would lead to the production of many proteins with an extended C-terminus that could result in misfolding and loss of protein function. However, a previous study found that human cells cultured in the presence of high concentrations of aminoglycosides exhibited only a small increase in the level of the Hsp70 molecular chaperone, suggesting that little protein misfolding occurred during aminoglycoside treatment.[70]

The apparent lack of global readthrough at normal stop codons could be explained in several ways. Evidence of an evolutionary bias toward natural stop codons and surrounding sequence context(s) that may represent the most efficient termination signals has been observed at the end of genes in many species.[42] No such selection for resistance to readthrough would occur at premature stop mutations. Thus, while the efficacy of aminoglycosides may be limited by the identity of the premature stop mutations and surrounding sequence context, these differences may also prevent readthrough at normal termination signals. In addition, multiple, in-frame stop codons are frequently found at the end of mRNAs.[97-100] The presence of multiple stop codons should dramatically reduce the ability of aminoglycosides to induce readthrough of normal termination signals. Furthermore, the termination complex formed at premature stop codons appears to differ from the complex formed at native stop codons at the end of an mRNA.[101,102] This intriguing finding suggests that the ribosome may terminate translation less efficiently at premature stop codons than native stop codons, possibly because the interactions between the termination complex at a premature stop codon and other factors bound in the 3' untranslated region of an mRNA cannot occur in a normal manner (see chapter by Amrani and Jacobson for further details).

Other Pharmacological Compounds That Suppress Nonsense Mutations

Since many of the toxic side effects caused by aminoglycosides are not directly associated with their ability to suppress stop mutations, another way to avoid these problems is to identify new, safer classes of compounds that suppress stop mutations. One such compound that has been investigated is negamycin. Negamycin is a dipeptide antibiotic that interacts with the ribosomal decoding site, much like aminoglycosides, even though it is structurally unrelated to aminoglycosides. Negamycin was shown to suppress the dystrophin premature stop mutation in the *mdx* mouse model, and was reported to be less toxic than aminoglycosides.[103] Another drug, PTC124, is a novel compound discovered by PTC Therapeutics, Inc. that has been shown to suppress nonsense mutations in cell culture and in animal models.[104] These results suggest that developing new pharmaceutical agents that suppress premature stop mutations without inducing the toxic side effects associated with aminoglycosides may have great potential for future therapeutic use.

Suppression of Nonsense Mutations Using Suppressor tRNAs

Another means of suppressing nonsense mutations involves expressing suppressor tRNAs. In this approach, DNA encoding a tRNA with an anticodon complementary to a stop codon is introduced into cells. This type of mutant tRNA, referred to as a suppressor tRNA, can compete with the termination factor eRF1 much more effectively than a near-cognate tRNA, resulting in a significant increase in stop codon suppression.

This approach has been shown to suppress premature stop mutations that cause β-thalassemia[105] and Duchenne muscular dystrophy[106] in mammalian cells. It has also been shown that the injection of DNA encoding a suppressor tRNA into the skeletal and heart muscles of a transgenic mouse expressing a reporter gene with a premature stop codon resulted in the suppression of the stop codon in vivo.[107]

Besides the potential suppression of natural stop codons that was discussed in a previous section, there are other significant drawbacks to this therapeutic approach. Although tRNA genes have strong RNA polymerase III promoters and the encoded tRNA molecules are generally stable after they are transcribed and undergo maturation,[108] the efficient introduction and maintenance of tRNA genes into cells by gene therapy methods remains a challenge. In addition, the termination signal to be suppressed and its surrounding context also affect the efficiency of suppressor tRNA-mediated suppression.[108,109]

Mutation Repair

Another novel approach that could be used to treat diseases caused by premature stop mutations is the repair of a mutation directly in the genome using a nucleotide exchange antisense oligonucleotide approach. Unlike nonsense suppression therapy, which requires that the mutation be in the correct open reading frame, the use of antisense oligonucleotides also has the potential to repair various point mutations, small deletions, insertions, or splicing defects. In this method, a mutant DNA sequence is replaced by the wild-type sequence. Although a number of approaches have been designed to target DNA for mutation repair, including the use of ribozymes, group II introns, and triplex-forming oligonucleotides, single-stranded DNA currently appears to generate the most robust and reproducible gene repair. This exchange is directed by a double-stranded DNA chimeric oligonucleotide that is typically 70-80 nucleotides long. This chimera is synthesized as a single-stranded molecule that is designed with sequence complementarity such that is folds into a double-hairpin structure. The double-hairpin structure prevents the molecule from nuclease digestion as well as concatenation.[110] When these molecules are introduced into cells, the chimeric oligonucleotide hybridizes to its complementary sequence in the target gene except for the region of mismatch sequence where the mutation lies. This mismatch is recognized by the cellular mismatch-repair system, which then catalyzes the exchange of the wild-type nucleotides for the mutant nucleotides, thus repairing the mutation in the DNA.

Gene repair of mutations has been demonstrated in yeast[111,112] and mammalian cell culture, where correction of mutations that caused sickle cell anemia[113] thalassemia,[114] alkaline phosphatase deficiency,[115] apolipoprotein A2-linked atherosclerosis,[116] and epidermolysis bullosa simplex[117] were accomplished. In addition, animal models of tyrosinemia,[118] muscular dystrophy,[119-122] hemophilia,[123] and renal tubular acidosis[124] have been used to demonstrate a partial correction of mutations and a restoration of some protein expression. The amount of normal protein levels restored among the various disease phenotypes varied from 0.5% to 20% of wild-type levels.[110] This level of functional protein could improve the phenotype of patients with these (and many other) genetic diseases.

Although this approach to correct mutations and restore wild-type protein production in mammalian cells is promising, its development for clinic application is still a work in progress. There are several problems that currently prevent the implementation of this approach.[125] First, the design of stable oligonucleotides capable of replacing the mutation efficiently can be difficult. Second, as with many gene therapy approaches, the introduction of oligonucleotides into the appropriate cell type is a major obstacle, and the efficiency of repair is cell-cycle dependent. Third, the percentage of corrected cells decreases with time. Finally, apoptosis has been observed in some cells that have undergone this form of targeted sequence alteration. Thus, further studies are required to determine the ultimate utility of this approach.

Suppression of NMD

Many strategies aimed at suppressing premature stop mutations could be compromised by the fact that mRNAs that contain premature stop mutations are often unstable, resulting in a severe reduction in their steady-state level. The reduced abundance of mRNAs that carry a stop mutation is due to the NMD pathway (see chapter by Maquat). Therefore, approaches that stabilize nonsense-containing mRNAs that are normally degraded by NMD should increase the steady-state amount of mRNA available for translation. This, in turn, could greatly enhance the level of protein produced by suppression therapy.

The NMD pathway and the NMD factors Upf1, Upf2, and Upf3 are conserved in eukaryotes ranging from yeast to humans (see chapters by Baker and Parker, Singh and Lykke-Andersen, Anderson, and Behm-Ansmant and Izaurralde). Several factors involved in mammalian NMD bind to mRNAs in the nucleus during transcription and the subsequent stages of mRNA processing. In particular, some NMD components assemble with the exon-junction complex (EJC), located ~20-24 nucleotides upstream of exon-exon junctions as a consequence of pre-mRNA splicing, and remain bound to the mRNA as it is exported to the cytoplasm[126] (see chapter by Maquat). In human cells, Upf3 binds to nuclear mRNA-protein (mRNP) complexes as a component of the EJC complex. Upf2 binds to these complexes as they leave the nucleus, while Upf1 is thought to associate with the complex in the cytoplasm.[127,128]

According to current models, once the mRNA reaches the cytoplasm these bound nuclear factors are removed as the ribosome translates the mRNA during the initial or "pioneer round" of translation[129,130] (see chapter by Maquat). If translation proceeds to the normal stop codon and all of the nuclear proteins are removed from the coding sequence during the pioneer round of translation, the transcript is remodeled to become steady-state mRNP and NMD can no longer occur. However, if a premature stop codon is present in the mRNA, the ribosome will not remove any nuclear proteins distal to the premature stop codon during this initial round of translation. This causes the mRNA to be identified as faulty and results in its rapid degradation by the NMD pathway. Generally, NMD occurs if an mRNA molecule carries a premature stop codon that is ≥50 nucleotides upstream of the 3'-most exon-exon junction.[131]

The destabilization of a nonsense-containing mRNA by NMD requires active translation of that mRNA. This conclusion is supported by the observations that several proteins that function in the NMD pathway associate with polysomes,[132-135] and NMD can be inhibited by translation elongation inhibitors such as cycloheximide and puromycin. In yeast, the NMD factors have been shown to associate with the translation termination factors eRF1 and eRF3[136] (see chapters by Baker and Parker, and Amrani and Jacobson). Of the Upf factors, at least Upf1 associates with eRF1 and eRF3 in mammals (see chapters by Maquat, and Singh and Lykke-Andersen), also indicating that NMD and the process of translation termination are likely to be tightly linked.

It has also been shown that suppression of a premature stop codon can stabilize the mRNA. Overexpression of a suppressor tRNA can inhibit the degradation of nonsense-containing mRNAs in yeast and in mammalian cells,[137,138] and suppression of premature stop codons by aminoglycosides has also been shown to stabilize nonsense-containing mRNAs in some cases. For example, a 5-fold increase in the abundance of *CFTR* mRNA containing a premature stop codon was detected in a human bronchial epithelial cell line derived from a CF patient that carried the *CFTR-W1282X* premature stop mutation after incubation with the aminoglycoside G418 for 24 hours[67] (Fig. 4). A 2-fold increase in mRNA was observed in human fibroblasts from a patient with Smith-Lemli-Opitz syndrome that carried a premature termination codon in the 7-dehydrocholesterol-delta 7-reductase (*DHCR7*) gene after G418 treatment.[139] Finally, a 2-fold increase in mRNA was observed in human fibroblasts derived from a patient with Hurler syndrome that carried nonsense mutations in the α-L-iduronidase (*IDUA*) gene after gentamicin treatment (K.M. Keeling and D.M. Bedwell, unpublished data). These results are surprising since aminoglycosides only induced readthrough of these premature stop mutations at a low frequency of 1-20%. While it is possible that premature stop mutations are more susceptible to aminoglycoside-mediated suppression during the pioneer round of translation than in subsequent rounds of translation, the mechanism by which such a modest level of readthrough can inhibit NMD is currently unknown.

Certain factors in the NMD pathway have also been identified as potential targets for pharmacological inhibition of mRNA degradation by the NMD pathway. One such target is Upf1, which is essential for the NMD process and is conserved in all eukaryotes. Upf1 is a phosphoprotein, and changes in its phosphorylation state regulate its function in the NMD pathway.[140,141] Several modifiers of the Upf1 phosphorylation state have been identified. For example, SMG1 is a kinase that phosphorylates at least two serine residues in Upf1 and stimulates the NMD process[142-144] (see chapter by Yamashita et al). SMG1 is a member of the phosphatidylinositol-4,5-bisphosphate (PIP2) family of protein kinases, which are frequently inhibited by caffeine and wortmannin. Accordingly, these compounds were found to abrogate the degradation of the collagen VI α2 subunit mRNA by NMD in human fibroblasts,

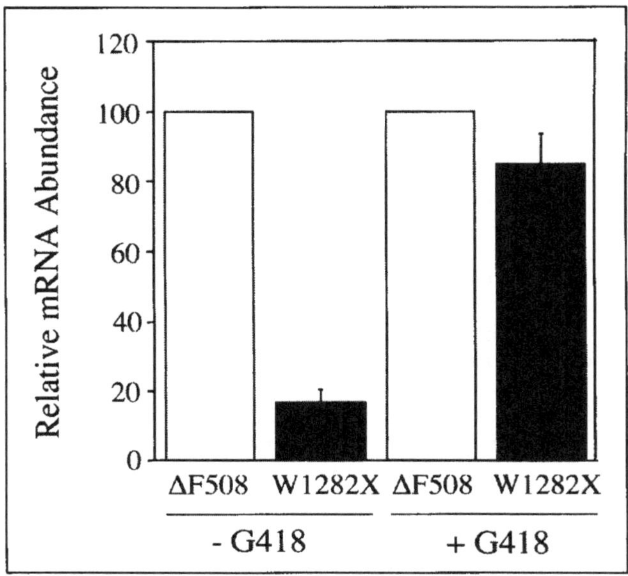

Figure 4. *CFTR*-W1282X mRNA abundance is increased in the IB3-1 bronchial epithelial cell line following G418 treatment. IB3-1 cells carry the W1282X premature stop mutation on one *CFTR* allele and the ΔF508 mutation on the other *CFTR* allele. The ΔF508 allele harbors a 3-nucleotide deletion of codon 508, so translation proceeds to the normal translation termination codon and a normal level of mRNA is produced.[160] In contrast, the W1282X allele contains a nonsense mutation at codon 1282 that results in NMD.[161,162] For this experiment, IB3-1 cells were grown in the presence or absence of G418, and the abundance of mRNA from each *CFTR* allele was determined by allele-specific oligonucleotide hybridization. The level of *CFTR* mRNA from the ΔF508 allele was defined as 100%, and the level of mRNA from the W1282X allele was normalized to the ΔF508 level. Each value represents the mean ± standard deviation of four independent measurements.

presumably by inhibiting SMG1 activity.[145] The phosphorylation status of Upf1 is also influenced by SMG5, SMG6 and SMG7, which participate in the dephosphorylation of Upf1 and thus complete its phosphorylation-dephosphorylation cycle.[146-149] These Upf1 modifiers also represent potential targets for pharmacological inhibition of NMD. Although the yeast Upf1 protein is also a phosphoprotein (K.M. Keeling and D.M. Bedwell, unpublished data), the kinase and phosphatase involved in its regulation have yet to be identified. Interestingly, a compound called diazoborine has been shown to stabilize aberrant mRNAs that are normally degraded by the NMD pathway in yeast.[150] While the target of this compound is not yet known, it is possible that this compound could also target a component of the Upf1 phosphorylation cycle or some other step of yeast NMD.

It is important to note that the NMD pathway has additional functions other than degrading mRNAs that carry premature stop mutations (see chapters by Abraham and Oliveira, Azzalin et al, Kaygun and Marzluff, and Kim and Maquat). Mutations in Upf1, Upf2, and Upf3 influence the abundance of many normal mRNA species in yeast (see chapter by He and Jacobson) and mammals (see chapters by Sharifi and Dietz, and Soergel et al).[151,152] Factors in the NMD pathway may also act as checkpoints for RNA processing and nuclear export.[153,154] Therefore, therapeutic approaches aimed at inhibition of the NMD pathway must be monitored carefully and optimized such that other vital cellular pathways are not adversely affected.

Future Development of Therapies for Nonsense-Associated Diseases

Even though the results of several clinical trials indicate that the suppression of nonsense mutations can partially restore protein function, none of these studies were designed in a manner that allowed the investigators to determine whether enough protein expression was restored to confer a therapeutic improvement in the disease phenotype. The threshold level of functional protein needed to alleviate a disease phenotype is unknown for most diseases. This is a complex issue, since the minimal level of functional protein needed to improve a particular disease phenotype will vary widely depending upon the structure and function of each protein, and the key tissue(s) in which each protein is required. For example, estimates of the amount of CFTR protein required to alleviate the CF disease phenotype range from 5%[155] to 30%[156] of wild-type CFTR levels. In contrast, as little as 1% of α-L-iduronidase can reduce the severity of the Hurler syndrome disease phenotype.[157,158] Therefore, suppression therapy may hold more potential for treating some disorders than others.

The development of new compounds or approaches that can suppress nonsense mutations without the side effects currently associated with aminoglycosides shows great potential for the future of nonsense suppression therapy. New pharmacological targets could include regions of rRNA, ribosomal proteins, translation termination factors, and components of the NMD machinery. Future therapies for the treatment of human diseases caused by premature stop mutations may include treatments designed to correct defects occurring at multiple levels simultaneously. For example, combining the suppression of premature stop mutations with the inhibition of NMD could restore a significantly higher level of functional protein than with either approach alone, thus providing a much greater opportunity to alleviate a disease phenotype. Currently, more research on the basic mechanisms of translation termination and NMD is needed to provide a better understanding of potential targets for therapies to treat nonsense-associated diseases.

References

1. Krawczak M, Ball EV, Fenton I et al. Human gene mutation database-a biomedical information and research resource. Hum Mutat 2000; 15(1):45-51.
2. Frischmeyer PA, Dietz HC. Nonsense-mediated mRNA decay in health and disease. Hum Mol Genet 1999; 8(10):1893-1900.
3. Kisselev L, Ehrenberg M, Frolova L. Termination of translation: Interplay of mRNA, rRNAs and release factors? EMBO J 2003; 22(2):175-182.
4. Bertram G, Innes S, Minella O et al. Endless possibilities: Translation termination and stop codon recognition. Microbiology 2001; 147(Pt 2):255-269.
5. Chavatte L, Frolova L, Kisselev L et al. The polypeptide chain release factor eRF1 specifically contacts the s(4)UGA stop codon located in the A site of eukaryotic ribosomes. Eur J Biochem 2001; 268(10):2896-2904.
6. Frolova L, Le Goff X, Rasmussen HH et al. A highly conserved eukaryotic protein family possessing properties of polypeptide chain release factor. Nature 1994; 372(6507):701-703.
7. Frolova L, Le Goff X, Zhouravleva G et al. Eukaryotic polypeptide chain release factor eRF3 is an eRF1- and ribosome-dependent guanosine triphosphatase. RNA 1996; 2(4):334-341.
8. Salas-Marco J, Bedwell DM. GTP Hydrolysis by eRF3 facilitates stop codon decoding during eukaryotic translation termination. Mol Cell Biol 2004; 24(17):7769-7778.
9. Stansfield I, Jones KM, Herbert P et al. Missense translation errors in Saccharomyces cerevisiae. J Mol Biol 1998; 282(1):13-24.
10. Mori N, Funatsu Y, Hiruta K et al. Analysis of translational fidelity of ribosomes with protamine messenger RNA as a template. Biochemistry 1985; 24(5):1231-1239.
11. Loftfield RB, Vanderjagt D. The frequency of errors in protein biosynthesis. Biochem J 1972; 128(5):1353-1356.
12. Bonetti B, Fu L, Moon J et al. The efficiency of translation termination is determined by a synergistic interplay between upstream and downstream sequences in Saccharomyces cerevisiae. J Mol Biol 1995; 251(3):334-345.
13. Fearon K, McClendon V, Bonetti B et al. Premature translation termination mutations are efficiently suppressed in a highly conserved region of yeast Ste6p, a member of the ATP-binding cassette (ABC) transporter family. J Biol Chem 1994; 269(27):17802-17808.

14. Zhang S, Ryden-Aulin M, Kirsebom LA et al. Genetic implication for an interaction between release factor one and ribosomal protein L7/L12 in vivo. J Mol Biol 1994; 242(5):614-618.
15. Van Dyke N, Xu W, Murgola EJ. Limitation of ribosomal protein L11 availability in vivo affects translation termination. J Mol Biol 2002; 319(2):329-339.
16. Torres M, Condon C, Balada JM et al. Ribosomal protein S4 is a transcription factor with properties remarkably similar to NusA, a protein involved in both nonribosomal and ribosomal RNA antitermination. EMBO J 2001; 20(14):3811-3820.
17. Tate WP, Dognin MJ, Noah M et al. The NH2-terminal domain of Escherichia coli ribosomal protein L11. Its three-dimensional location and its role in the binding of release factors 1 and 2. J Biol Chem 1984; 259(11):7317-7324.
18. Dahlgren A, Ryden-Aulin M. A novel mutation in ribosomal protein S4 that affects the function of a mutated RF1. Biochimie 2000; 82(8):683-691.
19. Brot N, Tate WP, Caskey CT et al. The requirement for ribosomal proteins L7 and L12 in peptide-chain termination. Proc Natl Acad Sci USA 1974; 71(1):89-92.
20. Velichutina IV, Hong JY, Mesecar AD et al. Genetic interaction between yeast Saccharomyces cerevisiae release factors and the decoding region of 18 S rRNA. J Mol Biol 2001; 305(4):715-727.
21. Velichutina IV, Dresios J, Hong JY et al. Mutations in helix 27 of the yeast Saccharomyces cerevisiae 18S rRNA affect the function of the decoding center of the ribosome. RNA 2000; 6(8):1174-1184.
22. Liu R, Liebman SW. A translational fidelity mutation in the universally conserved sarcin/ricin domain of 25S yeast ribosomal RNA. RNA 1996; 2(3):254-263.
23. Chernoff YO, Vincent A, Liebman SW. Mutations in eukaryotic 18S ribosomal RNA affect translational fidelity and resistance to aminoglycoside antibiotics. EMBO J 1994; 13(4):906-913.
24. Zadorskii SP, Borkhsenius AS, Sopova Iu V et al. Suppression of nonsense and frameshift mutations obtained by different methods for inactivating the translation termination factor eRF3 in yeast Saccharomyces cerevisiae. Genetika 2003; 39(4):489-494.
25. Wakem LP, Sherman F. Isolation and characterization of omnipotent suppressors in the yeast Saccharomyces cerevisiae. Genetics 1990; 124(3):515-522.
26. Vincent A, Newnam G, Liebman SW. The yeast translational allosuppressor, SAL6: A new member of the PP1-like phosphatase family with a long serine-rich N-terminal extension. Genetics 1994; 138(3):597-608.
27. Vincent A, Liebman SW. The yeast omnipotent suppressor SUP46 encodes a ribosomal protein which is a functional and structural homolog of the Escherichia coli S4 ram protein. Genetics 1992; 132(2):375-386.
28. Ono B, Tanaka M, Awano I et al. Two new loci that give rise to dominant omnipotent suppressors in Saccharomyces cerevisiae. Curr Genet 1989; 16(5-6):323-330.
29. Liebman SW, Cavenagh M. An antisuppressor that acts on omnipotent suppressors in yeast. Genetics 1980; 95(1):49-61.
30. Kulikov VN, Tikhodeev ON, Forafonov FS et al. Suppression of frameshift mutation as a result of partial inactivation of translation termination factors in Saccharomyces cerevisiae yeast. Genetika 2001; 37(5):602-609.
31. Carr-Schmid A, Valente L, Loik VI et al. Mutations in elongation factor 1beta, a guanine nucleotide exchange factor, enhance translational fidelity. Mol Cell Biol 1999; 19(8):5257-5266.
32. Valle RP, Morch MD, Haenni AL. Novel amber suppressor tRNAs of mammalian origin. EMBO J 1987; 6(10):3049-3055.
33. Vacher J, Grosjean H, de Henau S et al. Construction of a UGA suppressor tRNA by modification in vitro of yeast tRNACys. Eur J Biochem 1984; 138(1):77-81.
34. Kuchino Y, Beier H, Akita N et al. Natural UAG suppressor glutamine tRNA is elevated in mouse cells infected with Moloney murine leukemia virus. Proc Natl Acad Sci USA 1987; 84(9):2668-2672.
35. Beier H, Grimm M. Misreading of termination codons in eukaryotes by natural nonsense suppressor tRNAs. Nucleic Acids Res 2001; 29(23):4767-4782.
36. Tork S, Hatin I, Rousset JP et al. The major 5' determinant in stop codon read-through involves two adjacent adenines. Nucleic Acids Res 2004; 32(2):415-421.
37. Namy O, Hatin I, Rousset JP. Impact of the six nucleotides downstream of the stop codon on translation termination. EMBO Rep 2001; 2(9):787-793.
38. Mottagui-Tabar S, Tuite MF, Isaksson LA. The influence of 5' codon context on translation termination in Saccharomyces cerevisiae. Eur J Biochem 1998; 257(1):249-254.
39. McCaughan KK, Brown CM, Dalphin ME et al. Translational termination efficiency in mammals is influenced by the base following the stop codon. Proc Natl Acad Sci USA 1995; 92(12):5431-5435.
40. Harrell L, Melcher U, Atkins JF. Predominance of six different hexanucleotide recoding signals 3' of read-through stop codons. Nucleic Acids Res 2002; 30(9):2011-2017.

41. Cassan M, Rousset JP. UAG readthrough in mammalian cells: Effect of upstream and downstream stop codon contexts reveal different signals. BMC Mol Biol 2001; 2(1):3.
42. Brown CM, Stockwell PA, Trotman CN et al. Sequence analysis suggests that tetra-nucleotides signal the termination of protein synthesis in eukaryotes. Nucleic Acids Res 1990; 18(21):6339-6345.
43. Bulygin KN, Repkova MN, Ven'yaminova AG et al. Positioning of the mRNA stop signal with respect to polypeptide chain release factors and ribosomal proteins in 80S ribosomes. FEBS Lett 2002; 514(1):96-101.
44. Moazed D, Noller HF. Transfer RNA shields specific nucleotides in 16S ribosomal RNA from attack by chemical probes. Cell 1986; 47(6):985-994.
45. Yoshizawa S, Fourmy D, Puglisi JD. Structural origins of gentamicin antibiotic action. EMBO J 1998; 17(22):6437-6448.
46. Vicens Q, Westhof E. Crystal structure of paromomycin docked into the eubacterial ribosomal decoding A site. Structure (Camb) 2001; 9(8):647-658.
47. Recht MI, Fourmy D, Blanchard SC et al. RNA sequence determinants for aminoglycoside binding to an A-site rRNA model oligonucleotide. J Mol Biol 1996; 262(4):421-436.
48. Fourmy D, Yoshizawa S, Puglisi JD. Paromomycin binding induces a local conformational change in the A-site of 16 S rRNA. J Mol Biol 1998; 277(2):333-345.
49. Fourmy D, Recht MI, Puglisi JD. Binding of neomycin-class aminoglycoside antibiotics to the A-site of 16 S rRNA. J Mol Biol 1998; 277(2):347-362.
50. Fourmy D, Recht MI, Blanchard SC et al. Structure of the A site of Escherichia coli 16S ribosomal RNA complexed with an aminoglycoside antibiotic. Science 1996; 274(5291):1367-1371.
51. Van de Peer Y, Van den Broeck I, De Rijk P et al. Database on the structure of small ribosomal subunit RNA. Nucleic Acids Res 1994; 22(17):3488-3494.
52. Recht MI, Douthwaite S, Puglisi JD. Basis for prokaryotic specificity of action of aminoglycoside antibiotics. EMBO J 1999; 18(11):3133-3138.
53. Wilhelm JM, Pettitt SE, Jessop JJ. Aminoglycoside antibiotics and eukaryotic protein synthesis: Structure-function relationships in the stimulation of misreading with a wheat embryo system. Biochemistry 1978; 17(7):1143-1149.
54. Wilhelm JM, Jessop JJ, Pettitt SE. Aminoglycoside antibiotics and eukaryotic protein synthesis: Stimulation of errors in the translation of natural messengers in extracts of cultured human cells. Biochemistry 1978; 17(7):1149-1153.
55. Stahl G, Bidou L, Rousset JP et al. Versatile vectors to study recoding: Conservation of rules between yeast and mammalian cells. Nucleic Acids Res 1995; 23(9):1557-1560.
56. Singh A, Ursic D, Davies J. Phenotypic suppression and misreading Saccharomyces cerevisiae. Nature 1979; 277(5692):146-148.
57. Palmer E, Wilhelm JM, Sherman F. Phenotypic suppression of nonsense mutants in yeast by aminoglycoside antibiotics. Nature 1979; 277(5692):148-150.
58. Palmer E, Wilhelm JM. Mistranslation in a eucaryotic organism. Cell 1978; 13(2):329-334.
59. Manuvakhova M, Keeling K, Bedwell DM. Aminoglycoside antibiotics mediate context-dependent suppression of termination codons in a mammalian translation system. RNA 2000; 6(7):1044-1055.
60. Keeling KM, Bedwell DM. Clinically relevant aminoglycosides can suppress disease-associated premature stop mutations in the IDUA and p53 cDNAs in a mammalian translation system. J Mol Med 2002; 80(6):367-376.
61. Grentzmann G, Ingram JA, Kelly PJ et al. A dual-luciferase reporter system for studying recoding signals. RNA 1998; 4(4):479-486.
62. Firoozan M, Grant CM, Duarte JA et al. Quantitation of readthrough of termination codons in yeast using a novel gene fusion assay. Yeast 1991; 7(2):173-183.
63. Burke JF, Mogg AE. Suppression of a nonsense mutation in mammalian cells in vivo by the aminoglycoside antibiotics G-418 and paromomycin. Nucleic Acids Res 1985; 13(17):6265-6272.
64. Salas-Marco J, Bedwell DM. Discrimination between defects in elongation fidelity and termination efficiency provides mechanistic insights into translational readthrough. J Mol Biol 2005; 348(4):801-815.
65. Zsembery A, Jessner W, Sitter G et al. Correction of CFTR malfunction and stimulation of Ca-activated Cl channels restore HCO3- secretion in cystic fibrosis bile ductular cells. Hepatology 2002; 35(1):95-104.
66. Howard M, Frizzell RA, Bedwell DM. Aminoglycoside antibiotics restore CFTR function by overcoming premature stop mutations. Nat Med 1996; 2(4):467-469.
67. Bedwell DM, Kaenjak A, Benos DJ et al. Suppression of a CFTR premature stop mutation in a bronchial epithelial cell line. Nat Med 1997; 3(11):1280-1284.

68. Howard MT, Shirts BH, Petros LM et al. Sequence specificity of aminoglycoside-induced stop codon readthrough: Potential implications for treatment of Duchenne muscular dystrophy. Ann Neurol 2000; 48(2):164-169.
69. Barton-Davis ER, Cordier L, Shoturma DI et al. Aminoglycoside antibiotics restore dystrophin function to skeletal muscles of mdx mice. J Clin Invest 1999; 104(4):375-381.
70. Keeling KM, Brooks DA, Hopwood JJ et al. Gentamicin-mediated suppression of Hurler syndrome stop mutations restores a low level of alpha-L-iduronidase activity and reduces lysosomal glycosaminoglycan accumulation. Hum Mol Genet 2001; 10(3):291-299.
71. Hein LK, Bawden M, Muller VJ et al. alpha-L-iduronidase premature stop codons and potential read-through in mucopolysaccharidosis type I patients. J Mol Biol 2004; 338(3):453-462.
72. Sleat DE, Sohar I, Gin RM et al. Aminoglycoside-mediated suppression of nonsense mutations in late infantile neuronal ceroid lipofuscinosis. Eur J Paediatr Neurol 2001; 5(Suppl A):57-62.
73. Helip-Wooley A, Park MA, Lemons RM et al. Expression of CTNS alleles: Subcellular localization and aminoglycoside correction in vitro. Mol Genet Metab 2002; 75(2):128-133.
74. Schulz A, Sangkuhl K, Lennert T et al. Aminoglycoside pretreatment partially restores the function of truncated V(2) vasopressin receptors found in patients with nephrogenic diabetes insipidus. J Clin Endocrinol Metab 2002; 87(11):5247-5257.
75. Sossi V, Giuli A, Vitali T et al. Premature termination mutations in exon 3 of the SMN1 gene are associated with exon skipping and a relatively mild SMA phenotype. Eur J Hum Genet 2001; 9(2):113-120.
76. Aguiari G, Banzi M, Gessi S et al. Deficiency of polycystin-2 reduces Ca2+ channel activity and cell proliferation in ADPKD lymphoblastoid cells. FASEB J 2004; 18(7):884-886.
77. Lai CH, Chun HH, Nahas SA et al. Correction of ATM gene function by aminoglycoside-induced read-through of premature termination codons. Proc Natl Acad Sci USA 2004; 101(44):15676-15681.
78. Du M, Jones JR, Lanier J et al. Aminoglycoside suppression of a premature stop mutation in a Cftr-/- mouse carrying a human CFTR-G542X transgene. J Mol Med 2002; 80(9):595-604.
79. Wilschanski M, Yahav Y, Yaacov Y et al. Gentamicin-induced correction of CFTR function in patients with cystic fibrosis and CFTR stop mutations. N Engl J Med 2003; 349(15):1433-1441.
80. Wilschanski M, Famini C, Blau H et al. A pilot study of the effect of gentamicin on nasal potential difference measurements in cystic fibrosis patients carrying stop mutations. Am J Respir Crit Care Med 2000; 161(3 Pt 1):860-865.
81. Clancy JP, Bebok Z, Ruiz F et al. Evidence that systemic gentamicin suppresses premature stop mutations in patients with cystic fibrosis. Am J Respir Crit Care Med 2001; 163(7):1683-1692.
82. Wagner KR, Hamed S, Hadley DW et al. Gentamicin treatment of Duchenne and Becker muscular dystrophy due to nonsense mutations. Ann Neurol 2001; 49(6):706-711.
83. Politano L, Nigro G, Nigro V et al. Gentamicin administration in Duchenne patients with premature stop codon. Preliminary results. Acta Myol 2003; 22(1):15-21.
84. Howard MT, Anderson CB, Fass U et al. Readthrough of dystrophin stop codon mutations induced by aminoglycosides. Ann Neurol 2004; 55(3):422-426.
85. Nagai J, Takano M. Molecular aspects of renal handling of aminoglycosides and strategies for preventing the nephrotoxicity. Drug Metab Pharmacokinet 2004; 19(3):159-170.
86. Mingeot-Leclercq MP, Tulkens PM. Aminoglycosides: Nephrotoxicity. Antimicrob Agents Chemother 1999; 43(5):1003-1012.
87. Beauchamp D, Labrecque G. Aminoglycoside nephrotoxicity: Do time and frequency of administration matter? Curr Opin Crit Care 2001; 7(6):401-408.
88. Bartal C, Danon A, Schlaeffer F et al. Pharmacokinetic dosing of aminoglycosides: A controlled trial. Am J Med 2003; 114(3):194-198.
89. Sener G, Sehirli AO, Altunbas HZ et al. Melatonin protects against gentamicin-induced nephrotoxicity in rats. J Pineal Res 2002; 32(4):231-236.
90. Nakashima T, Teranishi M, Hibi T et al. Vestibular and cochlear toxicity of aminoglycosides - A review. Acta Otolaryngol 2000; 120(8):904-911.
91. Mazzon E, Britti D, De Sarro A et al. Effect of N-acetylcysteine on gentamicin-mediated nephropathy in rats. Eur J Pharmacol 2001; 424(1):75-83.
92. Kawamoto K, Sha SH, Minoda R et al. Antioxidant gene therapy can protect hearing and hair cells from ototoxicity. Mol Ther 2004; 9(2):173-181.
93. Gilbert DN, Wood CA, Kohlhepp SJ et al. Polyaspartic acid prevents experimental aminoglycoside nephrotoxicity. J Infect Dis 1989; 159(5):945-953.
94. Beauchamp D, Laurent G, Maldague P et al. Protection against gentamicin-induced early renal alterations (phospholipidosis and increased DNA synthesis) by coadministration of poly-L-aspartic acid. J Pharmacol Exp Ther 1990; 255(2):858-866.

95. Thibault N, Grenier L, Simard M et al. Attenuation by daptomycin of gentamicin-induced experimental nephrotoxicity. Antimicrob Agents Chemother 1994; 38(5):1027-1035.
96. Thibault N, Grenier L, Simard M et al. Protection against gentamicin nephrotoxicity by daptomycin in nephrectomized rats. Life Sci 1995; 56(22):1877-1887.
97. Major LL, Edgar TD, Yee Yip P et al. Tandem termination signals: Myth or reality? FEBS Lett 2002; 514(1):84-89.
98. Dalphin ME, Stockwell PA, Tate WP et al. TransTerm, the translational signal database, extended to include full coding sequences and untranslated regions. Nucleic Acids Res 1999; 27(1):293-294.
99. Dalphin ME, Brown CM, Stockwell PA et al. The translational signal database, TransTerm, is now a relational database. Nucleic Acids Res 1998; 26(1):335-337.
100. Brown CM, Dalphin ME, Stockwell PA et al. The translational termination signal database. Nucleic Acids Res 1993; 21(13):3119-3123.
101. Sachs MS, Wang Z, Gaba A et al. Toeprint analysis of the positioning of translation apparatus components at initiation and termination codons of fungal mRNAs. Methods 2002; 26(2):105-114.
102. Amrani N, Ganesan R, Kervestin S et al. A faux 3'-UTR promotes aberrant termination and triggers nonsense-mediated mRNA decay. Nature 2004; 432(7013):112-118.
103. Arakawa M, Shiozuka M, Nakayama Y et al. Negamycin restores dystrophin expression in skeletal and cardiac muscles of mdx mice. J Biochem (Tokyo) 2003; 134(5):751-758.
104. Hirawat S, Northcutt VJ, Welch EM et al. Phase 1 safety and PK study of PTC124 for nonsense-mutation suppression therapy of cystic fibrosis. Pediatric Pulmonology 2004; 38(S27):248.
105. Temple GF, Dozy AM, Roy KL et al. Construction of a functional human suppressor tRNA gene: An approach to gene therapy for beta-thalassaemia. Nature 1982; 296(5857):537-540.
106. Kiselev AV, Ostapenko OV, Rogozhkina EV et al. Suppression of nonsense mutations in the Dystrophin gene by a suppressor tRNA gene. Mol Biol (Mosk) 2002; 36(1):43-47.
107. Buvoli M, Buvoli A, Leinwand LA. Suppression of nonsense mutations in cell culture and mice by multimerized suppressor tRNA genes. Mol Cell Biol 2000; 20(9):3116-3124.
108. Atkinson J, Martin R. Mutations to nonsense codons in human genetic disease: Implications for gene therapy by nonsense suppressor tRNAs. Nucleic Acids Res 1994; 22(8):1327-1334.
109. Phillips-Jones MK, Watson FJ et al. The 3' codon context effect on UAG suppressor tRNA is different in Escherichia coli and human cells. J Mol Biol 1993; 233(1):1-6.
110. Kmiec EB. Targeted gene repair - In the arena. J Clin Invest 2003; 112(5):632-636.
111. Rice MC, Czymmek K, Kmiec EB. The potential of nucleic acid repair in functional genomics. Nat Biotechnol 2001; 19(4):321-326.
112. Liu L, Rice MC, Kmiec EB. In vivo gene repair of point and frameshift mutations directed by chimeric RNA/DNA oligonucleotides and modified single-stranded oligonucleotides. Nucleic Acids Res 2001; 29(20):4238-4250.
113. Cole-Strauss A, Yoon K, Xiang Y et al. Correction of the mutation responsible for sickle cell anemia by an RNA-DNA oligonucleotide. Science 1996; 273(5280):1386-1389.
114. Dominski Z, Kole R. Restoration of correct splicing in thalassemic pre-mRNA by antisense oligonucleotides. Proc Natl Acad Sci USA 1993; 90(18):8673-8677.
115. Kren BT, Cole-Strauss A, Kmiec EB et al. Targeted nucleotide exchange in the alkaline phosphatase gene of HuH-7 cells mediated by a chimeric RNA/DNA oligonucleotide. Hepatology 1997; 25(6):1462-1468.
116. Tagalakis AD, Graham IR, Riddell DR et al. Gene correction of the apolipoprotein (Apo) E2 phenotype to wild-type ApoE3 by in situ chimeraplasty. J Biol Chem 2001; 276(16):13226-13230.
117. D'Alessandro M, Morley SM, Ogden PH et al. Functional improvement of mutant keratin cells on addition of desmin: An alternative approach to gene therapy for dominant diseases. Gene Ther 2004; 11(16):1290-1295.
118. Alexeev V, Yoon K. Stable and inheritable changes in genotype and phenotype of albino melanocytes induced by an RNA-DNA oligonucleotide. Nat Biotechnol 1998; 16(13):1343-1346.
119. Rando TA, Disatnik MH, Zhou LZ. Rescue of dystrophin expression in mdx mouse muscle by RNA/DNA oligonucleotides. Proc Natl Acad Sci USA 2000; 97(10):5363-5368.
120. Mann CJ, Honeyman K, Cheng AJ et al. Antisense-induced exon skipping and synthesis of dystrophin in the mdx mouse. Proc Natl Acad Sci USA 2001; 98(1):42-47.
121. Lu QL, Mann CJ, Lou F et al. Functional amounts of dystrophin produced by skipping the mutated exon in the mdx dystrophic mouse. Nat Med 2003; 9(8):1009-1014.
122. Bertoni C, Rando TA. Dystrophin gene repair in mdx muscle precursor cells in vitro and in vivo mediated by RNA-DNA chimeric oligonucleotides. Hum Gene Ther 2002; 13(6):707-718.
123. Kren BT, Bandyopadhyay P, Steer CJ. In vivo site-directed mutagenesis of the factor IX gene by chimeric RNA/DNA oligonucleotides. Nat Med 1998; 4(3):285-290.

124. Lai LW, Chan DM, Erickson RP et al. Correction of renal tubular acidosis in carbonic anhydrase II-deficient mice with gene therapy. J Clin Invest 1998; 101(7):1320-1325.
125. Parekh-Olmedo H, Ferrara L, Brachman E et al. Gene therapy progress and prospects: Targeted gene repair. Gene Ther 2005; 12(8):639-646.
126. Le Hir H, Moore MJ, Maquat LE. Pre-mRNA splicing alters mRNP composition: Evidence for stable association of proteins at exon-exon junctions. Genes Dev 2000; 14(9):1098-1108.
127. Serin G, Gersappe A, Black JD et al. Identification and characterization of human orthologues to Saccharomyces cerevisiae Upf2 protein and Upf3 protein (Caenorhabditis elegans SMG-4). Mol Cell Biol 2001; 21(1):209-223.
128. Lykke-Andersen J, Shu MD, Steitz JA. Human Upf proteins target an mRNA for nonsense-mediated decay when bound downstream of a termination codon. Cell 2000; 103(7):1121-1131.
129. Lejeune F, Ishigaki Y, Li X et al. The exon junction complex is detected on CBP80-bound but not eIF4E-bound mRNA in mammalian cells: Dynamics of mRNP remodeling. EMBO J 2002; 21(13):3536-3545.
130. Ishigaki Y, Li X, Serin G et al. Evidence for a pioneer round of mRNA translation: mRNAs subject to nonsense-mediated decay in mammalian cells are bound by CBP80 and CBP20. Cell 2001; 106(5):607-617.
131. Nagy E, Maquat LE. A rule for termination-codon position within intron-containing genes: When nonsense affects RNA abundance. Trends Biochem Sci 1998; 23(6):198-199.
132. Zhang J, Maquat LE. Evidence that translation reinitiation abrogates nonsense-mediated mRNA decay in mammalian cells. EMBO J 1997; 16(4):826-833.
133. Lim SK, Sigmund CD, Gross KW et al. Nonsense codons in human beta-globin mRNA result in the production of mRNA degradation products. Mol Cell Biol 1992; 12(3):1149-1161.
134. Atkin AL, Schenkman LR, Eastham M et al. Relationship between yeast polyribosomes and Upf proteins required for nonsense mRNA decay. J Biol Chem 1997; 272(35):22163-22172.
135. Atkin AL, Altamura N, Leeds P et al. The majority of yeast UPF1 colocalizes with polyribosomes in the cytoplasm. Mol Biol Cell 1995; 6(5):611-625.
136. Czaplinski K, Ruiz-Echevarria MJ, Paushkin SV et al. The surveillance complex interacts with the translation release factors to enhance termination and degrade aberrant mRNAs. Genes Dev 1998; 12(11):1665-1677.
137. Gozalbo D, Hohmann S. Nonsense suppressors partially revert the decrease of the mRNA level of a nonsense mutant allele in yeast. Curr Genet 1990; 17(1):77-79.
138. Belgrader P, Cheng J, Maquat LE. Evidence to implicate translation by ribosomes in the mechanism by which nonsense codons reduce the nuclear level of human triosephosphate isomerase mRNA. Proc Natl Acad Sci USA 1993; 90(2):482-486.
139. Correa-Cerro LS, Wassif CA, Waye JS et al. DHCR7 nonsense mutations and characterisation of mRNA nonsense mediated decay in Smith-Lemli-Opitz syndrome. J Med Genet 2005; 42(4):350-357.
140. Page MF, Carr B, Anders KR et al. SMG-2 is a phosphorylated protein required for mRNA surveillance in Caenorhabditis elegans and related to Upf1p of yeast. Mol Cell Biol 1999; 19(9):5943-5951.
141. Bhattacharya A, Czaplinski K, Trifillis P et al. Characterization of the biochemical properties of the human Upf1 gene product that is involved in nonsense-mediated mRNA decay. RNA 2000; 6(9):1226-1235.
142. Yamashita A, Ohnishi T, Kashima I et al. Human SMG-1, a novel phosphatidylinositol 3-kinase-related protein kinase, associates with components of the mRNA surveillance complex and is involved in the regulation of nonsense-mediated mRNA decay. Genes Dev 2001; 15(17):2215-2228.
143. Pal M, Ishigaki Y, Nagy E et al. Evidence that phosphorylation of human Upf1 protein varies with intracellular location and is mediated by a wortmannin-sensitive and rapamycin-sensitive PI 3-kinase-related kinase signaling pathway. RNA 2001; 7(1):5-15.
144. Grimson A, O'Connor S, Newman CL et al. SMG-1 Is a Phosphatidylinositol kinase-related protein kinase required for nonsense-mediated mRNA decay in Caenorhabditis elegans. Mol Cell Biol 2004; 24(17):7483-7490.
145. Usuki F, Yamashita A, Higuchi I et al. Inhibition of nonsense-mediated mRNA decay rescues the phenotype in Ullrich's disease. Ann Neurol 2004; 55(5):740-744.
146. Ohnishi T, Yamashita A, Kashima I et al. Phosphorylation of hUPF1 induces formation of mRNA surveillance complexes containing hSMG-5 and hSMG-7. Mol Cell 2003; 12(5):1187-1200.
147. Chiu SY, Serin G, Ohara O et al. Characterization of human Smg5/7a: A protein with similarities to Caenorhabditis elegans SMG5 and SMG7 that functions in the dephosphorylation of Upf1. RNA 2003; 9(1):77-87.

148. Cali BM, Kuchma SL, Latham J et al. smg-7 is required for mRNA surveillance in Caenorhabditis elegans. Genetics 1999; 151(2):605-616.
149. Anders KR, Grimson A, Anderson P. SMG-5, required for C. elegans nonsense-mediated mRNA decay, associates with SMG-2 and protein phosphatase 2A. EMBO J 2003; 22(3):641-650.
150. Jungwirth H, Bergler H, Hogenauer G. Diazaborine treatment of Baker's yeast results in stabilization of aberrant mRNAs. J Biol Chem 2001; 276(39):36419-36424.
151. Lelivelt MJ, Culbertson MR. Yeast Upf proteins required for RNA surveillance affect global expression of the yeast transcriptome. Mol Cell Biol 1999; 19(10):6710-6719.
152. He F, Li X, Spatrick P et al. Genome-wide analysis of mRNAs regulated by the nonsense-mediated and 5' to 3' mRNA decay pathways in yeast. Mol Cell 2003; 12(6):1439-1452.
153. Reed R, Hurt E. A conserved mRNA export machinery coupled to pre-mRNA splicing. Cell 2002; 108(4):523-531.
154. Galy V, Gadal O, Fromont-Racine M et al. Nuclear retention of unspliced mRNAs in yeast is mediated by perinuclear Mlp1. Cell 2004; 116(1):63-73.
155. Ramalho AS, Beck S, Meyer M et al. Five percent of normal cystic fibrosis transmembrane conductance regulator mRNA ameliorates the severity of pulmonary disease in cystic fibrosis. Am J Respir Cell Mol Biol 2002; 27(5):619-627.
156. Kerem E. Pharmacologic therapy for stop mutations: How much CFTR activity is enough? Curr Opin Pulm Med 2004; 10(6):547-552.
157. Ashton LJ, Brooks DA, McCourt PA et al. Immunoquantification and enzyme kinetics of alpha-L-iduronidase in cultured fibroblasts from normal controls and mucopolysaccharidosis type I patients. Am J Hum Genet 1992; 50(4):787-794.
158. Aronovich EL, Pan D, Whitley CB. Molecular genetic defect underlying alpha-L-iduronidase pseudodeficiency. Am J Hum Genet 1996; 58(1):75-85.
159. Purohit P, Stern S. Interactions of a small RNA with antibiotic and RNA ligands of the 30S subunit. Nature 1994; 370(6491):659-662.
160. Trapnell BC, Chu CS, Paakko PK et al. Expression of the cystic fibrosis transmembrane conductance regulator gene in the respiratory tract of normal individuals and individuals with cystic fibrosis. Proc Natl Acad Sci USA 1991; 88(15):6565-6569.
161. Hamosh A, Trapnell BC, Zeitlin PL et al. Severe deficiency of cystic fibrosis transmembrane conductance regulator messenger RNA carrying nonsense mutations R553X and W1316X in respiratory epithelial cells of patients with cystic fibrosis. J Clin Invest 1991; 88(6):1880-1885.
162. Hamosh A, Rosenstein BJ, Cutting GR. CFTR nonsense mutations G542X and W1282X associated with severe reduction of CFTR mRNA in nasal epithelial cells. Hum Mol Genet 1992; 1(7):542-544.

Section III

Caenorhabditis elegans

Chapter 11

NMD in *Caenorhabditis elegans*

Philip Anderson*

Abstract

Loss-of-function mutations affecting seven different genes eliminate nonsense-mediated mRNA decay (NMD) in *Caenorhabditis elegans*. Three of these genes (*smg-2, smg-3,* and *smg-4*) are orthologs of *Saccharomyces cerevisiae* NMD genes (*UPF1*, *UPF2*, and *UPF3*, respectively). SMG-2, a central regulator of NMD, undergoes cycles of reversible phosphorylation, and all other *smg* genes influence the state of its phosphorylation. SMG-1, SMG-3, and SMG-4 are required to phosphorylate SMG-2, and SMG-5, SMG-6, and SMG-7 are required to dephosphorylate SMG-2. Both the SMG-2 phosphorylation and dephosphorylation steps are required for NMD, but the functions of this posttranslational modification are poorly understood. *Smg* mutations are allele-specific but not gene-specific modifiers of phenotypes caused by mutations affecting many different genes. NMD lessens the deleterious effects of heterozygous and homozygous nonsense mutations by degrading the affected mRNA. Wild-type *C. elegans* expresses substantial quantities of mRNAs that contain premature termination codons and are degraded by NMD. These include mRNAs regulated posttranscriptionally by unproductive alternative splicing and mRNAs of expressed pseudogenes.

Discovery of *C. elegans* Mutants Defective for NMD

Mutations affecting any of seven different genes (*smg-1* through *smg-7*) reduce or eliminate NMD in *C. elegans*. Such mutations were independently and almost simultaneously isolated by three different labs in the late 1980s.[1] The Hodgkin lab identified *smg* mutations both as suppressors of an allele of *tra-2* (a sex determination gene) and in mutants exhibiting the Smg morphological phenotype (see below). The Ambrose lab identified *smg* mutations as phenotypic suppressors of an allele of *lin-29*, a heterochronic gene. The Anderson lab identified *smg* mutations as phenotypic suppressors of an allele of *unc-54*, a myosin heavy chain gene. Several years passed, however, before any of these investigators established that *smg* mutants are defective in NMD.[2] They were isolated quite fortuitously, without knowing in advance that mRNA turnover was being affected.

A large collection of *smg* mutations defines seven different genes.[1,3] Single mutations affecting any of the seven genes reduce or eliminate NMD, indicating that functions of all seven are required for NMD. *Smg* mutations are plentiful in traditional genetic screens, because they are loss-of-function alleles affecting nonessential genes. Their rather uninformative gene names ("*smg*" denotes "suppressor with morphogenetic effects on genitalia") derive from the fact that the genes were named at a time when the molecular basis of their effects was unknown. *Smg(-)* hermaphrodites exhibit protruding vulvae, which are the openings through which embryos are expelled to the environment. *Smg(-)* males have a swollen bursa, which is a copulatory organ. In keeping with nomenclature conventions, *smg* genes were named for their most obvious mutant phenotypes. *Smg* mutants have since been found to have additional phenotypes (see below).

*Philip Anderson—Department of Genetics, University of Wisconsin, Madison, Wisconsin 53706, U.S.A. Email: andersn@wisc.edu

Nonsense-Mediated mRNA Decay, edited by Lynne E. Maquat. ©2006 Eurekah.com.

The genetic properties of *smg* mutants demonstrate that NMD functions in all tissues of the animal and at all times of development. *Smg* mutations are allele-specific but not gene-specific modifiers of the phenotypes of mutations affecting many different genes. The phenotypes caused by certain mutations (for example, the alleles of *unc-54*, *lin-29* and *tra-2* discussed above) are phenotypically suppressed by *smg* mutations. The phenotypes caused by many other mutations are enhanced (see below). The tissues in which *smg*-affected genes are expressed represent, essentially, the whole of the animal.

NMD is not required for viability of *C. elegans*, a situation similar to that observed in *S. cerevisiae*[4] but likely different to that seen in mammals.[5] Molecular null alleles (those that express none of the encoded protein) have been identified for all *smg* genes except possibly *smg-6*. Affected mutants are viable, fertile, and lack most or all NMD. Mutations affecting *smg-6* and *smg-7* are often observed to incompletely eliminate NMD (for example, see ref. 2), but interpreting these observations is complicated. An unambiguous null allele of *smg-6* has not been identified, and the NMD defect of *smg-7* null alleles is temperature sensitive. *Smg-7* null mutants are NMD-proficient at 15°C and 20°C, but they are NMD-deficient at 25°C.[3]

Traditional genetic screens for nonessential, nonredundant genes in *C. elegans* are quite effective, and it is likely that all such genes required strongly for NMD have already been identified. However, genes having only modest effects on NMD, redundant genes, genes involved in both NMD and other processes, genes essential for viability, and genes that are negative regulators of NMD might easily have been missed in the genetic screens done to date.

Conclusions concerning NMD in *C. elegans* are usually made either by assessing in vivo suppression phenotypes or by quantifying the steady-state abundance of NMD-sensitive mRNAs. Unfortunately, methods for directly measuring mRNA half-lives have not been developed or applied to studies of *C. elegans*. The steady-state abundance of a nonsense-containing mRNA in *smg(-)* animals is usually approximately equal to that of the corresponding wild-type mRNA in *smg(+)* animals. An example is shown in Figure 1. Exceptions to this generalization have been noted and are discussed below.

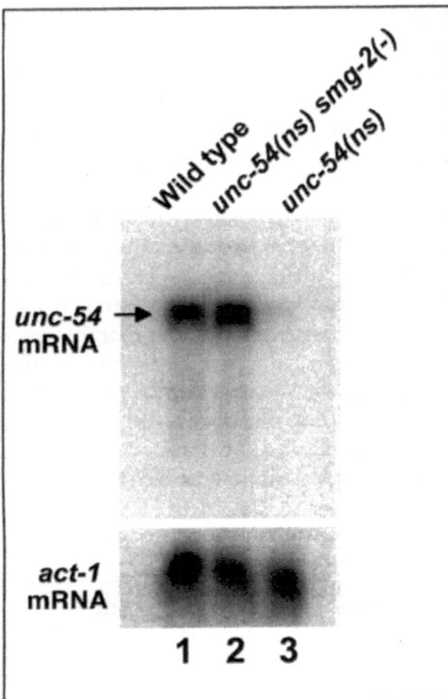

Figure 1. NMD in *C. elegans*. Northern blot of total *C. elegans* RNA from a wild-type strain (lane 1), an *unc-54(nonsense)* strain, a *smg-2(-)* strain (lane 2), and an *unc-54(nonsense)*; *smg(+)* strain (lane 3). The blot was probed for both *unc-54* and *act-1* mRNAs. *act-1* mRNA serves as a loading control. ns, nonsense.

Remarkably, *smg* mutants of *C. elegans* and *upf* mutants of *S. cerevisiae* were identified almost simultaneously about 100 yards apart at the University of Wisconsin. This occurred at a time when the existence of NMD[6] was not widely appreciated. Peter Leeds, a graduate student with Mike Culbertson, isolated *upf* ("up-frameshift") mutations as enhancers of tRNA-mediated suppression.[4] Rock Pulak, a graduate student with me, isolated *smg* mutations as phenotypic suppressors of a mutation deleting the polyadenylation signal sequence of the *unc-54* myosin gene.[1] When Leeds made the connection to NMD, it became apparent that NMD-defective mutations should suppress specific alleles of many genes. I was a member of Peter Leeds' thesis committee, and Culbertson was a member of Rock Pulak's committee. Annual meetings would usually conclude with speculation to the effect of, "Wouldn't it be remarkable if *upf* and *smg* genes proved to be related and have similar mechanisms of action". Cloning *C. elegans* genes was difficult at the time, and almost eight years elapsed before that speculation was shown to be true.[7]

SMG Proteins Are Conserved among Eukaryotes

NMD occurs in most, if not all, eukaryotes, and this is reflected in the conservation of sequence and function of NMD proteins (see chapters by Baker and Parker; Maquat; Singh and Lykke-Andersen; Yamashita et al; Behm-Ansmant and Izaurralde; van Hoof and Green). Table 1 summarizes yeast, nematode, and mammalian orthologs of to Upfl and SMG.

The seven SMG proteins fall into two groups based on conservation among sequenced eukaryotes. SMG-2, SMG-3, and SMG-4 are orthologs of yeast Upf1p, Upf2p, and Upf3p, respectively.[7-9] These proteins comprise a "core" group of factors that are required for NMD in all eukaryotes tested to date (reviewed in ref. 10). The remaining SMG proteins (SMG-1, SMG-5, SMG-6, and SMG-7) have orthologs in mammals,[11-16] and putative orthologs in many sequenced metazoa. No clear orthologs to SMG-1, SMG-5, SMG-6, and SMG-7, however, have been described in unicellular eukaryotes. Related proteins exist, but roles in NMD have not been described. For example, both SMG-1 and yeast Tor1p and Tor2p are members of the phosphatidylinositol kinase superfamily (see below), but functions for Tor1p and Tor2p in NMD have not been established.

Upf1/SMG-2 is the most highly conserved of the Upf and SMG proteins. A 781 amino acid central region of SMG-2 is 50% identical to a corresponding region of yeast Upf1p and 56% identical to mammalian RENT1/hUpf1.[17,18] Several lines of evidence identify Upf1/SMG-2 as a key regulator of NMD. It is required for NMD, and it elicits NMD when tethered downstream of stop codons. Such tethering can be engineered in reporter constructs[19] or can occur naturally on endogenous mRNAs.[20] Yeast Upf1p and mammalian Upf1 interact directly or indirectly with many of the major players of NMD, including translational release factors,[21,22] the decapping complex,[23,24] and, in the case of mammalian cells, components of the

Table 1. Proteins required for NMD

C. elegans				
Protein	**Length (aa)**	**Yeast**	**Human**	**References**
SMG-1	2322	-	hSMG-1	11, 12, 28
SMG-2	1069	Upf1p	hUpf1/RENT1	4, 17, 18, 7
SMG-3	1142	Upf2p/Nmd2p	hUpf2/RENT2	73, 23, 8, 74, 19, 75
SMG-4	368	Upf3p	hUpf3a, hUpf3b	76, 19, 75, 9
SMG-5	549	-	hSMG-5	33, 13, 15
SMG-6A	775	-	hSMG-6,	77, 14, 15
SMG-6B	1242		hSMG5/7a	
SMG-7	458	-	hSMG-7	3, 15

mammalian exon junction complex.[25-27] Understanding the regulated activities of Upf1/SMG-2 is fundamental to understanding the molecular mechanisms of NMD.

NMD Requires a Cycle of SMG-2 Phosphorylation

C. elegans SMG-2 undergoes cycles of phosphorylation and dephosphorylation, both of which are required for NMD.[7] Phosphorylated SMG-2 is detected in two ways: (1) it migrates more slowly in SDS-polyacrylamide gels, and (2) it is radiolabeled in crude extracts by its endogenous kinase, SMG-1.[28] The phosphorylated amino acid residue(s) of SMG-2 have not been defined, although the COOH-terminus of SMG-2 contains a cluster of serine-glutamine dipeptides similar to the COOH-terminus of human Upf1. These serines of human Upf1 are phosphorylated in vitro and in vivo by human SMG-1 (see chapters by Yamashita et al, and Abraham and Oliveira).[11]

The six remaining SMG proteins influence the state of SMG-2 phosphorylation (ref. 7; see Fig. 2). SMG-1, SMG-3 and SMG-4 are required to phosphorylate SMG-2, while SMG-5, SMG-6, and SMG-7 are required to efficiently dephosphorylate SMG-2 (see Fig. 2).

The genetic properties of *smg* mutants indicate that both phosphorylation and dephosphorylation steps are required for NMD. Single mutations eliminating SMG-1, SMG-3 or SMG-4 function eliminate NMD, even though such mutations do not affect SMG-2 dephosphorylation. Similarly, single mutations eliminating SMG-5, SMG-6 or SMG-7 function eliminate NMD, even though they are competent for SMG-2 phosphorylation. Thus, NMD fails either when SMG-2 cannot be phosphorylated or when it cannot be dephosphorylated. Possible roles for SMG-2/Upf1 phosphorylation and dephosphorylation are discussed below.

SMG-2 Phosphorylation

Four lines of evidence indicate that SMG-1 likely phosphorylates SMG-2 directly:[7,28] (1) SMG-1 is predicted to be a member of the phosphatidylinositol kinase (PIK) superfamily of protein kinases. (2) Both null and kinase-inactive alleles of *smg-1* are NMD defective. (3) SMG-2 is not phosphorylated in vivo or in vitro in *smg-1(-)* mutants. (4) Both SMG-1 and SMG-2 are immunoprecipitated as a complex from crude extracts using antibodies to either protein. Taken together, these observations suggest that SMG-1 directly phosphorylates SMG-2. Purified human SMG-1 directly phosphorylates purified human Upf1,[11,12] and it seems likely that *C. elegans* SMG-1 and SMG-2 behave similarly.

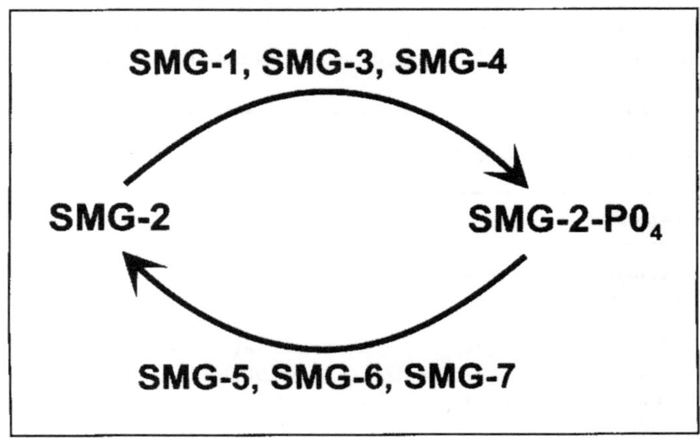

Figure 2. Cycle of SMG-2 phosphorylation. Activities of SMG-1, SMG-3 and SMG-4 are required to phosphorylate SMG-2, and activities of SMG-5, SMG-6 and SMG-7 are required to dephosphorylate SMG-2.

SMG-3 and SMG-4 are the *C. elegans* orthologs to yeast Upf2p and Upf3p, respectively.[8,9] Phosphorylation of SMG-2 is not detected in *smg-3(-)* or *smg-4(-)* mutants,[7,28] although the exact reasons for this are uncertain. SMG-3 and SMG-4 appear to be involved in recruiting SMG-2 to NMD complexes or in facilitating access of SMG-1 to its SMG-2 substrate. Based on both yeast two-hybrid and immunoprecipitation experiments, SMG-2, SMG-3, and SMG-4 form a trimeric complex[28] analogous to that of Upf1:Upf2:Upf3 complexes of yeast and mammalian cells.[23,29] SMG-2 interacts directly with SMG-3; SMG-3 interacts directly with SMG-4; and SMG-2 interacts indirectly with SMG-4 via a shared interaction with SMG-3 (A. Grimson et al, unpublished). All three proteins are coimmunoprecipitated from wild-type crude extracts. A detectable SMG-2:SMG-3 interaction requires the presence of SMG-4, but a detectable SMG-3:SMG-4 complex does not require the presence of SMG-2. Such interactions are analogous to described interactions of Upf1, Upf2, and Upf3 in mammals (reviewed in ref. 29,30). Human Upf3a and Upf3b interact with components of the exon junction complex and are believed to recruit Upf2 to messenger ribonucleoprotein particles (mRNPs). Upf2 may then recruit (or interact with previously recruited) Upf1, thus yielding mRNPs containing Upf1, Upf2, and Upf3.

Similar sequential recruitment in *C. elegans* might explain why SMG-2 is not phosphorylated in *smg-3(-)* and *smg-4(-)* mutants. Elements of the explanation include: (i) SMG-4 interacts with as-yet-unidentified components of mRNPs; (ii) SMG-4 recruits SMG-3 to those mRNPs, which, in turn, recruits (or interacts with) SMG-2; and (iii) SMG-1 phosphorylates SMG-2 only as part of such mRNP complexes. In either *smg-3(-)* or *smg-4(-)* mutants, SMG-2 fails to be phosphorylated because it fails to be recruited to mRNPs or to a complex that serves as the substrate of SMG-1 phosphorylation. SMG-1 is not required for formation or stability of the SMG-2:SMG-3:SMG-4 complex (A. Grimson et al, unpublished), implying that phosphorylation of SMG-2 occurs only after its recruitment to such complexes.

mRNP protein(s) with which SMG-4 might interact remain unidentified, and the mechanism(s) for "marking" premature termination codons in *C. elegans* has not been systematically investigated. Two lines of evidence indicate that introns are not necessary for *C. elegans* NMD, in contrast to the general situation in mammals (see chapter by Maquat). (1) Nonsense mutations within the final exon of the *unc-54* myosin heavy chain gene elicit NMD.[2] (2) The abundance of *smg-3* mRNA, which contains an intron within its 3' untranslated region (UTR), is not elevated in *smg(-)* mutants.[8] This makes *C. elegans* similar to *Drosophila melanogaster*, which also does not require downstream introns for NMD (ref. 15; see chapter by Behm-Ansmant and Izaurralde). mRNAs whose translation is repressed or are not exported to the cytoplasm do not elicit NMD,[31,32] demonstrating that *C. elegans* NMD likely occurs only during translation and only in the cytoplasm.

SMG-2 Dephosphorylation

SMG-5, SMG-6, and SMG-7 are all required for efficient SMG-2 dephosphorylation. Whereas phosphorylated SMG-2 is not detected in crude extracts of wild-type animals, approximately half of the SMG-2 contained in *smg-5(-)*, *smg-6(-)*, or *smg-7(-)* mutants is phosphorylated (ref. 7; see Fig. 3).

Thus, SMG-5, SMG-6, and SMG-7 either negatively regulate SMG-2 phosphorylation or positively regulate SMG-2 dephosphorylation. Single mutations affecting *smg-5*, *smg-6*, or *smg-7* inhibit NMD, demonstrating that SMG-5, SMG-6, and SMG-7 perform unique, nonredundant functions. *Smg-5(-)* mutants are consistently scored as being completely NMD defective, whereas *smg-6(-)* and *smg-7(-)* mutants are often observed to incompletely lack NMD (see above).

SMG-5 likely functions as a specificity determinant of protein phosphatase 2A (PP2A).[33] Eukaryotic PP2As are a family of ubiquitous, abundant, Ser/Thr protein phosphatases (reviewed in ref. 34). Active enzymes are usually trimeric complexes containing a catalytic subunit (PP2Ac), a subunit that scaffolds assembly (PR65), and one of several B-type regulatory subunits whose identity varies among different PP2A complexes. Identified B-type regulatory subunits are thought to regulate activity, subcellular localization, and substrate specificity of the catalytic complex.

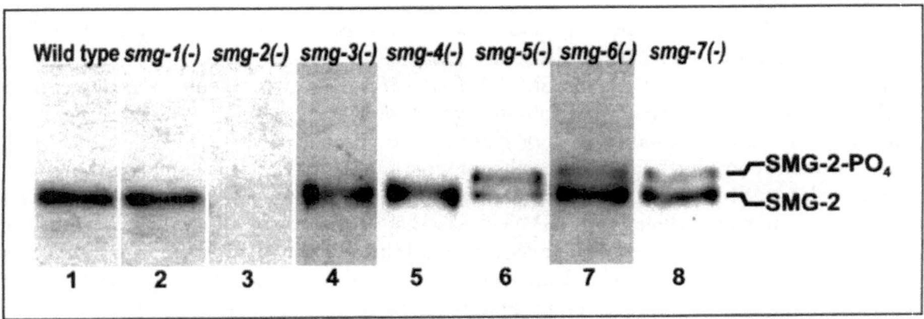

Figure 3. Phosphorylation of SMG-2. A Western blot of total *C. elegans* proteins was reacted with an anti-SMG-2 antibody. A *smg-2* null-allele (lane 3) demonstrates specificity of the antibody. Phosphorylated SMG-2 is not detected in a wild-type strain or in *smg-1(-), smg-3(-),* and *smg-4(-)* mutants (lanes 2, 4, and 5). Approximately half of total SMG-2 is phosphorylated in *smg-5(-)* mutants (lane 6), and slightly less than half is phosphorylated in *smg-6(-)* and *smg-7(-)* mutants (lanes 7 and 8).

SMG-5 interacts physically with both PP2A and SMG-2. SMG-5, SMG-2, PP2Ac, and PR65 copurify following immunoprecipitation of any of these four individual proteins. SMG-5 likely interacts directly with PR65, and SMG-2 likely interacts directly with PP2Ac based on yeast two-hybrid tests. SMG-5 additionally interacts with SMG-7, and the interaction is probably direct. Whether these interactions reflect the existence of a single, large, NMD complex or multiple subcomplexes is currently unknown. Based on work in other organisms, additional components of those complexes likely include translation release factors eRF1 and eRF3,[21,35] the decapping complex,[23,24,36] and the Lsm complex.[37] Most of these proteins are associated with polysomes,[35,38,39] presumably reflecting the fact that NMD occurs during or shortly after translation termination.

Possible Functions of SMG-2 Phosphorylation

The role of SMG-2 phosphorylation is poorly understood. Only a small fraction of total SMG-2 is phosphorylated in wild-type animals,[7,28] yet both SMG-2 phosphorylation and dephosphorylation are required for NMD. Thus, phosphorylated SMG-2 appears to be a short-lived species. Overexpression of wild-type human SMG-1 increases the efficiency of NMD,[11] suggesting that phosphorylation of Upf1/SMG-2 may be a rate-limiting step for NMD. But, what does Upf1/SMG-2 phosphorylation regulate? Several lines of evidence suggest that Upf1/SMG-2 phosphorylation influences dynamics of NMD-related protein-protein interactions. Phosphorylated human Upf1 is enriched on polysomes.[39] Human SMG-5 selectively associates with phosphorylated human Upf1.[13] Unphosphorylated and hyperphosphorylated human Upf1 form complexes containing differing isoforms of human Upf3a and differing amounts of human Upf2.[13] Almost any postulated step of NMD might be regulated by such interactions, including recruitment of SMG-2 to mRNPs, targeting of SMG-2 to a specific compartment of the cell, remodeling of mRNP structure before or after translation termination, activation of 5'-to-3', 3'-to-5', or endonucleolytic nucleases, or post-turnover disassembly of NMD complexes and recycling of its components.

It is presently difficult to assess whether the effects of various mutations and molecular manipulations of SMG-2/Upf1 phosphorylation are due to direct or indirect effects. Are changes in phosphorylation causing an observed effect, or are they reflecting more indirectly the effects of some other process? Two avenues of investigation seem needed: (i) an understanding of the differences in the biochemical properties or physical associations of phosphorylated and unphosphorylated SMG-2/Upf1, and (ii) a demonstration that those differences are instrumental to the mechanisms of NMD. It seems likely that not all phosphorylation events important for NMD have been discovered. For example, yeast Dcp2p, human Upf2, and overexpressed yeast

Upf1p are all phosphoproteins.[14,40,41] The significance of these phosphorylation events to NMD, however, is unknown. Our knowledge of the complexity of NMD-related protein phosphorylation is incomplete, as is our understanding of the biochemical and cell biological functions of NMD-related protein phosphorylation.

Biological Functions of NMD

NMD is likely an ancient process. NMD occurs in all tested eukaryotes, and proteins required for NMD are conserved among the sequenced eukaryotes. Its origins, therefore, predate the divergence of eukaryotes, and its functions have been selected during eukaryotic evolution. NMD may have evolved from factors involved more generally in translation, and NMD proteins may still retain a role in translation (for review, see ref. 42). But, its role in degrading nonsense-containing mRNAs is found throughout eukaryotes. Why do wild-type organisms need a system for degrading nonsense-containing mRNAs?

NMD appears to function in two broad areas. First, wild-type eukaryotes express a substantial amount of mRNA containing premature termination codons (PTCs). Some (perhaps many) PTC-containing mRNAs encode deleterious polypeptide fragments (see below). NMD minimizes the disruptive effects of such mRNAs by reducing their abundance, which prevents their translation. Thus, NMD improves the precision of gene expression by insuring that only mRNAs translatable through their full (or nearly full) length are stable. Second, NMD contributes to the normal posttranscriptional regulation of gene expression. A substantial and growing number of genes are regulated, in part, by NMD (see chapters by He and Jacobson, Saffari and Dietz, and Soergel et al). *C. elegans* examples of these broad functions of NMD are discussed below.

In both yeast and nematodes, NMD is a nonessential process. Both *upf* and *smg* mutants are viable and fertile. They are not, however, completely normal. *C. elegans smg* mutants exhibit numerous phenotypes, including the morphogenetic abnormalities discussed above, reduced brood sizes,[1] abnormal maintenance of RNA interference,[43] and mild embryonic patterning defects.[44] It is presently unknown why *smg* mutants exhibit any of these phenotypes. Presumably, PTC-containing mRNAs that fail to be degraded in *smg* mutants or global perturbations of gene expression (for examples in other organisms, see ref. 45-47) lead to the collection of Smg(-) phenotypes. The reduced brood size of *smg(-)* mutants alone is sufficient to insure long-term selection of NMD in *C. elegans*. The mammalian ortholog to Upf1 is essential for viability.[5] Perhaps NMD itself is essential in mammals, or perhaps functions of mammalian Upf1 in addition to its role in NMD are essential (see chatpers by Kim and Maquat, and Kaygun and Marzluff). If mammalian NMD per se is essential for viability while the NMD of lower eukaryotes is not, the reasons are likely to be very important.

Although *C. elegans smg* mutants exhibit only mild phenotypes, inactivating NMD influences the phenotypes caused by many other mutations. *Smg* mutations were isolated as extragenic suppressors of specific mutations (described above), and many other *smg*-suppressible alleles have been subsequently identified.[48-57] *Smg(-)* mutations also enhance (make more severe) the phenotypes of many known or suspected nonsense mutations. This was first observed with nonsense alleles of the *unc-54* myosin heavy chain gene.[2] Approximately one-fourth of *unc-54* nonsense mutations are conditionally dominant. In *smg(+)* genetic backgrounds, where PTC-containing mRNAs are unstable, such alleles are fully recessive, and *unc-54(nonsense)/+* heterozygotes exhibit no abnormal phenotypes. In *smg(-)* genetic backgrounds, where PTC-containing mRNAs are stable, such alleles are strongly dominant. *Smg(-); unc-54(nonsense)/+* heterozygotes are muscle-defective and paralyzed. Genetic tests indicate that the cause of dominance is expression of a myosin protein truncated by the nonsense mutation (rather than translation through the PTC).[58] Numerous additional mutations affecting a variety of genes have subsequently been shown to have phenotypes enhanced in *smg(-)* backgrounds.[58-61] In a relatively unbiased screen for dominant mutations isolated in *smg(-)* genetic backgrounds, approximately two-thirds of isolated mutations were recessive or less severely dominant in *smg(+)* genetic backgrounds.[58] These results demonstrate the potentially disruptive effects of truncated proteins and provide an attractive explanation for why NMD has been selected during eukaryotic evolution. NMD renders PTC-containing mRNAs less disruptive by insuring their rapid degradation.

The phenotypic effects of NMD discussed above pertain to inherited germline mutations. Such mutations are present in all cells and, when the affected gene is expressed, yield PTC-containing mRNA. It is likely that NMD has similar beneficial effects on phenotypes resulting from somatic mutations or from various errors of gene expression. For example, unspliced pre-mRNAs or aberrantly spliced mRNAs will usually elicit NMD. Such mRNAs were originally proposed to be biologically significant substrates of NMD, and "mRNA surveillance" was proposed as a biological role for NMD.[2] However, few mRNAs resulting unambiguously from errors of gene expression have been described. An unexpectedly large proportion of alternatively spliced human mRNAs contain PTCs,[62] but such mRNAs might represent those regulated by unproductive splicing and translation ("RUST"; see below; see chapter by Soergel et al). Nevertheless, degrading a heterogeneous collection of aberrant mRNAs resulting from errors of gene expression remains an attractive biological role for NMD.

Two classes of mRNAs expressed in wild-type animals and degraded by NMD have been described for *C. elegans*: (i) unproductively spliced mRNAs and (ii) mRNAs of expressed pseudogenes.

Unproductively Spliced mRNAs

Pre-mRNAs of four different *C. elegans* ribosomal proteins are alternatively spliced, such that certain of the resulting mRNAs are degraded by NMD.[63] "Productively spliced" mRNAs are normal mRNAs that are translated throughout their full length and encode the corresponding ribosomal protein. "Unproductively spliced" mRNAs retain a portion of a key intron. The retained intron sequences contain multiple PTCs, and such mRNAs are efficiently degraded by NMD (see Fig. 4).

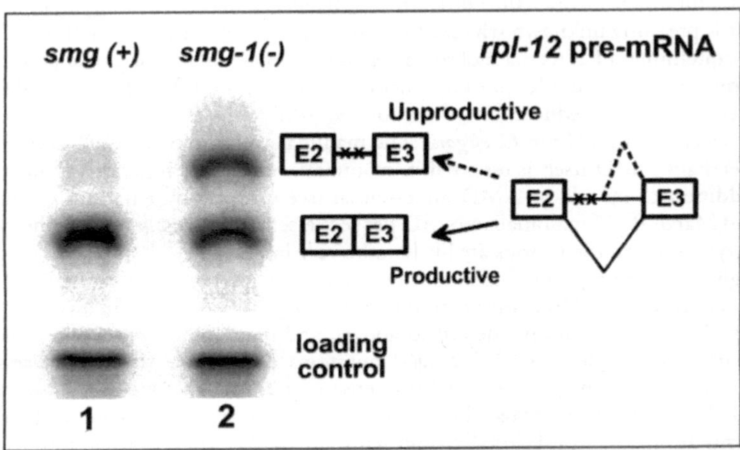

Figure 4. Alternative splicing of *rpl-12* pre-mRNA. A Northern blot of total *C. elegans* RNA from a wild-type strain and a *smg-1(-)* mutant was probed for *rpl-12* and *eft-3* (as a loading control) mRNAs. RNAs were first cleaved with an *rpl-12*–specific DNA oligonucleotide and RNase H to accentuate the size difference between productively and unproductively spliced mRNAs. RNase H cleaves the RNA strand of an RNA-DNA hybrid. Exon 2 (E2) and exon 3 (E3) of *rpl-12* pre-mRNA are diagrammed. The intron that separates these exons is alternatively spliced. The productively spliced mRNA is smaller and encodes full-length RPL-12 protein. An unproductively spliced mRNA is larger, retains a portion of intron 2, and contains multiple premature termination codons (indicated by "x"). Unproductively spliced *rpl-12* mRNA is barely detected in the wild-type strain (lane 1) but constitutes approximately half of total *rpl-12* mRNA in the *smg-1* mutant (lane 2).

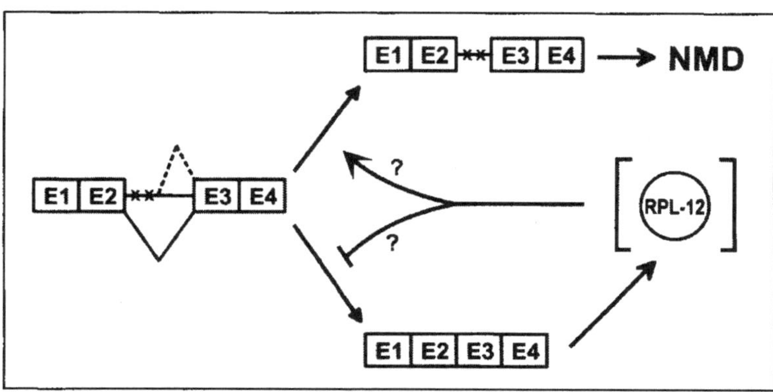

Figure 5. Posttranscriptional autoregulation of *rpl-12* mRNA abundance. RPL-12 abundance autoregulates *rpl-12* alternative splicing, either by activating unproductive splicing or by repressing productive splicing. Under conditions of RPL-12 excess, a greater proportion of total pre-mRNA is spliced unproductively. Unproductively spliced *rpl-12* mRNA contains premature termination codons (indicated by "x") and is degraded by NMD.

Unproductively spliced mRNAs are barely detected in wild-type animals, but they are remarkably abundant in *smg(-)* mutants. For example, unproductively spliced ribosomal protein (*rpl*) mRNAs can comprise 32% to 76% of total mRNA depending on the particular *rpl* gene. Productive and unproductive splicing of *rpl-12* pre-mRNA (and, presumably, the other *rpl* genes also) is autoregulated. Elevated levels of RPL-12 promote unproductive splicing, and reduced levels of RPL-12 promote productive splicing. This constitutes an autoregulatory circuit in which unneeded mRNA is destroyed by targeting it for NMD, presumably in response to metabolic needs (see Fig. 5).

Unproductive splicing followed by decay of the resulting PTC-containing mRNA appears to be a general pathway in *C. elegans* and other organisms. Six of the seven *C. elegans* SR-protein genes are both productively and unproductively spliced (ref. 64; D. Markwardt and P. Anderson, unpublished). Pre-mRNAs for at least three mammalian splicing factors are autoregulated by unproductive splicing followed by NMD.[65-67] Such regulated splicing has been termed "RUST", denoting "regulated unproductive splicing and translation".[62] RNA-binding proteins are enriched among a large number of *C. elegans* mRNAs whose abundance increases 2-fold or more in *smg(-)* mutants (D. Markwardt and P. Anderson, unpublished). Perhaps the inherent ability of RNA-binding proteins to bind RNA and alter splicing has been adopted, together with NMD, as a general mechanism to regulate their expression.

Expressed Pseudogene mRNAs

Pseudogenes are nonfunctional genes or gene fragments that accumulate in all genomes through duplication and genetic drift.[68] Pseudogenes are remarkably abundant[69,70] and can, on rare occasion, contribute sequence information to functional genes.[71] While pseudogene transcripts are usually not detectable, certain pseudogenes yield polyadenylated mRNAs. Such mRNAs encode truncated proteins having the potential to interfere with function of the genes from which they derive. mRNAs of expressed *C. elegans* pseudogenes are substrates of NMD. mRNAs of seven different tested pseudogenes are absent (or barely detected) in *smg(+)* strains but abundant in *smg(-)* mutants.[72] Although expressed pseudogenes are not under selection to produce functional proteins, they are likely under selection to avoid expressing disruptive proteins or protein fragments. NMD likely reduces the disruptive effects of expressed pseudogenes, thereby allowing them to persist in genomes for longer periods of times. By allowing pseudogenes to persist without deleterious effect, NMD may contribute to the evolution of pseudogenes and genomes.

References

1. Hodgkin J, Papp A, Pulak R et al. A new kind of informational suppression in the nematode Caenorhabditis elegans. Genetics 1989; 123:301-313.
2. Pulak R, Anderson P. mRNA surveillance by the Caenorhabditis elegans smg genes. Genes Dev 1993; 7:1885-1897.
3. Cali BM, Kuchma SL, Latham J et al. SMG -7 is required for mRNA surveillance in Caenorhabditis elegans. Genetics 1999; 151:605-616.
4. Leeds P, Peltz SW, Jacobson A et al. The product of the yeast UPF1 gene is required for rapid turnover of mRNAs containing a premature translational termination codon. Genes Dev 1991; 5:2303-2314.
5. Medghalchi SM, Frischmeyer PA, Mendell JT et al. Rent1, a trans-effector of nonsense-mediated mRNA decay, is essential for mammalian embryonic viability. Hum Mol Genet 2001; 10:99-105.
6. Losson R, Lacroute F. Interference of nonsense mutations with eukaryotic messenger RNA stability. Proc Natl Acad Sci USA 1979; 76:5134-5137.
7. Page MF, Carr B, Anders KR et al. SMG-2 is a phosphorylated protein required for mRNA surveillance in Caenorhabditis elegans and related to Upf1p of yeast. Mol Cell Biol 1999; 19:5943-5951.
8. Kuchma SL. Cloning and characterization of smg-3, a gene required for nonsense-mediated mRNA decay in Caenorhabditis elegans. PhD Thesis, University of Wisconsin 1999; 142.
9. Aronoff R, Baran R, Hodgkin J. Molecular identification of smg-4, required for mRNA surveillance in C. elegans. Gene 2001; 268:153-164.
10. Culbertson MR, Leeds PF. Looking at mRNA decay pathways through the window of molecular evolution. Curr Opin Genet Dev 2003; 13:207-214.
11. Yamashita A, Ohnishi T, Kashima I et al. Human SMG-1, a novel phosphatidylinositol 3-kinase-related protein kinase, associates with components of the mRNA surveillance complex and is involved in the regulation of nonsense-mediated mRNA decay. Genes Dev 2001; 15:2215-2228.
12. Denning G, Jamieson L, Maquat LE et al. Cloning of a novel phosphatidylinositol kinase-related kinase: Characterization of the human SMG-1 RNA surveillance protein. J Biol Chem 2001; 276:22709-22714.
13. Ohnishi T, Yamashita A, Kashima I et al. Phosphorylation of hUPF1 induces formation of mRNA surveillance complexes containing hSMG-5 and hSMG-7. Mol Cell 2003; 12:1187-1200.
14. Chiu SY, Serin G, Ohara O et al. Characterization of human Smg5/7a: A protein with similarities to Caenorhabditis elegans SMG5 and SMG7 that functions in the dephosphorylation of Upf1. RNA 2003; 9:77-87.
15. Gatfield D, Unterholzner L, Ciccarelli FD et al. Nonsense-mediated mRNA decay in Drosophila: At the intersection of the yeast and mammalian pathways. EMBO J 2003; 22:3960-3970.
16. Unterholzner L, Izaurralde E. SMG-7 acts as a molecular link between mRNA surveillance and mRNA decay. Mol Cell 2004; 16:587-596.
17. Perlick HA, Medghalchi SM, Spencer FA et al. Mammalian orthologues of a yeast regulator of nonsense transcript stability. Proc Natl Acad Sci USA 1996; 93:10928-10932.
18. Applequist SE, Selg M, Raman C et al. Cloning and characterization of HUPF1, a human homolog of the Saccharomyces cerevisiae nonsense mRNA-reducing UPF1 protein. Nucl Acids Res 1997; 25:814-821.
19. Lykke-Andersen J, Shu MD, Steitz JA. Human Upf proteins target an mRNA for nonsense-mediated decay when bound downstream of a termination codon. Cell 2000; 103:1121-1131.
20. Kim YK, Furic L, Desgroseillers L et al. Mammalian Staufen1 recruits Upf1 to specific mRNA 3'UTRs so as to elicit mRNA decay. Cell 2005; 120:195-208.
21. Czaplinski K, Ruiz-Echevarria MJ, Paushkin SV et al. The surveillance complex interacts with the translation release factors to enhance termination and degrade aberrant mRNAs. Genes Dev 1998; 12:1665-1677.
22. Wang W, Czaplinski K, Rao Y et al. The role of Upf proteins in modulating the translation read-through of nonsense-containing transcripts. EMBO J 2001; 20:880-890.
23. He F, Jacobson A. Identification of a novel component of the nonsense-mediated mRNA decay pathway by use of an interacting protein screen. Genes Dev 1995; 9:437-454.
24. Lykke-Andersen J. Identification of a human decapping complex associated with hUpf proteins in nonsense-mediated decay. Mol Cell Biol 2002; 22:8114-8121.
25. Kim VN, Kataoka N, Dreyfuss G. Role of the nonsense-mediated decay factor hUpf3 in the splicing-dependent exon-exon junction complex. Science 2001; 293:1832-1836.
26. Le Hir H, Gatfield D, Izaurralde E et al. The exon-exon junction complex provides a binding platform for factors involved in mRNA export and nonsense-mediated mRNA decay. EMBO J 2001; 20:4987-4997.

27. Lykke-Andersen J, Shu MD, Steitz JA. Communication of the position of exon-exon junctions to the mRNA surveillance machinery by the protein RNPS1. Science 2001; 293:1836-1839.
28. Grimson A, O'Connor S, Newman CL et al. SMG-1 is a phosphatidylinositol kinase-related protein kinase required for nonsense-mediated mRNA decay in Caenorhabditis elegans. Mol Cell Biol 2004; 24:7483-7490.
29. Singh G, Lykke-Andersen J. New insights into the formation of active nonsense-mediated decay complexes. Trends Biochem Sci 2003; 28:464-466.
30. Maquat LE. Nonsense-mediated mRNA decay: Splicing, translation and mRNP dynamics. Nature Rev Mol Cell Biol 2004; 5:89-99.
31. MacMorris MA, Zorio DAR, Blumenthal TE. An exon that prevents transpot of a mature mRNA. Proc Natl Acad Sci USA 1999; 96:3813-3818.
32. Lee MH, Schedl T. Translation repression of GLD-1 protects its mRNA targets from nonsense-mediated mRNA decay in C. elegans. Genes Dev 2004; 18:1047-1059.
33. Anders KR, Grimson A, Anderson P. SMG-5, required for C. elegans nonsense-mediated mRNA decay, associates with SMG-2 and protein phosphatase 2A. EMBO J 2003; 22:641-650.
34. Janssens V, Goris J. Protein phosphatase 2A: A highly regulated family of serine/threonine phosphatases implicated in cell growth and signaling. Biochemical J 2001; 353:417-439.
35. Andjelkovic N, Zolnierowicz S, Van Hoof C et al. The catalytic subunit of protein phosphatase 2A associates with the translation termination factor eRF1. EMBO J 1996; 15:7156-7167.
36. Dunckley T, Parker R. The DCP2 protein is required for mRNA decapping in Saccharomyces cerevisiae and contains a functional MutT motif. EMBO J 1999; 18:5411-5422.
37. Gavin AC, Bosche M, Krause R et al. Functional organization of the yeast proteome by systematic analysis of protein complexes. Nature 2002; 415:141-147.
38. Atkin AL, Schenkman LR, Easttham M et al. Relationship between yeast polyribosomes and Upf proteins required for nonsense mRNA decay. J Biol Chem 1997; 272:22163-22172.
39. Pal M, Ishigaki Y, Nagy E et al. Evidence that phosphorylation of human Upfl protein varies with intracellular location and is mediated by a wortmannin-sensitive and rapamycin-sensitive PI 3-kinase-related kinase signaling pathway. RNA 2001; 7:5-15.
40. LaGrandeur TE, Parker R. Isolation and characterization of Dcp1p, the yeast mRNA decapping enzyme. EMBO J 1998; 17:1487-1496.
41. de Pinto B, Lippolis R, Castaldo R et al. Overexpression of Upf1p compensates for mitochondrial splicing deficiency independently of its role in mRNA surveillance. Mol Microbiol 2004; 51:1129-1142.
42. Wilkinson MF. A new function for nonsense-mediated mRNA decay factors. Trends Genet 2005; 21:143-148.
43. Domeier ME, Morse DP, Knight SW et al. A link between RNA interference and nonsense-mediated decay in Caenorhabditis elegans. Science 2000; 289:1928-1931.
44. Hunter CP, Kenyon CJ. Spatial and temporal controls target pal-1 blastomere-specification activity to a single blastomere lineage in C. elegans embryos. Cell 1996; 87:217-226.
45. Lelivelt MJ, Culbertson MR. Yeast Upf proteins required for RNA surveillance affect global expression of the yeast transcriptome. Mol Cell Biol 1999; 19:6710-6719.
46. He F, Li X, Spatrick P et al. Genome-wide analysis of mRNAs regulated by the nonsense-mediated and 5' to 3' mRNA decay pathways in yeast. Mol Cell 2003; 12:1439-1452.
47. Mendell JT, Sharifi NA, Meyers JL et al. Nonsense surveillance regulates expression of diverse classes of mammalian transcripts and mutes genomic noise. Nature Gen 2004; 36:1073-1078.
48. Furuta T, Tuck SP, Kirchner JW et al. EMB-30: An APC4 homologue required for metaphase-to-anaphase transitions during meiosis and mitosis in Caenorhabditis elegans. Mol Biol Cell 2000; 11:1401-1419.
49. Schnabel H, Bauer G, Schnabel R. Suppressors of the organ-specific differentiation gene pha-1 of Caenorhabditis elegans. Genetics 1991; 129:69-77.
50. Emtage L, Gu G, Hartwieg E et al. Extacellular proteins organize the mechanosensory channel complex in C. elegans touch receptor neurons. Neuron 2004; 44:795-807.
51. Sun AY, Lambie EJ. Gon-2, a gene required for gonadogenesis in Caenorhabditis elegans. Genetics 1997; 147:1077-1089.
52. Lackner MR, Kim SK. Genetic analysis of the Caenorhabditis elegans MAP kinase gene mpk-1. Genetics 1998; 150:103-117.
53. Blelloch RH, Kimble JE. Control of organ shape by a secreted metalloprotease in the nematode Caenorhabditis elegans. Nature 1999; 399:586-690.
54. Ding M, Goncharov A, Jin Y et al. C. elegans ankyrin repeat protein VAB-19 is a component of epidermal attachment structures and is essential for epidermal morphogenesis. Development 2003; 130:5791-5801.
55. Bloom L, Horvitz HR. The Caenorhabditis elegans gene unc-76 and its human homologs define a new gene family involved in axonal outgrowth and fasciculation. Proc Natl Acad Sci USA 1997; 94:3414-3419.

56. Van Auken K, Weaver D, Robertson B et al. Roles of the Homothorax/Meis/Prep homolog UNC-62 and the Exd/Pbx homologs CEH-20 and CEH-40 in C. elegans embryogenesis. Development 2002; 129:5255-5238.
57. Tenenhaus C, Subramaniam K, Dunn MA et al. PIE-1 is a bifunctional protein that regulates maternal and zygotic gene expression in the embryonic germ line of Caenorhabditis elegans. Genes Dev 2001; 15:1031-1040.
58. Cali BM, Anderson P. mRNA surveillance mitigates genetic dominance in Caenorhabditis elegans. Mol Gen Genet 1998; 260:176-184.
59. Jacobs D, Beitel GJ, Clark SG et al. Gain-of-function mutations in the Caenorhabditis elegans lin-1 ETS gene identify a C-terminal regulatory domain phosphorylated by ERK MAP kinase. Genetics 1998; 149:1809-1822.
60. Mango SE, Maine EM, Kimble JE. Carboxy-terminal truncations activates glp-1 protein to specify vulval fates in Caenorhabditis elegans. Nature 1991; 352:811-815.
61. Lamitina ST, L'Hernault SW. Dominant mutations in the Caenorhabditis elegans Myt1 ortholog wee-1.3 reveal a novel domain that controls M-phase entry during spermatogenesis. Development 2002; 129:5009-5018.
62. Lewis BP, Green RE, Brenner SE. Evidence for the widespread coupling of alternative splicing and nonsense-mediated mRNA decay in humans. Proc Natl Acad Sci USA 2003; 100:189-192.
63. Mitrovich QM, Anderson P. Unproductively spliced ribosomal protein mRNAs are natural targets of mRNA surveillance in C. elegans. Genes Dev 2000; 14:2173-2184.
64. Morrison M, Harris KS, Roth MB. Smg mutants affect the expression of alternatively spliced SR protein mRNAs in Caenorhabditis elegans. Proc Natl Acad Sci USA 1997; 94:9782-9785.
65. Sureau A, Gattoni R, Dooghe Y et al. SC35 autoregulates its expression by promoting splicing events that destabilize its mRNAs. EMBO J 2001; 20:1785-1796.
66. Le Guiner C, Lejeune F, Galiana D et al. TIA-1 and TIAR activate splicing of alternative exons with weak 5' splice sites followed by a U-rich stretch on their own pre-mRNAs. J Biol Chem 2001; 276:40638-40646.
67. Wollerton MC, Gooding C, Wagner EJ et al. Autoregulation of polypyrimidine tract binding protein by alternative splicing leading to nonsense-mediated decay. Mol Cell 2004; 13:91-100.
68. Mighell AJ, Smith NR, Robinson PA et al. Vertebrate pseudogenes. FEBS Lett 2000; 468:109-114.
69. Harrison PM, Gerstein M. Studying genomes through the aeons: Protein families, pseudogenes and proteome evolution. J Mol Biol 2002; 318:1155-1174.
70. Torrents D, Suyama M, Zdobnov E et al. A genome-wide survey of human pseudogenes. Genome Res 2003; 13:2559-2567.
71. Trabesinger-Ruef N, Jermann T, Zankel T et al. Pseudogenes in ribonuclease evolution: A source of new biomacromolecular function? FEBS Lett 1996; 382:319-322.
72. Mitrovich QM, Anderson P. mRNA surveillance of expressed pseudogenes in C. elegans. Curr Biol 2005; 15:963-967.
73. Cui Y, Hagan KW, Zhang S et al. Identification and characterization of genes that are required for the accelerated degradation of mRNAs containing a premature translational termination codon. Genes Dev 1995; 9:423-436.
74. Mendell JT, Medghalchi SM, Lake RG et al. Novel Upf2p orthologues suggest a functional link between translation initiation and nonsense surveillance complexes. Mol Cell Biol 2000; 20:8944-8957.
75. Serin G, Gersappe A, Black JD et al. Identification and characterization of human orthologues to Saccharomyces cerevisiae Upf2 protein and Upf3 protein (Caenorhabditis elegans SMG-4). Mol Cell Biol 2001; 21:209-223.
76. Lee BS, Culbertson MR. Identification of an additional gene required for eukaryotic nonsense mRNA turnover. Proc Natl Acad Sci USA 1995; 92:10354-10358.
77. Shang H. Molecular analysis of smg-6, a gene required for nonsense-mediated mRNA decay in Caenorhabditis elegans. PhD Thesis. University of Wisconsin, 2004:122.

SECTION IV
Drosophila melanogaster

CHAPTER 12

NMD in *Drosophila*:
A Snapshot into the Evolution of a Conserved mRNA Surveillance Pathway

Isabelle Behm-Ansmant and Elisa Izaurralde*

Abstract

Eukaryotic gene expression involves a number of interlinked post-transcriptional steps that are subject to surveillance or quality control mechanisms to ensure that only fully processed and error-free mRNAs are translated. Among these, the nonsense-mediated mRNA decay (NMD) pathway recognizes and targets for degradation mRNAs containing premature translation termination codons (PTCs), which could give rise to truncated and potentially harmful proteins. The NMD pathway not only prevents the accumulation of malfunctioning proteins but also modulates the clinical manifestations of many human genetic disorders (see chapters by Holbrook et al and Keeling et al). A cross-species analysis of this pathway has revealed important conserved key components, and has provided the basis for elucidating the NMD network in humans. This chapter focuses on the NMD pathway in the fruitfly *Drosophila melanogaster*, emphasizing how studies in this model organism have provided new insights into the mechanisms underlying NMD and its evolution.

Introduction

The gene expression pathway involves a number of interlinked steps that begin with gene transcription and end with protein synthesis, wherein mRNA is a key intermediate. mRNA is transcribed from DNA as pre-mRNA, which is generally processed by the removal of introns and the addition of the poly(A) tail. The resulting mature mRNA is exported to the cytoplasm where it is translated and finally degraded. It is now established that modulation of most, if not all, steps in the gene expression pathway can contribute to the regulation of gene expression.

While the complexity of eukaryotic gene expression allows for the control of protein production at multiple levels, it also makes the process vulnerable to errors. Eukaryotic cells have evolved elaborated mRNA quality control (also called surveillance) mechanisms that ensure the fidelity of gene expression by detecting and degrading aberrant transcripts. These quality control or surveillance mechanisms operate in both the nucleus and the cytoplasm. For instance, mechanisms in the nucleus degrade improperly processed mRNAs before they are exported to the cytoplasm (reviewed by Fasken and Corbett in ref. 1). In the cytoplasm, surveillance pathways assess the translatability of the mRNA and degrade any that have no translation termination codons (nonstop-mediated mRNA decay, NSD) or that have PTCs (NMD) (reviewed by Fasken and Corbett in ref. 1).

*Corresponding Author: Elisa Izaurralde—European Molecular Biology Laboratory, Meyerhofstrasse 1, 69117 Heidelberg, Germany. Email: izaurralde@embl-heidelberg.de

Nonsense-Mediated mRNA Decay, edited by Lynne E. Maquat. ©2006 Eurekah.com.

The NMD pathway is one of the best characterized mRNA surveillance mechanisms. It not only eliminates mRNAs containing frameshift or nonsense mutations, but also regulates the expression of naturally occurring transcripts having features recognized by the NMD machinery (see chapters by He and Jacobson, Sharifi and Dietz, and Soergel et al). In this way, the NMD pathway contributes to the post-transcriptional regulation of about 10% of the transcriptome in yeast, fruitfly and human cells.[2-5]

Studies using model organisms have revealed that despite conservation of this pathway, different species have evolved different mechanisms to discriminate natural from premature translation termination codons and to degrade transcripts identified as NMD substrates (reviewed by Conti and Izaurralde in ref. 6). In this chapter, we discuss the information gathered from studies of the NMD pathway in the model organism *Drosophila melanogaster* (fruitfly) and the insights gained from these studies into the molecular mechanism underlying NMD and its evolution.

The NMD Protein Interaction Network: An Overview

The RNA helicase Upf1 is a key molecular component of the NMD machinery. Upf1 was originally identified in genetic screens of yeast (*Saccharomyces cerevisiae*) and worms (*Caenorhabditis elegans*) and shown to be essential for NMD in these organisms (see chapter by Baker and Parker and Anderson). Subsequently, it was shown that silencing of the Upf1 gene by RNA interference (RNAi) in both human and fruitfly cells results in the stabilization of PTC-containing mRNAs indicating that these Upf1 orthologs are also essential for NMD.[7,8] Upf1 interacts with two additional proteins, Upf2 and Upf3, to form the so-called 'surveillance complex'. Surveillance complex function in NMD is conserved in all organisms that have been studied. Indeed data suggests that for all eukaryotes recognition of a PTC by a translating ribosome leads to the assembly of the surveillance complex, which in turn targets the mRNA for decay.

The role of the surveillance complex is therefore to couple a premature translation termination event to mRNA degradation. Consistent with this, NMD factors have genetic, functional and/or physical links with translation factors and mRNA decay enzymes (Fig. 1). A link between NMD and translation termination is provided by the association of Upf1, Upf2 and Upf3 with the eukaryotic translation termination factors (eRF)1 and eRF3.[9-11] As the actual substrate of NMD is not a naked mRNA, but a ribonucleoprotein particle (mRNP), many RNA binding proteins play roles in NMD. For yeast, these proteins include the poly(A)-binding protein 1 (Pab1p) (see chapter by Amrani and Jacobson). For mammals, they include the exon junction complex (EJC), which is a multi-protein assembly deposited ~20-24 nucleotides upstream of exon-exon junctions (see chapter by Maquat).

As mentioned above, although the core of the NMD machinery (Upf1-3) is conserved, the mechanisms by which PTCs are recognized and mRNA targets are degraded differ between species (reviewed by Conti and Izaurralde in ref. 6). These mechanistic differences correlate with changes in the NMD protein interaction network (Fig. 1). Moreover, the complexity of the network underlying this pathway increases from simpler organisms such as the budding yeast *S. cerevisiae* to the mammal *Homo sapiens*. For instance, additional factors have been integrated into the network in multicellular organisms, including the proteins Smg1, Smg5, Smg6 and Smg7, which regulate Upf1 function.

Upf1 is regulated by phosphorylation and dephosphorylation. In man and worms, phosphorylation of Upf1 is catalyzed by Smg1, a phosphoinositide-3-kinase-related protein kinase, and requires Upf2 and Upf3[12-17] (see chapters by Yamashita et al and Anderson). This suggests that it is the assembly of the surveillance complex that triggers Upf1-phosphorylation. The dephosphorylation of Upf1 is mediated by Smg5, Smg6 and Smg7, three similar but not functionally redundant proteins.[14,18-20] Smg5-7 are not phosphatases themselves, but they are thought to trigger Upf1 dephosphorylation by recruiting protein phosphatase 2A (PP2A). This model is based on the observations that Smg5 and Smg7 interact with each

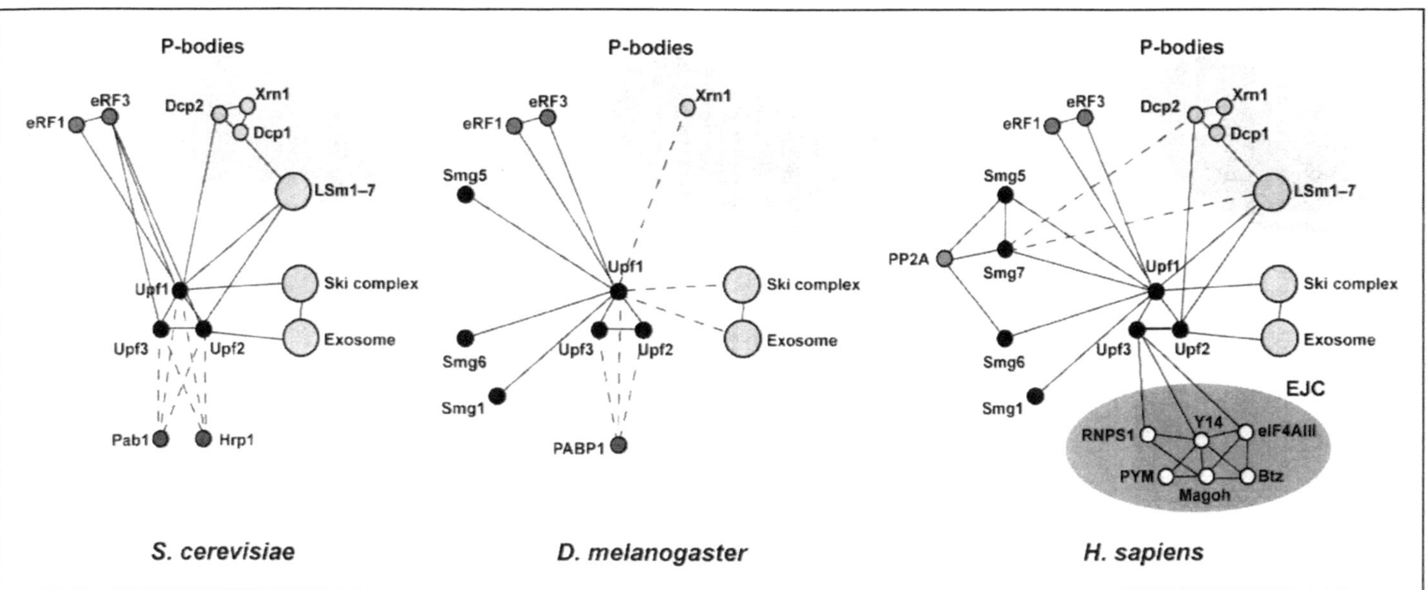

Figure 1. The NMD protein interaction network. Proteins for which experimental evidence for a role in NMD is available are shown. Links do not necessarily represent direct interactions. Dashed lines represent functional interactions. Individual proteins are denoted as small balls, and multimeric protein complexes (i.e., the exosome, the Ski complex, and the LSm1-7 complex) are represented as large balls. RNPS1, Y14, Magoh, eIF4AIII, Barentsz (Btz) and PYM, which are components of the exon junction complex (EJC), are part of the NMD network in *H. sapiens* but not *D. melanogaster*. Components of the EJC and proteins localizing in P-bodies are shadowed in, respectively, dark gray and light gray ovals.

other and are part of a larger complex comprising phosphatase PP2A and phosphorylated Upf1.[19,20] Similarly, Smg6 is part of a protein complex comprising PP2A and phosphorylated Upf1.[21] With the exception of Smg7, *Drosophila* orthologs to core components of the NMD machinery have been identified, including Upf1, Upf2, Upf3, Smg1, Smg5 and Smg6.[7] Depletion of these proteins by RNAi in *Drosophila* Schneider cells stabilizes various PTC-containing mRNAs, providing evidence for a role in NMD.[7] Intriguingly, mutations in the *smg1* gene do not appear to affect NMD in *Drosophila* embryos.[22] These observations suggest that different and/or redundant kinases phosphorylate Upf1 in embryos.

Regardless of whether Upf1 is phosphorylated by another redundant kinase in embryos, the role of *Drosophila* Smg1 in NMD is well established. Firstly, depletion of this protein stabilizes different PTC-containing reporters in Schneider cells to a similar extent as does depletion of Upf3.[7] Secondly, depletion of Smg1 or depletion of any of the other known NMD factors leads to similar mRNA expression profiles.[5] Thus, in addition of Upf1-3, NMD in *Drosophila* requires the Smg1, Smg5 and Smg6 proteins, which are a specific feature of higher eukaryotic NMD.

Mechanism of PTC Definition in *Drosophila*

In mammals, PTC recognition relies on splicing. In this case, stop codons are defined as premature if they are located ~50 nucleotides or more upstream of an exon-exon junction[23] (see chapter by Maquat). The positions of exon-exon junctions are communicated to translating ribosomes by the EJC (see chapter by Maquat) (Fig. 2). In particular, the EJC components Y14, Magoh, eIF4AIII, Barentsz and RNPS1 elicit NMD when bound at least ~20-25 nucleotides downstream of a termination codon[24-27] (Fig. 2).

Mammalian Upf3 interacts with components of the EJC and is loaded onto mRNAs during splicing, while Upf2 is thought to join the complex in the cytoplasm after export.[26,28-31] During a pioneer round of translation, Upf2, Upf3 and the additional EJC components are displaced by the ribosomes as they traverse the mRNA.[32] If translating ribosomes were to encounter a stop codon upstream of an EJC, this would lead to the incomplete removal of Upf2 and Upf3 proteins from downstream mRNA sequences and to the recruitment of Upf1, probably via interactions with the eRF1:eRF3 complex. The recruitment of Upf1 creates an opportunity for the assembly of the Upf1:Upf2:Upf3 complex, which targets the mRNA for rapid degradation (reviewed by Wagner and Lykke-Andersen,[33] see chapters by Maquat, and Singh and Lykke-Andersen).

In *S. cerevisiae*, most genes lack introns, and most EJC components do not exist. Consistent with this, exon-exon junctions are not required to distinguish natural from premature stop codons in this organism. Two models have been proposed to explain the mechanism of PTC recognition in yeast. In one model, PTCs are defined relative to a downstream sequence

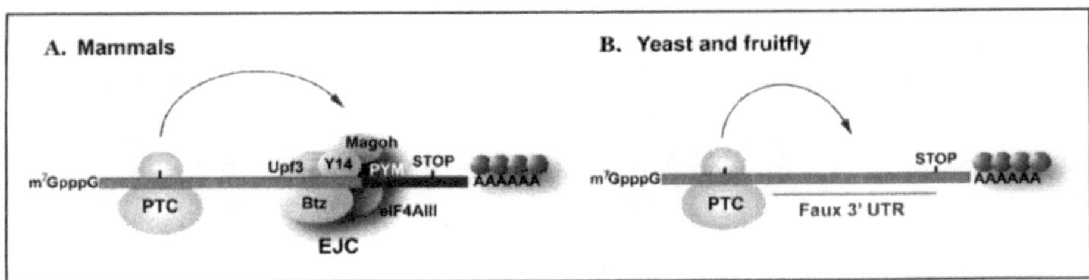

Figure 2. Mechanisms of PTC recognition. A premature termination codon (PTC) is recognized relative to the position of a downstream cis-acting signal that varies across species. In mammals (A), the cis-acting signal is an exon-exon junction marked by a group of proteins that constitute the exon junction complex (EJC). In yeast and fruitfly (B), the cis-acting signal is a faux 3'-UTR. Natural termination codons (STOP), in contrast to PTCs, are flanked by proper 3'-untranslated region (UTR) marked by a specific protein or set of proteins, including the poly(A)-binding protein shown over the poly(A) tail.

element (DSE), which might function analogously to a mammalian exon-exon junction (see chapters by Baker and Parker, and Amrani and Jacobson). The protein Hrp1p has been implicated in binding to at least one DSE, thereby providing positional information for PTC recognition.

An alternative, the faux 3'-untranslated region (UTR) model, posits that the process of premature translation termination is intrinsically aberrant, because the stop codon is not in the appropriate context (i.e., not flanked by a bona fide 3'-UTR [see chapter by Amrani and Jacobson]). According to this model, bona fide 3'-UTRs would be marked by a specific protein or set of proteins (e.g., Pab1p). If a terminating ribosome is able to interact with these proteins, proper termination can occur. In contrast, translation termination at a PTC would be impaired or too slow, because of the inability of the terminating ribosome to establish interactions with bona fide 3'-UTR-associated proteins. In this case the NMD complex could be assembled, leading to rapid degradation of the mRNA.

Several lines of evidence support the faux 3'-UTR model in yeast. Firstly, translation termination is indeed aberrant at PTCs compared to normal termination codons, and prematurely terminating ribosomes fail to release efficiently.[34] Secondly, this effect is abolished when the PTC is flanked by a normal 3'-UTR.[34] Moreover, tethering Pab1p downstream of the PTC apparently mimics a normal 3'-UTR since it leads to efficient translation termination at the PTC and suppresses NMD.[34]

In *Drosophila*, as in yeast, PTC recognition occurs independently of exon-exon junctions, and although *Drosophila* orthologs of vertebrate EJC proteins do exist (i.e., Y14, Mago, eIF4AIII, Barentsz and RNPS1), they are dispensable for NMD.[7] This concurs with the observation that PTC-containing mRNAs that are transcribed from intronless genes are subject to NMD in *Drosophila*,[7] in contrast to the situation in mammals (see chapter by Maquat).

What differentiates a PTC from a normal termination codon in *Drosophila*? One possibility is that PTCs are defined relative to a DSE, as may be the case for some transcripts in yeast. However, the observation that transcripts derived from heterologous genes such as bacterial chloramphenicol-acetyl transferase (CAT) or green fluorescent protein (GFP) can be recognized by the NMD machinery when they carry a PTC makes unlikely the possibility that specific sequence elements mark *Drosophila* transcripts. Analogously to the results reported for yeast, we have recently found that tethering PABPC1 downstream of a PTC abolishes NMD in *Drosophila* cells (I. Behm-Ansmant and E. Izaurralde, unpublished). These findings strongly suggest that the faux 3'-UTR model provides a mechanism for PTC recognition in *Drosophila*, too.

The changes in the mechanisms of PTC definition described above are reflected in the NMD protein interaction network (Fig. 1). For instance, the proteins Y14, Magoh, eIF4AIII and Barentsz are part of a protein interaction module conserved in multicellular organisms, but not in budding yeast. In *Drosophila*, this module plays a role in mRNA localization, while in mammals, it interacts with the NMD machinery, reflecting the requirement of exon-exon junctions for PTC-recognition (see chapter by Maquat). In yeast and *Drosophila*, in contrast, poly(A)-binding proteins (Pab1p and its ortholog PABPC1, respectively) play a role in PTC definition and are therefore functionally linked to the NMD network in these organisms (Fig. 1).

Decay of NMD Targets in *Drosophila*

In eukaryotic cells, mRNAs are generally degraded via alternative pathways, each of which is initiated by removal of the poly(A)-tail by deadenylases (reviewed by Parker and Song in ref. 35). Following this first rate-limiting step, the cap structure is removed by the decapping enzymes Dcp1/Dcp2 and decapping coactivators such as the LSm1-7 complex, which makes the mRNA susceptible to digestion by the major cytoplasmic 5'-to-3' exonuclease Xrn1. Alternatively, following deadenylation, mRNAs can be degraded from their 3'-ends. Decay through the 3'-to-5' pathway requires the exosome, a multimeric complex of 3'-to-5' exonucleases, and the Ski complex, a trimeric protein complex that regulates exosome activity (reviewed by Parker and Song in ref. 35).

The enzymes involved in general mRNA decay also function in NMD. In yeast, the major decay pathway for NMD substrates involves deadenylation-independent decapping and 5'-to-3' degradation by Xrn1p[36-38] (Fig. 3). This implies that one function of the surveillance complex is to bypass deadenylation, the rate-limiting step in mRNA decay, and to promote decapping directly. An alternative pathway, which also contributes to the decay of PTC-containing mRNAs, relies on accelerated deadenylation and 3'-to-5' degradation by the exosome and the Ski complex[37,38] (Fig. 3). The mechanism by which NMD substrates are degraded in mammals has not been investigated in as much detail, but has been shown to occur by a mechanism similar to that described in yeast.[39-41]

In *Drosophila*, in contrast, degradation of nonsense transcripts is initiated by endonucleolytic cleavage near the PTC (Fig. 3). The resulting 5' fragment is degraded by the exosome, while the 3' fragment is degraded by Xrn1.[42] Thus, the mRNA fragments are degraded from the newly generated ends without undergoing decapping or deadenylation, suggesting that, in contrast to yeast and mammals, the decapping enzymes, the LSm1-7 complex and deadenylases are not required for NMD in *Drosophila*, and are not part of the NMD protein interaction network (Fig. 1).

Recently, the decapping enzymes Dcp1 and Dcp2, the LSm1-7 complex and Xrn1 have been shown to reside in specialized cytoplasmic bodies or mRNA decay foci.[43-47] These foci have been named mRNA processing bodies (P-bodies) or GW bodies, due to the accumulation of the RNA binding protein GW182 in these bodies.[44,48] These findings have raised the question of whether decay of NMD substrates occurs in the cytoplasm independently of P-bodies or whether the entire surveillance complex escorts the mRNA to P-bodies, where it undergoes rapid 5'-3' decay (see chapters by Baker and Parker, and Singh and Lykke-Andersen).

Smg7 provides a molecular link between NMD and decay enzymes localized in P-bodies.[49,50] When over-expressed, human Smg7 accumulates in P-bodies and also causes the accumulation of Smg5 and Upf1 in these bodies.[49,50] These observations suggest a mechanism that couples the assembly of the surveillance complex to the degradation of PTC-containing transcripts. Accordingly, the assembly of the surveillance complex on PTC-containing mRNAs leads to the phosphorylation of Upf1. Phosphorylated Upf1 recruits Smg7 (most likely in a complex with Smg5 and PP2A).[19,20,49] Smg7 would then recruit the decay enzymes in the cytoplasm or target the PTC-containing transcript for degradation in P-bodies.[49] The association of Smg7, Smg5 and PP2A would also trigger the dephosphorylation of Upf1, which might be involved in recycling of the NMD factors for another round of targeting.

The mechanism for connecting the surveillance complex to the mRNA decay enzymes in yeast and *Drosophila* cells (which appear to lack a Smg7 ortholog) has not been fully elucidated. Moreover, the identity of the *Drosophila* endonuclease and how it is recruited to nonsense transcripts are unknown.

Evolution of the Physiological Role of NMD

NMD components are not essential in yeast.[51] Similarly, inhibition of the NMD pathway in *C. elegans* leads to viable worms with defects in the male bursa and the hermaphrodite vulva[16,52] (see chapter by Anderson). In contrast, depletion of Upf1 or Upf2 induces a G2/M cell cycle arrest and impairs cell proliferation in *Drosophila*.[5] Upf1 is also required for cell viability in mice, and deletion of the UPF1 gene leads to early embryonic death.[53]

At least two mechanisms can account for the different phenotypes observed across species after depletion of NMD factors. One mechanism could be the acquisition of novel functions by NMD components during evolution. Consistent with this possibility, it has been reported that human Upf1 can be recruited to the 3' UTR of specific transcripts via interactions with Staufen1 or stem-loop binding protein and elicits mRNA decay by a mechanism distinct from but related to NMD[54] (see chapters by Kim and Maquat, and Kaygun and Marzluff). Moreover, recent reports have implicated a subset of NMD factors in other cellular processes including telomere maintenance and DNA repair (see chapters by Abraham and Oliveira, and Azzalin et al).

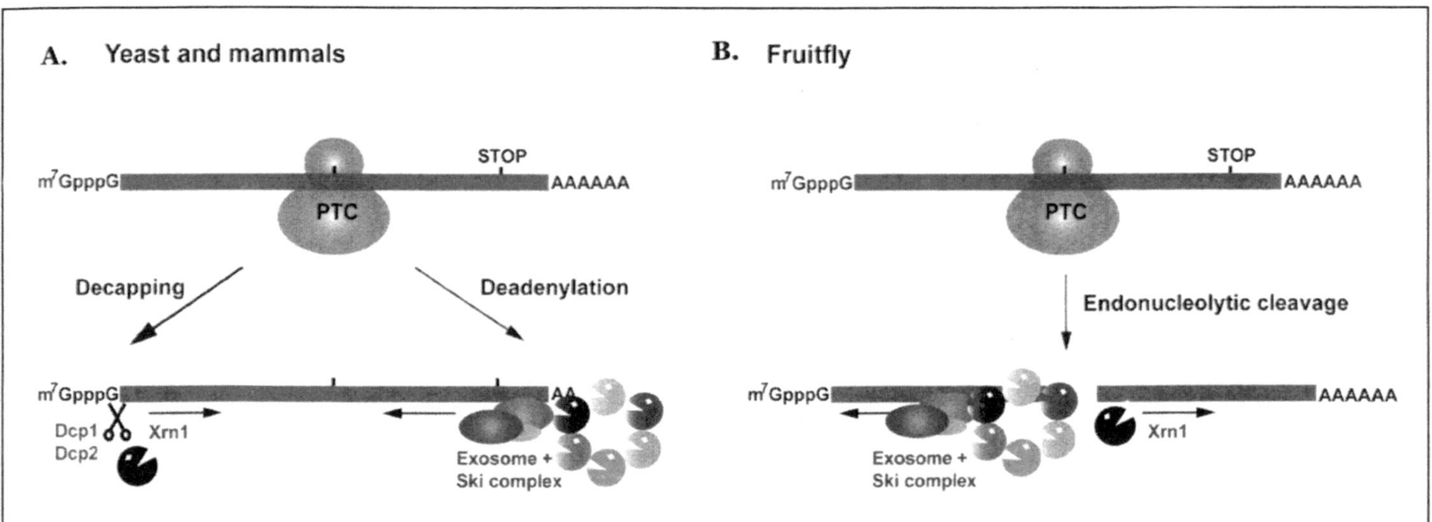

Figure 3. Decay of NMD substrates. In yeast and mammals (A), decay of NMD targets is initiated by removal of the cap structure by the decapping enzymes Dcp1 and Dcp2. This exposes the mRNA to exonucleolytic digestion by the 5'-to-3' exonuclease Xrn1. An alternative pathway involves deadenylation followed by 3'-to-5' exosome-mediated decay. In Fruitfly (B), decay of NMD targets is initiated by endonucleolytic cleavage near the PTC. The resulting 5' and 3' mRNA fragments are degraded from the newly generated ends by the exosome and Xrn1, respectively. This decay pathway is independent of deadenylation and decapping. Exosome activity requires the Ski complex in all organisms.

Another mechanism that can lead to phenotypic differences is changes in selected targets. Evidence already exists that this is indeed the case. For instance, rearrangements of the immunoglobulin and T-cell receptor genes in vertebrates result in frameshifted genes at high frequency (~66% of the recombination events), and transcripts from these genes are degraded by NMD (see chapter by Gudikote and Wilkinson). More recently gene expression profiling of yeast, fruitfly or human cells defective in NMD have allowed the identification of endogenous targets of the pathway (see ref. 5 and chapters by He and Jacobson, Sharafi and Dietz, and Soergel et al). These studies revealed that although NMD regulates the expression of ~10% of the transcriptome in these organism,[2-5] with few exceptions, NMD targets in different species do not represent orthologous genes.[5]

The exceptions include transcripts involved in DNA repair and telomere maintenance. SMG5 mRNA is regulated by NMD in both human and *Drosophila* cells, and in human cells it has been implicated in telomerase function.[4,5,55,56] Similarly, the telomerase-associated protein Est1p (a close relative of Smg5-7) is regulated by NMD in yeast.[3] In addition to EST1, NMD regulates six genes involved in telomere maintenance in yeast, including ESB1, EST2, EST3, STN1, YKu70 and TEL1.[3,57] This regulation has functional implications, as inhibition of NMD in yeast leads to telomere shortening and de-repression of silenced telomeric loci.[3,57,58] Finally, the gene for *Drosophila* ATM, a relative of TEL1 involved in DNA repair and telomere function, is also a target of NMD.[5] All together, this suggests a conserved role for NMD in regulating the expression of genes involved in telomere maintenance.

The analysis of gene expression profiles in cells lacking individual NMD factors also revealed that in both yeast and fruitfly cells the major role of these factors is to regulate in concert the expression of a common set of genes. Indeed, yeast strains lacking Upf1p, Upf2p or Upf3p exhibit similar expression profiles.[3] Moreover, depletion of Upf1, Upf2, Upf3, Smg1, Smg5 or Smg6 from fruitfly cells, results in correlated changes in gene expression.[5] This and the observation that only a few transcripts are regulated exclusively by the individual proteins suggest that most NMD factors have not acquired specialized functions in RNA turnover in these organisms. These findings do not rule out the possibility that individual NMD factors have acquired additional functions that do not affect steady-state mRNA levels. Consequently, differences in the cellular role of NMD between yeast and fruitfly largely reflect changes in selected targets in addition to a potential functional diversification of NMD components. An important goal for the future is to understand how changes in regulated targets lead to the different phenotypes observed at the cellular level across species.

Perspectives

NMD is linked to other post-transcriptional processes by the use of common factors. For instance, many RNA binding proteins play roles at multiple steps of gene expression. As one example, poly(A)-binding proteins have roles in translation, NMD and mRNA turnover. In addition, enzymes involved in general mRNA degradation are recruited by the NMD and RNA silencing machineries to degrade targeted mRNAs. Understanding how the interplay of these regulatory circuits determines the expression pattern of many genes is an important goal for the future.

The NMD pathway contributes to the phenotypic manifestations of human genetic diseases, and therapeutic strategies based on the modulation of this pathway are under evaluation (see chapters by Holbrook et al and Keeling et al). However, the effect of such strategies on the whole patient cannot be assessed in the absence of information on the physiological role of this pathway. A deeper understanding of the mechanisms underlying NMD and the comprehensive identification of its components are also required. Studies in model organisms such as yeast, worms and fruitfly have begun to provide a detailed understanding of the mechanisms of NMD, and they have led to the identification of the core components of the machinery. These studies are likely to continue to unveil deep insights into the role and regulatory mechanisms of NMD, and they will contribute to establishing the NMD network in humans.

Acknowledgements

We gratefully acknowledge David Thomas for critical reading of the manuscript. Both authors are supported by the European Molecular Biology Organization (EMBO) and the Human Frontier Science Program Organization (HFSPO).

References

1. Fasken MB, Corbett AH. Process or perish: Quality control in mRNA biogenesis. Nat Struct Mol Biol 2005; 12(6):482-488.
2. Lelivelt MJ, Culbertson MR. Yeast Upf proteins required for RNA surveillance affect global expression of the yeast transcriptome. Mol Cell Biol 1999; 19(10):6710-6719.
3. He F, Li X, Spatrick P et al. Genome-wide analysis of mRNAs regulated by the nonsense-mediated and 5' to 3' mRNA decay pathways in yeast. Mol Cell 2003; 12(6):1439-1452.
4. Mendell JT, Sharifi NA, Meyers JL et al. Nonsense surveillance regulates expression of diverse classes of mammalian transcripts and mutes genomic noise. Nat Genet 2004; 36(10):1073-1078.
5. Rehwinkel J, Letunic I, Raes J et al. Nonsense-mediated mRNA decay factors act in concert to regulate common mRNA targets. RNA 2005; 11(10):1530-1544.
6. Conti E, Izaurralde E. Nonsense-mediated mRNA decay: Molecular insights and mechanistic variations across species. Curr Opin Cell Biol 2005; 17(3):316-325.
7. Gatfield D, Unterholzner L, Ciccarelli FD et al. Nonsense-mediated mRNA decay in Drosophila: At the intersection of the yeast and mammalian pathways. EMBO J 2003; 22(15):3960-3970.
8. Mendell JT, ap Rhys CM, Dietz HC. Separable roles for rent1/hUpf1 in altered splicing and decay of nonsense transcripts. Science 2002; 298(5592):419-422.
9. Weng Y, Czaplinski K, Peltz SW. Identification and characterization of mutations in the UPF1 gene that affect nonsense suppression and the formation of the Upf protein complex but not mRNA turnover. Mol Cell Biol 1996; 16(10):5491-5506.
10. Wang W, Czaplinski K, Rao Y et al. The role of Upf proteins in modulating the translation read-through of nonsense-containing transcripts. EMBO J 2001; 20(4):880-890.
11. Czaplinski K, Ruiz-Echevarria MJ, Paushkin SV et al. The surveillance complex interacts with the translation release factors to enhance termination and degrade aberrant mRNAs. Genes Dev 1998; 12(11):1665-1677.
12. Denning G, Jamieson L, Maquat LE et al. Cloning of a novel phosphatidylinositol kinase-related kinase: Characterization of the human Smg-1 RNA surveillance protein. J Biol Chem 2001; 276(25):22709-22714.
13. Grimson A, O'Connor S, Newman CL et al. Smg-1 is a phosphatidylinositol kinase-related protein kinase required for nonsense-mediated mRNA Decay in Caenorhabditis elegans. Mol Cell Biol 2004; 24(17):7483-7490.
14. Page MF, Carr B, Anders KR et al. Smg-2 is a phosphorylated protein required for mRNA surveillance in Caenorhabditis elegans and related to Upf1p of yeast. Mol Cell Biol 1999; 19(9):5943-5951.
15. Pal M, Ishigaki Y, Nagy E et al. Evidence that phosphorylation of human Upfl protein varies with intracellular location and is mediated by a wortmannin-sensitive and rapamycin-sensitive PI 3-kinase-related kinase signaling pathway. RNA 2001; 7(1):5-15.
16. Pulak R, Anderson P. mRNA surveillance by the Caenorhabditis elegans smg genes. Genes Dev 1993; 7(10):1885-1897.
17. Yamashita A, Ohnishi T, Kashima I et al. Human Smg-1, a novel phosphatidylinositol 3-kinase-related protein kinase, associates with components of the mRNA surveillance complex and is involved in the regulation of nonsense-mediated mRNA decay. Genes Dev 2001; 15(17):2215-2228.
18. Cali BM, Kuchma SL, Latham J et al. smg-7 is required for mRNA surveillance in Caenorhabditis elegans. Genetics 1999; 151(2):605-616.
19. Anders KR, Grimson A, Anderson P. Smg-5, required for C. elegans nonsense-mediated mRNA decay, associates with Smg-2 and protein phosphatase 2A. EMBO J 2003; 22(3):641-650.
20. Ohnishi T, Yamashita A, Kashima I et al. Phosphorylation of hUPF1 induces formation of mRNA surveillance complexes containing hSMG-5 and hSMG-7. Mol Cell 2003; 12(5):1187-1200.
21. Chiu SY, Serin G, Ohara O et al. Characterization of human Smg5/7a: A protein with similarities to Caenorhabditis elegans SMG5 and SMG7 that functions in the dephosphorylation of Upf1. RNA 2003; 9(1):77-87.
22. Chen Z, Smith KR, Batterham P et al. Smg1 nonsense mutations do not abolish nonsense-mediated mRNA decay in Drosophila melanogaster. Genetics 2005.
23. Nagy E, Maquat LE. A rule for termination-codon position within intron-containing genes: When nonsense affects RNA abundance. Trends Biochem Sci 1998; 23(6):198-199.

24. Fribourg S, Gatfield D, Izaurralde E et al. A novel mode of RBD-protein recognition in the Y14-Mago complex. Nat Struct Biol 2003; 10(6):433-439.
25. Gehring NH, Neu-Yilik G, Schell T et al. Y14 and hUpf3b form an NMD-activating complex. Mol Cell 2003; 11(4):939-949.
26. Lykke-Andersen J, Shu MD, Steitz JA. Communication of the position of exon-exon junctions to the mRNA surveillance machinery by the protein RNPS1. Science 2001; 293(5536):1836-1839.
27. Palacios IM, Gatfield D, St Johnston D et al. An eIF4AIII-containing complex required for mRNA localization and nonsense-mediated mRNA decay. Nature 2004; 427(6976):753-757.
28. Kim VN, Kataoka N, Dreyfuss G. Role of the nonsense-mediated decay factor hUpf3 in the splicing-dependent exon-exon junction complex. Science 2001; 293(5536):1832-1836.
29. Le Hir H, Gatfield D, Izaurralde E et al. The exon-exon junction complex provides a binding platform for factors involved in mRNA export and nonsense-mediated mRNA decay. EMBO J 2001; 20(17):4987-4997.
30. Lykke-Andersen J, Shu MD, Steitz JA. Human Upf proteins target an mRNA for nonsense-mediated decay when bound downstream of a termination codon. Cell 2000; 103(7):1121-1131.
31. Serin G, Gersappe A, Black JD et al. Identification and characterization of human orthologues to Saccharomyces cerevisiae Upf2 protein and Upf3 protein (Caenorhabditis elegans SMG-4). Mol Cell Biol 2001; 21(1):209-223.
32. Ishigaki Y, Li X, Serin G et al. Evidence for a pioneer round of mRNA translation: mRNAs subject to nonsense-mediated decay in mammalian cells are bound by CBP80 and CBP20. Cell 2001; 106(5):607-617.
33. Wagner E, Lykke-Andersen J. mRNA surveillance: The perfect persist. J Cell Sci 2002; 115(Pt 15):3033-3038.
34. Amrani N, Ganesan R, Kervestin S et al. A faux 3'-UTR promotes aberrant termination and triggers nonsense-mediated mRNA decay. Nature 2004; 432(7013):112-118.
35. Parker R, Song H. The enzymes and control of eukaryotic mRNA turnover. Nat Struct Mol Biol 2004; 11(2):121-127.
36. Muhlrad D, Parker R. Aberrant mRNAs with extended 3' UTRs are substrates for rapid degradation by mRNA surveillance. RNA 1999; 5(10):1299-1307.
37. Cao D, Parker R. Computational modeling and experimental analysis of nonsense-mediated decay in yeast. Cell 2003; 113(4):533-545.
38. Mitchell P, Tollervey D. An NMD pathway in yeast involving accelerated deadenylation and exosome-mediated 3'—>5' degradation. Mol Cell 2003; 11(5):1405-1413.
39. Chen CY, Shyu AB. Rapid deadenylation triggered by a nonsense codon precedes decay of the RNA body in a mammalian cytoplasmic nonsense-mediated decay pathway. Mol Cell Biol 2003; 23(14):4805-4813.
40. Couttet P, Grange T. Premature termination codons enhance mRNA decapping in human cells. Nucleic Acids Res 2004; 32(2):488-494.
41. Lejeune F, Li X, Maquat LE. Nonsense-mediated mRNA decay in mammalian cells involves decapping, deadenylating, and exonucleolytic activities. Mol Cell 2003; 12(3):675-687.
42. Gatfield D, Izaurralde E. Nonsense-mediated messenger RNA decay is initiated by endonucleolytic cleavage in Drosophila. Nature 2004; 429(6991):575-578.
43. Cougot N, Babajko S, Seraphin B. Cytoplasmic foci are sites of mRNA decay in human cells. J Cell Biol 2004; 165(1):31-40.
44. Eystathioy T, Jakymiw A, Chan EK et al. The GW182 protein colocalizes with mRNA degradation associated proteins hDcp1 and hLSm4 in cytoplasmic GW bodies. RNA 2003; 9(10):1171-1173.
45. Ingelfinger D, Arndt-Jovin DJ, Luhrmann R et al. The human LSm1-7 proteins colocalize with the mRNA-degrading enzymes Dcp1/2 and Xrnl in distinct cytoplasmic foci. RNA 2002; 8(12):1489-1501.
46. Sheth U, Parker R. Decapping and decay of messenger RNA occur in cytoplasmic processing bodies. Science 2003; 300(5620):805-808.
47. van Dijk E, Cougot N, Meyer S et al. Human Dcp2: A catalytically active mRNA decapping enzyme located in specific cytoplasmic structures. EMBO J 2002; 21(24):6915-6924.
48. Eystathioy T, Chan EK, Tenenbaum SA et al. A phosphorylated cytoplasmic autoantigen, GW182, associates with a unique population of human mRNAs within novel cytoplasmic speckles. Mol Biol Cell 2002; 13(4):1338-1351.
49. Unterholzner L, Izaurralde E. SMG7 acts as a molecular link between mRNA surveillance and mRNA decay. Mol Cell 2004; 16(4):587-596.
50. Fukuhara N, Ebert J, Unterholzner L et al. SMG7 is a 14-3-3-like adaptor in the nonsense-mediated mRNA decay pathway. Mol Cell 2005; 17(4):537-547.

51. Leeds P, Peltz SW, Jacobson A et al. The product of the yeast UPF1 gene is required for rapid turnover of mRNAs containing a premature translational termination codon. Genes Dev 1991; 5(12A):2303-2314.
52. Hodgkin J, Papp A, Pulak R et al. A new kind of informational suppression in the nematode Caenorhabditis elegans. Genetics 1989; 123(2):301-313.
53. Medghalchi SM, Frischmeyer PA, Mendell JT et al. Rent1, a trans-effector of nonsense-mediated mRNA decay, is essential for mammalian embryonic viability. Hum Mol Genet 2001; 10(2):99-105.
54. Kim YK, Furic L, Desgroseillers L et al. Mammalian Staufen1 recruits Upf1 to specific mRNA 3'UTRs so as to elicit mRNA decay. Cell 2005; 120(2):195-208.
55. Reichenbach P, Hoss M, Azzalin CM et al. A human homolog of yeast Est1 associates with telomerase and uncaps chromosome ends when overexpressed. Curr Biol 2003; 13(7):568-574.
56. Snow BE, Erdmann N, Cruickshank J et al. Functional conservation of the telomerase protein Est1p in humans. Curr Biol 2003; 13(8):698-704.
57. Dahlseid JN, Lew-Smith J, Lelivelt MJ et al. mRNAs encoding telomerase components and regulators are controlled by UPF genes in Saccharomyces cerevisiae. Eukaryot Cell 2003; 2(1):134-142.
58. Lew JE, Enomoto S, Berman J. Telomere length regulation and telomeric chromatin require the nonsense-mediated mRNA decay pathway. Mol Cell Biol 1998; 18(10):6121-6130.

SECTION V
Plants

Chapter 13

NMD in Plants

Ambro van Hoof* and Pamela J. Green

NMD Is Widespread in Plants

Nonsense-mediated mRNA decay (NMD) appears to be widespread in the plant kingdom. Numerous studies have documented a correlation between the presence of premature termination codons (PTCs) and a decrease in mRNA abundance for a variety of genes and in a large number of flowering plants. Examples include the Kunitz trypsin inhibitor gene in soybean, the phytohemagglutinin (PHA) gene in common bean, the waxy gene in rice and wheat, the AUX1 and FAD7 genes in *Arabidopsis*, the Chs gene in petunia, the Xantha-f gene in barley, and the Fed-1 gene in tobacco.[1-10] Therefore, NMD appears to be a general feature of not only animals (see chapter by Maquat) and fungi (see chapter by Baker and Parker) but also plants. Animals and fungi diverged from each other after they diverged from plants. Thus, any features in common between plants and either animals or fungi are likely ancestral features. Where animals and fungi differ, plants can provide insight into the evolutionary origin of these differences. In this review we will summarize the evidence for NMD in plants, and will highlight some of the cases where plants provide additional insight not available through studies of fungi and animals.

Three studies provide evidence for NMD in plants that supplements the correlation of a PTC with a low level of mRNA. Each of these studies grew from an initial characterization of alleles found in important crop plants. The first evidence that PTCs may trigger rapid mRNA degradation came from characterization of a naturally occurring allele of a Kunitz trypsin inhibitor gene of soybean that produced an abnormally low level of mRNA. This allele contained a point mutation approximately 70% of the way into the coding region that introduced a PTC.[5] Transcription rates of the wild type and mutant alleles, as measured by nuclear run-on analysis, were not significantly different. These results suggested that the effect of the PTC was post-transcriptional, possibly at the level of mRNA stability.[5] More direct evidence for NMD was found in a second case where PTCs in the 5' part of the PHA gene of common bean triggered rapid mRNA decay, but a more distal PTC had no effect. Initially, a naturally occurring allele of the PHA gene was found to produce a greatly reduced level of mRNA.[7] Sequencing showed that this allele contained a frameshift mutation that led to a PTC in the new reading frame approximately 20% of the way through the PHA coding region. Replacing the wild-type coding region with the frameshift-containing coding region decreased gene expression in transgenic plants, and correcting the mutation restored the level of gene expression to normal.[6] Further evidence for NMD was obtained using both natural and artificial alleles of the PHA gene that contained a PTC.[11] PTCs within the first 60% of the PHA coding region decreased mRNA stability in cultured tobacco cells and mRNA levels in whole tobacco or *Arabidopsis* plants. However, a PTC at 80% of the coding region had no effect on the mRNA level. Therefore, the fact that NMD appears to be triggered by early but not late

*Corresponding Author: Ambro van Hoof—Department of Microbiology and Molecular Genetics, University of Texas Health Science Center - Houston, 6431 Fannin MSB 1.212, Houston, Texas 77030, U.S.A. Email: ambro.van.hoof@uth.tmc.edu

PTCs appears to be a conserved feature between plants, yeast, and mammals. A third case showed that NMD also occurs in monocot plants (the other main group of plants, which includes the cereals). Specifically, Ishiki et al studied two waxy alleles in rice that produced reduced levels of mRNA but normal levels of partially spliced mRNA.[4] This decrease in mRNA abundance was caused by an increase in mRNA degradation rates. In summary, these three studies firmly establish that a PTC can increase the rate of mRNA degradation in a variety of flowering plants. This conclusion suggests that the basic mechanism of NMD evolved before plants diverged from animals and fungi.

To Be Premature or Not Premature, That Is the Question

Given the fact that the sequence of a PTC is the same as that of a normal termination codon, all cells must use some second signal to distinguish termination codons that elicit NMD from those that do not. Several models have been proposed for what this second signal may be and how it may act, and these models appear to fall in three distinct groups. First, elegant experiments in mammalian cells have established that this signal can be a downstream exon junction complex (EJC) (see chapter by Maquat). EJCs are deposited upstream of mRNA exon-exon junctions during splicing. The net effect is that termination codons that are situated more than ~50- to 55-nucleotides upstream of the last exon-exon junction, i.e., the last intron within pre-mRNA, are recognized as premature and trigger NMD, while more distal termination codons generally do not trigger NMD.

In plants, NMD is not dependent on the position of exon-exon junctions, and thus the signal that distinguishes termination codons that elicit NMD from those that do not must be some other feature of the mRNA. The conclusion that intron position within pre-mRNA is irrelevant to NMD in plants is based on three pieces of evidence. First, many plant genes do not have introns, yet they are susceptible to NMD. For example, the Kunitz trypsin inhibitor and PHA genes mentioned above do not contain introns, yet they are susceptible to NMD.[5,6,11] In contrast, humans have very few genes that are naturally intronless, and these few genes are insensitive to NMD.[12] Second, while no systematic effort has been made to compare the effects of termination codons before and after the last intron, a mutation that introduces a termination codon within the last exon of the barley Xantha-f gene reduces mRNA abundance. This observation suggests that even for intron-containing genes, the position of an intron is not important for NMD.[3] Third, insertion of an intron within the 3' untranslated region (UTR) of a reporter gene so that it resides more than 50-55 nucleotides downstream of the normal termination codon fails to reduce the level of product mRNA.[13-15] From all of these data, it appears that plants use some feature other than exon-exon junction position to distinguish termination codons that elicit NMD from those that do not. Similar conclusions have been reached for yeast and *Drosophila*[16] (see chapters by Amrani and Jacobson, and Behm-Ansmant and Izaurralde).

Interestingly, one of the initial hints that introns downstream of a termination codons were disadvantageous derived from the observation that introns are usually found within coding regions but seldom occur within the 3' UTR in mammals, *Drosophila* and plants.[17,18] This observation was made many years ago based on the limited number of gene sequences available at the time. It would be interesting to revisit the frequency of introns within 3' UTRs based on whole *Arabidopsis*, rice, and *Drosophila* genome sequences. If introns are truly rare within the 3'UTRs of plant and *Drosophila* genes, then there may be more ancient reasons for the rarity of introns in 3' UTRs than the function in NMD that has been established for mammals.

A second proposed mechanism to distinguish termination codons that trigger NMD from those that do not relies on a "downstream sequence element".[19] It has been proposed that in yeast PTCs are recognized by the presence of a loosely defined downstream sequence. This downstream region would normally be present in most or all coding regions, and thus be upstream of the normal termination codon. It appears unlikely that such a specific sequence exists in plants, since a synthetic Bt-toxin gene is also susceptible to NMD (E.J. De Rocher and

P.J. Green, unpublished data). Similarly, a PTC in an exogenous CAT gene triggers NMD in *Drosophila*.[16] It therefore seems unlikely that any specific downstream sequence that normally resides within the coding region is required for NMD in plants.

A third proposed mechanism to distinguish termination codons that elicit NMD from those that do not relies on the spacing between the termination codon and other sequences normally present within the 3'UTR (e.g., see ref. 20). In this mechanism, distal termination codons are recognized as normal because of their close proximity to proper 3' UTR sequences, which may include the context of the termination codon, the proximity of the termination codon to the poly(A) tail, or both. PTCs would fail to be recognized as normal because, although a proper 3' UTR is present, the 3' UTR or some component of the 3' UTR is situated too far from the PTC. To our knowledge there are no experiments that address this mechanism in plants. However, by the process of elimination, it currently is the most attractive mechanism. Importantly, if this is indeed the correct mechanism in plants, yeast (see chapter by Amrani and Jacobson) and *Drosophila* (see chapter by Behm-Ansmant and Izaurralde), then it must have evolved earlier than, and may be ancestral to, the mechanism that typifies termination codon classification in mammals.

Conserved trans-Acting Factors in NMD

There have been no detailed functional studies of potential trans-acting factors that are required for NMD in plants. However, many proteins required for NMD in other species have homologs in plants (see below). Since this section is limited to such homologs, it excludes any putative factors that may be required for NMD in plants but are not required in other organisms.

In all eukaryotes where this has been studied, NMD requires homologs to yeast Upf1p, Upf2p, and Upf3p (see chapters by Baker and Parker, Singh and Lykke-Andersen, Anderson, and Behm-Ansmant and Izaurralde). The two plant genomes that have been completely sequenced (*Arabidopsis* and rice) each contain a single gene that appears to encode a Upf1 ortholog (Table 1). Although *upf1* mutants in various organisms are defective in NMD, other important phenotypic differences have been observed. Yeast and *C. elegans upf1* mutants are relatively normal, with only a few subtle phenotypes.[21,22] In contrast, the mouse UPF1 gene is essential to organismal development;[23] see chapter by Sharifi and Dietz). Thus, while the core molecular mechanism of NMD appears to be conserved in all eukaryotes, the importance of the process to the normal organismal survival and development varies.

Several insertion mutations within the *Arabidopsis* UPF1 gene have been generated as part of a large scale effort to create mutants in all *Arabidopsis* genes (Table 1; refs. 24-26). While the resulting mutants are freely available as heterozygotes, information about their homozygous phenotype is currently not forthcoming. Analysis of these mutants would shed light on whether the essentiality of mouse Upf1 is limited to vertebrates or more widespread.

The *Arabidopsis* and rice genomes also encode a single ortholog to Upf2 and to Upf3. This is similar to the genomes of yeast, *Drosophila*, and *C. elegans*, but it is different from the genome of humans, which encodes two Upf3 homologs. Very recently published data support the conclusion that *Arabidopsis* Upf3 is required for NMD. Hori and Watanabe (2003) analyzed endogenous alternatively spliced mRNAs, where one splicing product harbored a PTC, and the other did not. Results indicated that an *Arabidopsis upf3* gene disruption mutation is viable, but it is typified by an increased level of the alternatively spliced mRNA that contains the PTC relative to the level of mRNA that lacks the PTC. Furthermore the difference in transcript levels depends on ongoing translation. Given that Upf homologs are present in plants, and that at least Upf3 is implicated in NMD, it seems likely that the basic NMD machinery that is conserved between yeast and animals is also conserved in plants.

NMD in humans, *Drosophila* and *C. elegans* requires additional "Smg" proteins that are not present in yeast (see chapters by Anderson, and Behm-Ansmant and Izaurralde). The Smg proteins include the protein kinase Smg1 that phosphorylates Upf1 in vitro, targeting SQ sequences. It is thought that this phosphorylation plays a key role in NMD.[27,28] There

Table 1. *The* Arabidopsis *and rice genomes encode homologs to proteins required for NMD*[a]

Yeast Gene	Human Gene	Function	*Arabidopsis* Gene	a.k.a.[b]	*Arabidopsis* Mutant[c]	Rice Gene
UPF1	UPF1	NMD factor	At5g47010		Y	BAD30435
UPF2	UPF2	NMD factor	At2g39260		Y	BAD26479
UPF3	UPF3a, Upf3b	NMD factor	At1g33980		Y	CAE03133
	SMG1	NMD regulator	none			AAR07071
	SMG5, SMG6, SMG7	NMD regulator	At5g19400, At1g28260		Y Y	AAQ56472
	Y14	EJC component	At1g51510		Y	AAT77901
	Magoh	EJC component	At1g02140	HAP1	pollen defect	AAK08102
	eIF4AIII	EJC component	At3g19760		Y	BAB63781
	RNPS1	EJC component	At1g16610		Y	BAD87117, AAU44172
	SRm160	EJC component	At2g29210		Y	AAP06864
	REF/ALY	EJC component	At5g59950 At5g02530 At1g66260 At5g37720	 DIP1 DIP2	Y N Y Y	BAD01466, BAD46300, BAC80073

[a] *Arabidopsis* and rice homologs were identified using BLAST and the indicated human and/or yeast genes. [b] a.k.a.: also known as. DIP1 and DIP2 have been isolated in a yeast two-hybrid screen as proteins interacting with Tomato Bushy Stunt Virus protein P19 by Uhrig et al (2004) Plant physiol 135:2411-2423. [c] Y indicates that the mutant is available as a heterozygote (there is often allelic redundancy). N indicates that no mutant is available. EJC, exon function complex.

are 22 copies of this SQ motif in human Upf1, but only four in yeast Upf1. *Arabidopsis* Upf1 has 18 SQ motifs, making it more similar in this respect to the human protein. The Smg1 protein kinase is a member of a larger family of kinases, and although members of this family are encoded in the *Arabidopsis* genome, there does not appear to be a Smg1 ortholog. In contrast, the rice genome does contain a Smg1 ortholog. The remaining three Smg proteins, Smg5, Smg6, and Smg7, are related to each other and have a TPR and a PIN domain. The *Arabidopsis* genome contains two proteins that have a similar organization, but we were unable to find a third protein (Table 1). The rice genome only appears to have one homolog to Smg5/6/7. The evolutionary relationship between the plant proteins and the three metazoan proteins is unclear.

As explained above, exon-exon junctions do not appear to play a role in NMD in plants and *Drosophila*, but orthologs of EJC components are encoded in the *Drosophila*, *Arabidopsis* and rice genomes (Table 1; ref. 16). Interestingly, it has recently been reported that many EJC components are mainly localized to the nucleolus of *Arabidopsis*.[29] One possible explanation is that the EJC carries out some other role, possibly in the nucleolus, and was recruited to the NMD pathway in the mammalian ancestry relatively recently.

In human cells, a major effect of the EJC appears to be an increase in the amount of protein that is produced from a given amount of mRNA.[30] A similar mechanism may be operating in plants. It has been shown that the presence of an intron in a reporter gene can increase the abundance of the encoded mRNA and protein.[13,31] To facilitate this effect, the intron has to be near the 5' end of the gene. This suggests that the role of the EJC in stimulating expression may predate a role in NMD.

While some EJC functions may be shared by mammals and plants, it can be expected that plants have exploited the EJC for aspects unique to plant biology. As one example, *Arabidopsis* has a single ortholog of Magoh that is essential for proper pollen function. An insertion in the *Arabidopsis* Magoh gene was isolated in a screen for mutants with specific defects in pollen function. The insertion mutant abnormality is surprisingly specific in that pollen germination and pollen tube growth are relatively normal, but the pollen tube fails to grow towards the ovule and thus the plants are male sterile.[32] It is unclear whether this phenotype is unique to Magoh, or whether defects in other EJC components have a similar phenotype.

Yeast contains two pathways for mRNA degradation. In one pathway, the mRNA is decapped by the Dcp2p pyrophosphatase, and degraded by the 5' to 3' exoribonuclease Xrn1p. In the alternative pathway, the mRNA is degraded from the 3' end by the exosome, which is a complex of multiple 3' to 5' exoribonucleases. NMD targets are also mainly degraded by these two pathways. Orthologs to all of these yeast genes are present in most other eukaryotes, including *Arabidopsis* and rice, although it is not clear whether either one of these pathways is responsible for NMD in any plant. As with genes encoding the Upf proteins and EJC components, insertions within genes encoding *Arabidopsis* Dcp2, Xrn1 and exosome components are readily available. The *Arabidopsis* Xrn1 homolog, AtXrn4, is not essential and has been shown to function in mRNA decay, and its substrates include selected micro RNA targets.[33] However, the role of plant mRNA degradative activities in NMD has not been investigated.

Concluding Remarks

Although NMD is conserved in plants, yeast and metazoans, there are some mechanistic differences. Plants, as an "outgroup", offer the opportunity to study whether yeast or humans reflect an ancestral state. *Arabidopsis* offers many tools, including insertion mutants in most of the genes discussed here, which make answering these important questions possible. We expect that future studies of plants will yield insights into the evolutionary origins of NMD, and will yield some surprises, such as the pollen-specific defect that typifies Magoh mutants.

Acknowledgements

We are grateful to Jarred Lyons for assistance in generating Table 1 and Michelle Trosclair for editorial assistance. Work in the authors' laboratories was supported by NSF grant MCB0228144 and DOE grant #DE-FG01-04ER04-01 (P.J. Green), and by NIH grant GM069900 and a PEW Scholarship in the Biomedical Sciences (A. van Hoof).

References

1. Petracek ME, Nuygen T, Thompson WF et al. Premature termination codons destabilize ferredoxin-1 mRNA when ferredoxin-1 is translated. Plant J 2000; 21(6):563-9.
2. Saito M, Nakamura T. Two point mutations identified in emmer wheat generate null Wx-A1 alleles. Theor Appl Genet 2005; 110(2):276-82.
3. Gadjieva R, Axelsson E, Olsson U et al. Nonsense-mediated mRNA decay in barley mutants allows the cloning of mutated genes by a microarray approach. Plant Physiol Biochem 2004; 42(7-8):681-5.
4. Isshiki M, Yamamoto Y, Satoh H et al. Nonsense-mediated decay of mutant waxy mRNA in rice. Plant Physiol 2001; 125(3):1388-95.
5. Jofuku KD, Schipper RD, Goldberg RB. A frameshift mutation prevents Kunitz trypsin inhibitor mRNA accumulation in soybean embryos. Plant Cell 1989; 1(4):427-35.
6. Voelker TA, Moreno J, Chrispeels MJ. Expression analysis of a pseudogene in transgenic tobacco: A frameshift mutation prevents mRNA accumulation. Plant Cell 1990; 2:255-261.

7. Voelker TA, Staswick P, Chrispeels MJ. Molecular analysis of two phytohemagglutinin genes and their expression in Phaseolus vulgaris cv. Pinto, a lectin-deficient cultivar of the bean. EMBO J 1986; 5:3075-3082.
8. Que Q, Wang HY, English JJ et al. The frequency and degree of cosuppression by sense chalcone synthase transgenes are dependent on transgene promoter strength and are reduced by premature nonsense codons in the transgene coding sequence. Plant Cell 1997; 9(8):1357-1368.
9. Kusumi J, Iba K. Characterization of a nonsense mutation in FAD7, the gene which encodes w-3 desaturase in Arabidopsis thaliana. J Plant Res 1998; 111:87-91.
10. Marchant A, Bennett MJ. The Arabidopsis AUX1 gene: A model system to study mRNA processing in plants. Plant Mol Biol 1998; 36(3):463-71.
11. van Hoof A, Green PJ. Premature nonsense codons decrease the stability of phytohemagglutinin mRNA in a position-dependent manner. Plant J 1996; 10(3):415-24.
12. Maquat LE, Li X. Mammalian heat shock p70 and histone H4 transcripts, which derive from naturally intronless genes, are immune to nonsense-mediated decay. RNA 2001; 7(3):445-56.
13. Callis J, Fromm M, Walbot V. Introns increase gene expression in cultured maize cells. Genes Dev 1987; 1(10):1183-200.
14. Rose AB. Requirements for intron-mediated enhancement of gene expression in Arabidopsis. RNA 2002; 8(11):1444-53.
15. Snowden KC, Buchhholz WG, Hall TC. Intron position affects expression from the tpi promoter in rice. Plant Mol Biol 1996; 31(3):689-92.
16. Gatfield D, Unterholzner L, Ciccarelli FD et al. Nonsense-mediated mRNA decay in Drosophila: At the intersection of the yeast and mammalian pathways. EMBO J 2003; 22(15):3960-70.
17. Hawkins JD. A survey on intron and exon lengths. Nucleic Acids Res 1988; 16(21):9893-908.
18. Nagy E, Maquat LE. A rule for termination-codon position within intron-containing genes: When nonsense affects RNA abundance. Trends Biochem Sci 1998; 23(6):198-9.
19. Zhang S, Ruiz-Echevarria MJ, Quan Y et al. Identification and characterization of a sequence motif involved in nonsense-mediated mRNA decay. Mol Cell Biol 1995; 15(4):2231-44.
20. Hilleren P, Parker R. mRNA surveillance in eukaryotes: Kinetic proofreading of proper translation termination as assessed by mRNP domain organization? RNA 1999; 5(6):711-9.
21. Leeds P, Peltz SW, Jacobson A et al. The product of the yeast UPF1 gene is required for rapid turnover of mRNAs containing a premature translational termination codon. Genes Dev 1991; 5(12A):2303-14.
22. Pulak R, Anderson P. mRNA surveillance by the Caenorhabditis elegans smg genes. Genes Dev 1993; 7(10):1885-97.
23. Medghalchi SM, Frischmeyer PA, Mendell JT et al. Rent1, a trans-effector of nonsense-mediated mRNA decay, is essential for mammalian embryonic viability. Hum Mol Genet 2001; 10(2):99-105.
24. Alonso JM, Stepanova AN, Leisse TJ et al. Genome-wide insertional mutagenesis of Arabidopsis thaliana. Science 2003; 301(5633):653-7.
25. Li Y, Rosso MG, Strizhov N et al. GABI-Kat SimpleSearch: A flanking sequence tag (FST) database for the identification of T-DNA insertion mutants in Arabidopsis thaliana. Bioinformatics 2003; 19(11):1441-2.
26. Sessions A, Burke E, Presting G et al. A high-throughput Arabidopsis reverse genetics system. Plant Cell 2002; 14(12):2985-94.
27. Grimson A, O'Connor S, Newman CL et al. SMG-1 is a phosphatidylinositol kinase-related protein kinase required for nonsense-mediated mRNA Decay in Caenorhabditis elegans. Mol Cell Biol 2004; 24(17):7483-90.
28. Yamashita A, Ohnishi T, Kashima I et al. Human SMG-1, a novel phosphatidylinositol 3-kinase-related protein kinase, associates with components of the mRNA surveillance complex and is involved in the regulation of nonsense-mediated mRNA decay. Genes Dev 2001; 15(17):2215-28.
29. Yamashita A, Ohnishi T, Kashima I et al. Proteomic analysis of the Arabidopsis nucleolus suggests novel nucleolar functions. Mol Biol Cell 2005; 16(1):260-9.
30. Nott A, Le Hir H, Moore MJ. Splicing enhances translation in mammalian cells: An additional function of the exon junction complex. Genes Dev 2004; 18(2):210-22.
31. Rose AB. The effect of intron location on intron-mediated enhancement of gene expression in Arabidopsis. Plant J 2004; 40(5):744-51.
32. Johnson MA, von Besser K, Zhou Q et al. Arabidopsis hapless mutations define essential gametophytic functions. Genetics 2004; 168(2):971-82.
33. Souret FF, Kastenmayer JP, Green PJ. AtXRN4 degrades mRNA in Arabidopsis and its substrates include selected miRNA targets. Mol Cell 2004; 15(2):173-83.

SECTION VI

Evolutionary Aspects of NMD

Chapter 14

Regulation of Gene Expression by Coupling of Alternative Splicing and NMD

David A.W. Soergel, Liana F. Lareau and Steven E. Brenner*

Abstract

Most human genes exhibit alternative splicing, but not all alternatively spliced transcripts produce functional proteins. Computational and experimental results indicate that roughly a third of reliably inferred alternative splicing events in humans result in mRNA isoforms that harbor a premature termination codon (PTC). These transcripts are predicted to be degraded by the NMD pathway. One potential explanation for this startling observation is that cells routinely link alternative splicing and NMD to regulate the abundance of mRNA transcripts. This mechanism, which we call "Regulated Unproductive Splicing and Translation" (RUST), has been experimentally shown to regulate the expression of a wide variety of genes in many organisms from yeast to humans. It is frequently employed to autoregulate proteins that affect the splicing process itself. Thus, alternative splicing and NMD, acting together, play an important and widespread role in regulating gene expression.

Introduction

One major result of the large-scale sequencing projects of the last decade has been an appreciation of the extent of alternative splicing of mammalian transcripts. Estimates vary, but most reports agree that over half of human genes generate transcripts that are alternatively spliced.[1,2] What is the biological function of this extensive alternative splicing? Many claim it is a mechanism of proteome expansion,[3] but relatively few alternative forms encode truly distinct proteins. More often, alternative splicing seems to modulate gene function by adding or removing protein domains, affecting protein activity, or altering the stability of the transcript or the resulting protein.[4-6]

In the last few years, it has become clear that many alternative splice forms previously thought to encode truncated proteins are actually targets of NMD (Fig. 1). In mammals, a termination codon located more than about 50 nucleotides upstream of an exon-exon junction is generally recognized as premature, eliciting NMD[7,8] (see also chapter by Maquat). An understanding of this rule allowed identification of numerous transcripts that are predicted to be degraded rather than translated into protein. Such transcripts can arise through various patterns of alternative splicing (Fig. 2), which may introduce an in-frame termination codon, may induce a frameshift which gives rise to a downstream termination codon, or may introduce an exon-exon junction downstream of the original stop codon. The prevalence of these NMD-targeted transcripts calls for a reconsideration of the roles of alternative splicing and NMD.

*Corresponding Author: Steven E. Brenner—Department of Molecular and Cell Biology, and Department of Plant and Microbial Biology, University of California, Berkeley, California 94707-3102, U.S.A. Email: brenner@compbio.berkeley.edu

Nonsense-Mediated mRNA Decay, edited by Lynne E. Maquat. ©2006 Eurekah.com.

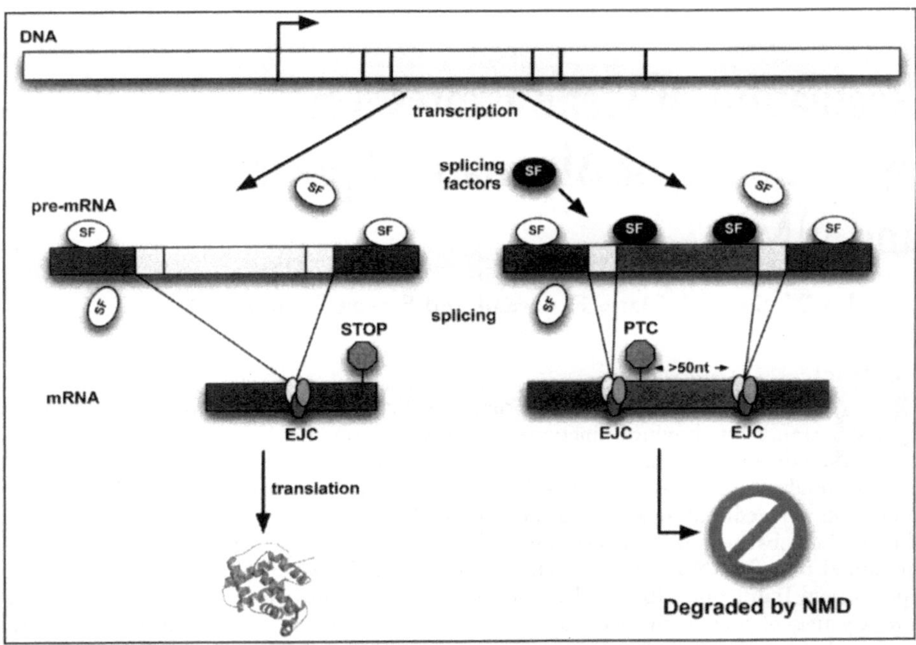

Figure 1. Some alternatively spliced transcripts are degraded by NMD. The spliceosome deposits an exon junction complex (EJC) on the mRNA ~20-24 nucleotides upstream of the splice junction, thereby marking the former location of the excised intron.[8] During the pioneering round of translation,[18] any in-frame stop codon found more than 50 nucleotides upstream of the splice junction triggers NMD.[7,8] Alternative splicing can lead to the inclusion of a premature termination codon (PTC) on an alternatively spliced region, or may introduce a downstream PTC due to a frameshift. Thus, alternative splicing can give rise to unproductive transcripts. A splicing factor (SF) can alter the ratio of productive transcripts to transcripts that contain a PTC, targeting the latter for degradation. In this example, the dark SF induces the inclusion of an alternative exon with a PTC, thereby decreasing the abundance of the productive isoform and downregulating protein expression. Components of the splicing machinery such as small nuclear (sn)RNA U2-associated factor (U2AF) 35 and polypyrimidine tract binding protein (PTB) can similarly regulate isoform proportions.

NMD was originally considered to be a quality control mechanism, protecting cells from the potentially toxic effects of nonsense mutations, errors in transcription, and errors in splicing.[9,10] We now know that there are many natural targets of NMD (see chapters by He and Jacobson, and Sharifi and Dietz), including transcripts with upstream open translational reading frames (uORFs), products of alternative splicing, byproducts of V(D)J recombination, and transcripts arising from transposons and retroviruses.[11] Indeed, it now seems that a major effect of NMD is to downregulate physiological transcripts, in addition to clearing cells of erroneous transcripts.

Alternative Splice Forms Are Frequently Targets of NMD

While it was long known that alternative splicing may produce isoforms that are degraded by NMD, this was not appreciated as a pervasive process until 2002. At that point, genome-wide studies provided the first indication that a substantial fraction of human genes are routinely spliced to produce isoforms that are targeted for NMD.

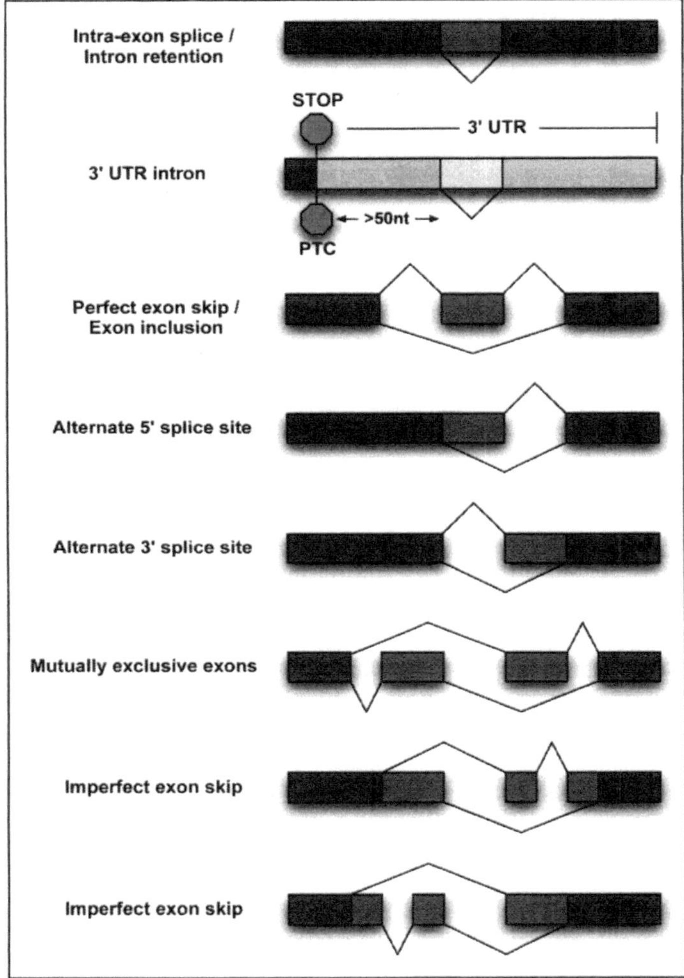

Figure 2. Patterns of alternative splicing. Alternative selection of 5' and 3' splice sites can lead to various patterns of included exons. Any exon that is included in an alternative form may harbor a premature termination codon (PTC). Also, whenever an exon whose length is not a multiple of 3 nucleotides is included or removed, the concomitant frameshift may result in a downstream PTC. Finally, splicing out an intron in the 3' untranslated region (UTR) can cause the normal stop codon to trigger NMD.

In the first study to show widespread predicted NMD of alternative splice forms, Lewis et al used human mRNA and expressed sequence tags (ESTs) available from public databases to infer alternative splice forms and identify PTCs.[12] Although the data set considered was limited, it revealed that at least 12% of genes for which RefSeq mRNA sequences are available generate a PTC^+ isoform. The actual prevalence of such genes may be substantially higher than this lower bound. This study considered 16,780 human mRNA sequences from the reviewed category of RefSeq,[13] a set of well-characterized, experimentally confirmed transcript sequences. Alignment of the RefSeq mRNAs to their genomic loci showed that 617 of these curated mRNA sequences,

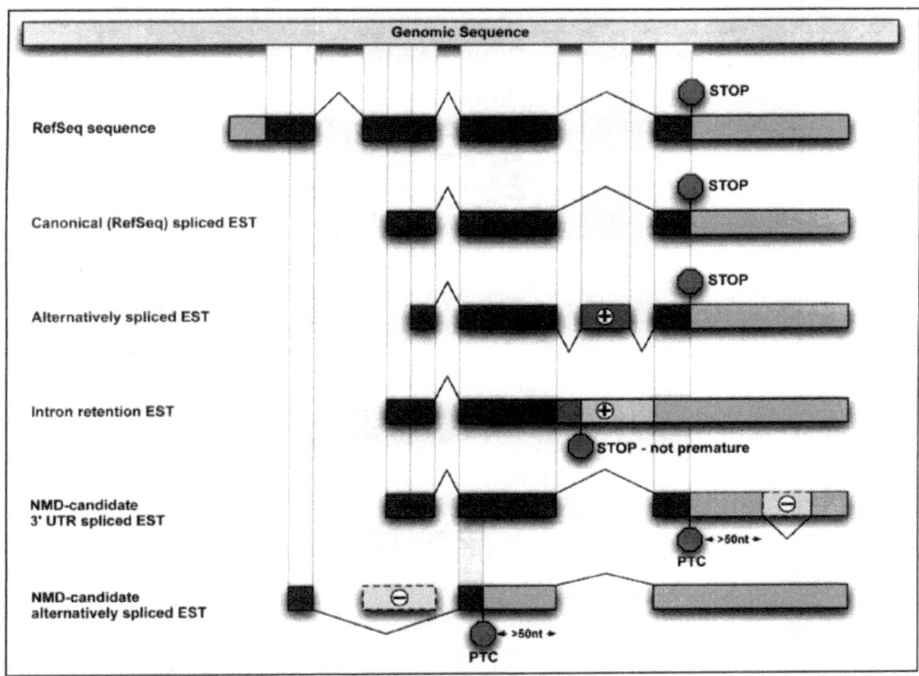

Figure 3. Inference of alternative splice forms and PTCs from RefSeq and EST data. Lewis et al. aligned coding regions of RefSeq mRNAs to the human genome to determine canonical splicing patterns.[12] EST alignments to the genomic sequence confirmed the canonical splices and indicated alternative splices. Canonical (RefSeq) splices are indicated above the exons, whereas alternative splices are indicated below the exons. When an alternative splice introduced a stop codon >50 nucleotides upstream of the final exon-exon junction of an inferred mRNA isoform, the stop codon was classified as a PTC and the corresponding mRNA isoform was labeled an NMD candidate.

or 3.7%, contained PTCs. The alternative splice forms inferred by aligning EST sequences from dbEST[14] to the remaining 16,163 genomic loci (Fig. 3) substantially increased the estimated fraction of genes with PTC⁺ isoforms. Based on the EST data, over 3000 of the RefSeq genes had alternative splice forms, and 45% of these alternatively spliced genes had at least one form that was predicted to be a target of NMD.[12]

These results have been confirmed and strengthened by later studies. An analysis of the isoforms contained in SWISS-PROT[15] showed that even this reliable, curated database contained presumed translation products of mRNA sequences that were more likely to have been degraded by NMD. Alignment of the mRNA sequence of each protein isoform reported in SWISS-PROT to the human genome identified reliable exon-intron structures for 2483 isoforms from 1363 genes. The 50-nucleotide rule predicts that 144 isoforms (5.8% of 2483) from 107 genes (7.9% of 1363) contain a PTC of the type that should elicit NMD.[16]

A new study by Baek and Green has extended the analysis of PTC⁺ alternative splicing to consider conservation of splice forms between human and mouse.[17] This approach helps distinguish aberrant splice events from rare but functional variants. Starting from a large set of cDNA and EST sequences, Baek and Green identified about 1500 pairs of exon inclusion/exclusion splice forms found in both human and mouse. A quarter of the conserved alternative forms contain a conserved PTC,[17] suggesting that these isoforms play a functional role, and that the PTC is important to their function.

Direct experimental evidence from human cells supports these computational results. Mendell and coworkers made HeLa cells depleted of Upf1, an essential component of the NMD pathway, and used microarrays to compare the abundance of mRNAs in these cells to the abundance of mRNAs in unmodified cells.[11] They found that 4.9% of the ~4000 transcripts tested showed significantly higher abundance in cells deficient in NMD, suggesting that NMD normally downregulates those transcripts. They confirmed that their observations were largely due to the direct action of NMD, rather than being a downstream regulatory consequence, by showing that several of the putative NMD-targeted transcripts decayed faster in normal cells than in cells depleted of Upf1. They also confirmed that the effect they observed was due to NMD and not to the action of Upf1 in some other pathway, by showing that the abundance of PTC⁺ transcripts was similarly upregulated upon the depletion of Upf2, another protein essential for NMD. Finally, Mendell et al also observed that 4.3% of transcripts were decreased in abundance in NMD-deficient cells. The stability of those transcripts was not altered by NMD deficiency, showing that the change in their abundance was an indirect effect.

The striking prevalence of PTC⁺ alternative splice forms begs for an explanation. Why would cells routinely expend energy and resources transcribing, splicing, and degrading PTC⁺ isoforms? Are these isoforms often translated after all, contravening the 50-nucleotide rule? Are the observed PTC⁺ isoforms all due to transcriptional or splicing noise? Are they an unavoidable side effect of productive alternative splicing, which itself is conserved as an important mechanism for producing a diversity of proteins? Or, does the combination of alternative splicing and NMD constitute a novel system for regulating gene expression? We will consider each of these potential explanations in turn.

Do the Observed PTC⁺ mRNA Isoforms Evade NMD to Produce Functional Protein?

The existence of numerous PTC⁺ isoforms was first inferred from EST data.[12] One may wonder why EST evidence exists at all for isoforms that are expected to be degraded by NMD. As observed in numerous experiments (Table 1), NMD substantially reduces the abundance of PTC⁺ transcripts, but it does not eliminate the transcripts entirely. One explanation is that NMD surveillance may not be completely effective. Furthermore, PTC⁺ isoforms are not degraded instantly upon being spliced; rather, their degradation occurs only as a consequence of a pioneer round of translation,[18] which for most mRNAs might occur near the nuclear pore during or soon after export of the mRNA from the nucleus (reviewed in ref. 19). Thus, we expect some steady-state abundance of PTC⁺ isoforms that have not yet been degraded, especially inside the nucleus, or that have escaped decay, particularly in the cytoplasm. A series of elegant experiments and computational modeling in yeast[20] suggests that the dominant reason for the presence of PTC⁺ mRNAs in the yeast cell is the temporal lag between splicing and degradation, rather than incomplete surveillance. A similar temporal lag may occur in mammals, despite differences in NMD dynamics between mammals and yeast. Evidently, the resulting abundance of PTC⁺ isoforms is in many cases high enough for ESTs deriving from those isoforms to be observed and deposited in dbEST. Indeed, many of the alternative splice junctions that generate a PTC are supported by two or more ESTs.

EST libraries are nonetheless biased against less stable isoforms. Using sequence features such as splice-site strength, Baek and Green modeled the predicted inclusion rates of alternative exons.[17] They showed that PTC⁺ isoforms are probably produced at a higher rate than they are observed in EST data, but are degraded before they can be sequenced. Thus, the EST data underestimate the fraction of PTC⁺ mRNA deriving from a given gene, and also underestimate the number of genes with PTC⁺ alternative splicing. For this reason, and also because the quality filters used in the above studies excluded many genes and isoforms, reports offer a lower bound on the number of PTC⁺ isoforms; the true prevalence of alternative splicing and of PTC⁺ isoforms may be substantially higher.

Some PTC⁺ transcripts may evade NMD, increasing their likelihood of being observed and deposited in sequence databases. There are a few known examples in which a transcript that

Table 1. Experimentally confirmed examples of unproductive splicing

Name	Organism	AS→PTC	AS Regulated	PTC→Low Abundance	NMD	Notes	Refs.
a) Unproductive Splicing							
FGFR2	Rat	•	•	•	•	Side effect. Productive forms are tissue-specific.	33,67
Calpain-10	Human	•		•	•		16,34,35
TCR-beta	Human	•		•	•	V(D)J cleanup. NMD strength boosted by sequence elements.	68-70
ABCC4	Human, Monkey, Mouse	•		•	•	High conservation of PTC-producing exons from mouse to human suggests that they are under ESE control, or that translation is reinitiated downstream of the PTC.	59
HPRT	Human	•		•	•	Unproductive transcripts are likely noise.	71
b) Regulated Unproductive Splicing							
MID-1	Human, Mouse, Fugu	•	•	•	•		38
POLB	Human	•	•	•	•		71
FAH	Human	•	•	•	•	PTC^+ transcript is productive.	27
Nicastrin	Rat	•	•	•	•		72
U2AF35	Human	•	•	•	•	Mutually exclusive exons; PTC^+ isoforms are an apparent side effect.	73-77
MER2	Yeast	•	•	•	•	Splicing is regulated by MER1, which is produced only in meiotic cells. As a result, MER2 transcripts are productively spliced only during meiosis. In mitotic cells, a PTC^+ form is produced and degraded.	45,78
ClC-1	Human, Mouse	•	•	•	•		42,43
NDUFS4	Human	•		•	•		79
ITSN-1	Human, Mouse	•	•	•		Productive and unproductive isoforms are tissue-specific.	80

continued on next page

Table 1. Continued

Name	Organism	AS→PTC	AS Regulated	PTC→Low Abundance	NMD	Notes	Refs.
b) Regulated Unproductive Splicing							
LARD	Human	•	•			PTC^+ isoforms are abundant.	81
ARD-1, NIPP-1	Human	•		•		Translation reinitiation. ARD1 is downregulated by NMD, but is nonetheless expressed. Also, ARD1 and NIPP1 may influence splicing via PP1.	21
c) Autoregulatory Unproductive Splicing Not Affecting Splicing Factors							
RPL30	Yeast	•	•	•	•		44
RPL12	Worm	•	•	•	•		46
AtGRP7	Arabidopsis	•	•	•			55
d) Autoregulatory Unproductive Splicing Affecting Splicing Factors							
Sc35	Human	•	•	•	•		60
CLKs	Human, Mouse, Ciona	•	•	•	•		16,53, 54,82,83
TIAR /TIA-1	Human	•	•	•	•		84
PTB	Human	•	•	•	•		49,51,52
SRp20, SRp30b	Worm	•		•	•		85
AUF1	Human	•	•	•			86,87
TRA2-beta	Human	•	•			PTC^+ forms are abundant, but are not translated, perhaps due to sequestration.	88

Dots indicate direct experimental confirmation; lack of a dot means only that the experiment has not been performed to our knowledge. For instance, while it seems certain that ITSN-1 PTC^+ isoforms are degraded by NMD, this has not been directly observed using Upf1 knockdown or another NMD assay. The few cases where an experiment was performed but yielded a negative result are noted. AS, alternative splicing.

should be degraded according to the 50-nucleotide rule is in fact stable and translated to make protein. These include polycistronic transcripts on which translation is reinitiated downstream of a PTC;[21-23] apolipoprotein B, which is protected from NMD by an RNA editing complex;[24] some transcripts with a PTC near the initiation codon;[25] and an aberrant beta-globin transcript which is protected from NMD by an unknown mechanism.[26] Although NMD does not prevent protein production entirely in such cases, it may nonetheless limit expression from PTC⁺ transcripts substantially, as was shown for an alternative transcript of fumarylacetoacetate hydrolase (FAH) and for activator of RNA decay (ARD-1).[21,27]

Nonetheless, documented exceptions to the 50-nucleotide rule are rare, and there are many more known cases in which the rule is honored; indeed, the microarray results described above are consistent with the rule, as are diverse experiments on individual transcripts (Table 1).

Even for transcripts that are degraded by NMD, the possibility remains that the truncated protein products of the pioneer round of translation are functionally significant, since some regulatory proteins can have an effect even at very low copy number.[28] Also, to the extent that NMD is not completely effective at detecting and degrading PTC⁺ transcripts, the overlooked transcripts may be translated to produce truncated proteins. However, these proteins will frequently lack critical domains, rendering them inactive or even harmful. In any case, it is hard to imagine that functional roles of truncated proteins could explain the high prevalence of genes with PTC⁺ isoforms, especially given the wide functional diversity of those genes, and no data exist to support such a view.

While there may be exceptions, it seems unlikely that many PTC⁺ isoforms produce functional protein due to incomplete surveillance or by otherwise evading NMD.

Do the Observed PTC⁺ mRNA Isoforms Represent Missplicing or Cellular Noise?

NMD was originally described as a means of clearing erroneous transcripts from the cell.[9,10] In keeping with this role, some alternative splice forms that are degraded by NMD could represent splicing errors. Such errors could arise from mutations disrupting splice sites or regulatory sequences, including mutations in intronic regions that are invisible after intron removal. Also, the splicing machinery itself could recognize incorrect splice sites. The spliceosome distinguishes true splice sites from nearby cryptic sites with impressive fidelity, but splice site recognition is a complex process and errors must occur at some low rate. There are at present no clear data on the extent of missplicing. However, since EST libraries contain millions of transcript sequences, even extremely rare events may be represented.

In EST-based computational analyses, splicing errors can be distinguished from genuine alternative splicing to some extent by filtering out splicing events that are seen only in a few ESTs. However, filtering will certainly exclude some legitimate rare splice forms as well. With multiple mammalian genomes available, recent work has focused on evolutionary conservation to suggest purifying selection and, perhaps, functional roles for conserved alternative forms.[29]

Conservation of alternative splice forms between closely related organisms can be used to distinguish functional alternative splicing from probable splicing errors. Minor isoforms, i.e., those that occur only a fraction of the time, are less often conserved than major isoforms,[30] and they may sometimes represent recent mutations or splicing errors. Those minor isoforms that are conserved, including PTC⁺ isoforms, are more likely to be functional than minor isoforms that are seen only in one species.[31]

As described above, Baek and Green identified PTC⁺ alternative splice forms that were conserved between human and mouse, to filter out aberrant splicing. They noted that the inclusion of the same "accidental" alternative exon is unlikely to happen by chance in both species, because accidental recognition of the same position in two species—a position that is not under selective pressure to be recognized as a splice site—is unlikely. On the other hand, occasional accidental skipping of the same exon could easily be seen in both human and mouse; if the spliceosome were to miss bone fide splice sites at some low frequency, the same accident might be found by chance in homologous transcripts in two species. To reduce the influence of these conserved but aberrant splices on their data set, they designed a statistical method to discriminate between splicing errors and functional alternative splicing. Using this method, 80% of the conserved PTC-producing splice events were legitimate, compared to 20% that appeared aberrant.[17] Thus, most of the conserved PTC-producing splice events were not due to missplicing.

Are the Observed PTC⁺ mRNA Isoforms a Side Effect of Productive Alternative Splicing?

In the particular situation of mutually exclusive exon usage, NMD may be a mechanism for removing transcripts that erroneously include both exons or neither exon. In some instances there are physical constraints that force the splicing machinery to include one exon or the other but not both.[32] In other cases, including both exons or neither exon would introduce a frameshift resulting in a PTC, targeting the mRNA for degradation. In the case of fibroblast growth factor receptor (FGFR)2 RNA, an isoform including exon IIIb while skipping exon IIIc is productive; similarly, the isoform including exon IIIc but excluding exon IIIb is productive. However, the spliceosome may instead pair the same splice sites differently such that both exons are included, or such that neither is included. Each of these latter possibilities introduces a PTC (Fig. 4).

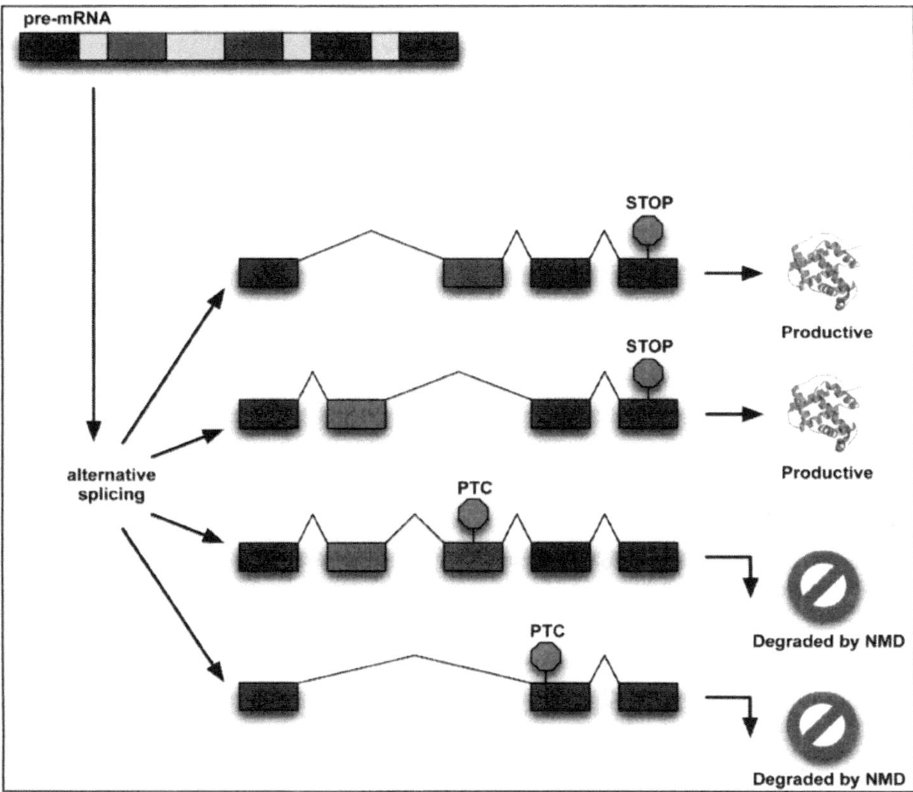

Figure 4. NMD can be employed to remove "side effect" isoforms in the case of mutually exclusive exons. Alternative splicing may generate two productive isoforms including one or the other of a pair of mutually exclusive exons. By choosing different pairings from the same set of 5′ and 3′ splice sites, the spliceosome may also generate isoforms that include both exons or neither exon. Frameshifts can give rise to PTCs in the undesired isoforms so that they will be targeted for NMD.

Each of the splice sites involved in the removal of exons IIIb and IIIc is required for the production of at least one productive isoform; the unproductive isoforms arise simply from alternate pairings of these otherwise productive splice sites. Given that the spliceosome is prone to such alternate pairings, there may be evolutionary pressure to ensure that the undesired isoforms include a PTC. This case differs from the quality control scenario described above, in that the degraded isoforms result not from random noise but as an inevitable side effect of the mechanism for productive alternative splicing. NMD is used as a filter to remove these "side-effect" isoforms, which may comprise a substantial fraction of the transcripts produced (up to 50% in the case of FGFR2 RNA).[33]

We examined the alternative isoforms inferred from human dbEST data (above), and found that PTC^+ isoforms could be explained as this kind of side effect for 34% of the genes that produce them. That is, 66% of the genes that generate a PTC^+ isoform have a splice site that is specific to PTC^+ isoforms and that is responsible for introducing the PTC (D.A.W. Soergel, unpublished data). If these unproductive isoforms were on the whole detrimental to the cell, then we would expect evolution to have eliminated the PTC^+-specific splice sites long ago; but in fact many of them are strikingly conserved, as we discuss below. Thus, while the contribution of "side-effect" isoforms may be significant, they alone cannot explain the high prevalence of PTC^+ isoforms.

Are the Observed PTC^+ mRNA Isoforms Part of a Mechanism for Regulating Gene Expression?

None of the phenomena discussed above—NMD evasion, noise, and "side-effect" splicing—are sufficient to explain the high prevalence of PTC^+ isoforms that are observed. A remaining explanation is that the cell commonly produces a substantial fraction of NMD-targeted isoforms in a functional or regulated manner. This process may provide an additional level of regulatory circuitry to help the cell achieve the proper level of expression for a given protein. The cell could change the level of productive mRNA after transcription by shunting some fraction of the already-transcribed pre-mRNA into an unproductive splice form and thence to the decay pathway (Fig. 1).

The literature contains numerous examples in which a regulatory process involving alternative splicing and NMD has been experimentally confirmed (Table 1), and many more examples in which experiments are consistent with and suggestive of this mode of regulation (Table 2). We propose that gene regulation through the coupled action of alternative splicing and NMD is widespread, and that this is a major explanation for the large number of observed PTC^+ isoforms. We have termed this process "regulated unproductive splicing and translation," or RUST.

RUST can be used to regulate protein levels, and the process is itself regulated by changes in the splicing environment. In the simplest case, some constant fraction of pre-mRNA transcribed from a given gene is spliced into an unproductive, NMD-targeted form. In other cases, the proportion of transcripts targeted for degradation is regulated by an external input. Finally, autoregulatory loops can arise in which a protein affects the splicing pattern of its own pre-mRNA.

Constitutive Unproductive Splicing

The simplest type of coupled alternative splicing and NMD is one in which the ratio of productive to unproductive splice forms is not significantly variable. In this case, the combined effect of alternative splicing with NMD reduces mRNA abundance by a more or less constant factor. An apparent example of this is provided by the Calpain-10 transcript, which encodes a ubiquitously expressed protease. This transcript is alternatively spliced to produce eight mRNA isoforms.[16,34] Of these, our analysis of SWISS-PROT and genomic sequences showed that four contain PTCs. An expression study by Horikawa et al showed that these very isoforms were "less abundant" in vivo than the other four.[34] An experiment in our lab later showed that the PTC^+ isoforms increased in abundance relative to the PTC^- isoforms when cells were treated with cycloheximide, which blocks translation and thereby inactivates the NMD pathway.[35] This result is consistent with the idea that all eight mRNA isoforms are produced but that the four PTC⁺ isoforms are degraded by NMD. Numerous transcripts in the literature are similarly processed (Table 1a). Of course, in each case, there may be as-yet-unknown regulatory inputs to splicing that do alter the isoform proportions.

Regulated Unproductive Splicing

Many examples of regulated alternative splicing leading to NMD are also known (Table 1b). In addition to changing the relative abundance of different functions, changes in the splicing environment may increase or decrease the production of functional isoforms relative to PTC⁺ isoforms that are degraded by NMD (Fig. 1).

The spliceosome recognizes a range of related sequence signals as 5' and 3' splice sites, with a range of "strengths" or binding affinities. Selection of splice sites is also under the control of a host of regulatory splicing factors, which bind to specific sequence signals on the pre-mRNA. These sequences may be exonic or intronic, and may be associated with enhancement or suppression of splicing at nearby (and sometimes at distant) splice sites. Exonic splicing enhancers (ESEs), intronic splicing enhancers (ISEs), exonic splicing silencers (ESSs), and intronic splicing silencers (ISSs) are frequently found in clusters, suggesting combinatorial regulation of splicing by complexes of splicing factors.[36,37]

Through the selection of alternative splice sites, splicing can give rise to a PTC in various ways. Inclusion of an alternative exon can introduce a PTC directly or shift the frame of the downstream exons to cause a downstream PTC. Similarly, exclusion of an exon can result in a frameshift and PTC. Finally, removal of an intron from the 3' untranslated region (UTR) may cause the normal stop codon to trigger NMD.

A change in the abundance of splicing factors can shift the balance of splicing patterns towards the production of NMD-targeted isoforms, thereby reducing the abundance of productive transcripts and hence the rate of protein production. In this way, splicing factors can alter gene expression, analogous to transcription factors.

This intriguing mechanism is used to regulate expression of MID1, a microtubule-associated protein involved in triggering degradation of phosphatase 2A.[38] This gene is ubiquitously transcribed, but its transcripts are spliced in a tissue- and development-specific manner. Winter and coworkers observed numerous alternatively spliced transcripts that included alternative exons, in addition to the nine previously known exons. Most of these novel exons contained in-frame stop codons. Some of these stop codons were followed by alternative poly(A) tails, allowing translation of a C-terminally truncated protein. A second class of alternative transcripts contained stop codons closely followed by an in-frame start codon, suggesting the possibility of translation reinitiation and production of N-terminally truncated protein. A third class of alternative transcripts contained PTCs that were associated neither with an alternate poly(A) signal nor with an alternate translation start site. These transcripts should be subject to NMD according to the 50-nucleotide rule. Consistent with this prediction, Winter et al found that the abundance of human MID1 transcripts including exon 1c (an alternative exon that introduces a PTC) increased in the presence of the NMD inhibitor cycloheximide.[38] Finally, Winter et al used RT-PCR to observe that different MID1 isoforms are produced in different tissues and at different developmental stages in both human and mouse. For instance, the PTC-introducing exon 1a was observed in five distinct transcripts in human fetal brain cells, two transcripts in fetal liver cells, and none in fetal fibroblasts. These results strongly suggest that alternative splicing and NMD are employed to regulate the overall abundance of productive MID1 transcripts.

Defects in regulating unproductive splicing can lead to disease. Myotonic dystrophy (DM), an autosomal dominant disease, is the most common form of adult-onset muscular dystrophy. DM has been shown to be caused by one of two repeat expansions, whose presence in mRNA sequesters several splicing factors[39] including muscleblind, and thus induces splicing

Table 2. Putative examples of unproductive splicing

Name	Organism	AS→PTC	AS Regulated	PTC→Low Abundance	NMD	Notes	Refs.
GFAP	Rat	•	•	•		Isoform proportions are tissue-regulated.	89
TR4	Human, Rat	•	•	•			90
HLA-G	Human	•	•	•		Exception to the 50-nucleotide rule; does not undergo NMD.	7,91
Alpha2(XI)	Human, Mouse	•	•	•			92
FIBP	*Drosophila*	•	•				93
Dystrophin from Purkinje promoter	Human	•	•			Regulated tissue-specific splicing in healthy individuals. The PTC+ isoform is abundant.	94
SRp20	Mouse	•	•			A truncated protein is observed.	95
mdm2	Human, Dog	•	•			Translation reinitiation.	22
SmHSF	*Schistosoma mansoni*	•	•			Developmental regulation. The protein product is a transcription factor controlling expression of heat shock proteins, including HSP70. Thus, at a developmental stage when splicing generates primarily unproductive SmHSF isoforms, HSP70 mRNAs are not observed.	96
TM	Rat		•			A PTC+ isoform is believed to exist.	49,52,97
MPZ	Human	•		•			98

continued on next page

Table 2. Continued

Name	Organism	AS→PTC	AS Regulated	PTC→Low Abundance	NMD	Notes	Refs.
NRAMP	Human	•		•		The PTC-containing exon is an Alu element.	99
ACF/ASP	Mouse, Rat	•		•			100
Scgb3a1/ Ugrp2	Mouse	•		•		Translation reinitiation is likely.	101
WDNM1	Rat	•		•			102
U1-70k snRNP	Human, *Arabidopsis*	•				Vertebrates share a conserved PTC-inducing included exon. In Arabidopsis, a different PTC-inducing exon is included.	103-105
Scorpion venom	Scorpion	•				Trans-splicing creates some PTC⁺ isoforms.	106
Cds1/CHK2	Yeast, Human	•					107
hMSH2	Human	•					108
ABCA13	Mouse	•					109
snRNP B and B′	Human	•				A truncated protein is observed.	110
RPL28, RPS17B	Yeast				•		45
RPL3, RPL7a, RPL10a	Worm				•		46

In these examples taken from the literature, available data are consistent with or suggestive of regulated unproductive splicing, but they do not yet conclusively demonstrate the phenomenon.

changes in several genes.[40,41] Patients develop myotonia from lack of muscle-specific chloride channel 1 (ClC-1), which is misspliced in DM tissue.[42] The normal developmental splice pattern for the ClC-1 transcript has a PTC in embryos but no PTC in adult cells. In DM tissue, ClC-1 transcript splicing reverts to its embryonic, PTC-containing splicing pattern. Consequently, ClC-1 mRNA is greatly reduced in abundance, likely due to the action of NMD.[43] Thus, it appears that normal ClC-1 gene expression is governed by RUST, and that the DM disease is caused when this regulation is undermined by sequestration of splicing factors.

Autoregulatory Unproductive Splicing

There is abundant evidence that RUST is used for autoregulation. The autoregulated gene often, but not always, encodes a protein that is part of the splicing machinery. In some fascinating cases, proteins that are not generally involved in mRNA processing bind specifically to their own transcripts to affect splicing and elicit NMD. The clearest example of this is found not in a human gene but in yeast. Yeast genes are generally unspliced, but in the few intron-containing genes, intron inclusion can introduce an in-frame stop codon and target the transcript for NMD. The yeast ribosomal protein RPL30 binds to its own pre-mRNA to prevent the removal of a PTC-containing intron, which in turn triggers NMD.[44] The mRNAs of other ribosomal protein genes, including RPL28 (CYH2) and RPS17B (RP51B), also sometimes retain their introns and become natural NMD targets, leaving open the possibility that their splicing is also regulated to elicit NMD.[45]

Some ribosomal proteins in *Caenorhabditis elegans* are similarly autoregulated. A screen for natural targets of NMD identified L3, L7a, L10a, and L12 ribosomal protein transcripts. Each of these transcripts can be alternatively spliced to generate either a productive isoform or an unproductive isoform that contains a PTC and is therefore degraded by NMD. The ratio of productive to unproductive alternative splicing of *rpl-12* RNA is affected by levels of RPL-12 protein, indicating that unproductive splicing of *rpl-12* RNA is under feedback control.[46]

A striking number of splicing factors and elements of the splicing machinery are autoregulated through RUST (Fig. 5 and Table 1c,d). One such example is polypyrimidine tract binding protein (PTB), which inhibits splicing by competing with small nuclear (sn)RNA U2 associated factor (U2AF) for the polypyrimidine tract and perhaps through other mechanisms (reviewed in refs. 36 and 47). PTB RNA is alternatively spliced to produce two major productive isoforms (one of which lacks exon 9),[48,49] one minor productive isoform lacking exons 3-9,[49,50] and two unproductive isoforms lacking exon 11. Removing exon 11 causes a frameshift leading to a downstream PTC. PTB protein has been found to promote the removal of exon 11 from its own transcripts.[49] Consequently, when PTB levels are high, PTB production is slowed by targeting PTB transcripts for NMD; and when PTB levels are low, production is accelerated by reducing the proportion of transcripts that are degraded.[49,51,52]

The CDC-like kinases (CLKs), which regulate an important family of splicing factors known as SR proteins, seem to be affected by RUST as well.[16] RUST appears to regulate CLK1 levels through an indirect feedback mechanism. CLK1 has been shown to modify splicing of its own transcript indirectly, most likely through phosphorylation of SR proteins.[53] Thus, as a variation of the autoregulatory circuit described above, increased CLK1 activity results in changes in the activity of one or more SR proteins. These SR proteins in turn affect the splicing of CLK1 pre-mRNA to favor a PTC$^+$ transcript that is predicted to undergo NMD. This PTC$^+$ transcript is stabilized by cycloheximide, consistent with its being normally degraded by NMD.[54]

RUST regulation of Clk1 levels may have a downstream effect on numerous SR proteins, and in turn on the splicing of many pre-mRNAs that are substrates of those SR proteins. Thus, alternative splicing can regulate factors that control splicing of other gene products.

Such a regulatory cascade of alternative splicing seems likely in the case of neuronal (n) PTB, a paralog of PTB that is regulated by RUST. There are reports that PTB affects the

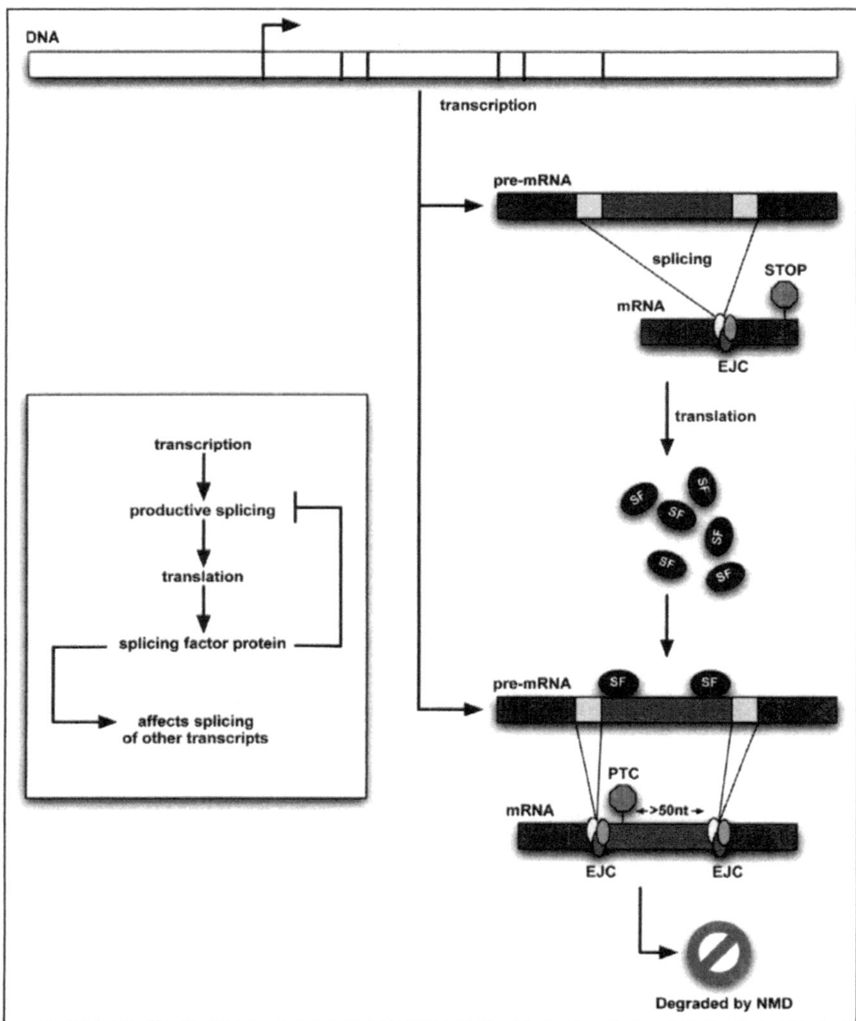

Figure 5. Autoregulatory unproductive splicing. Some splicing factors (SFs), such as PTB and SC35, regulate the splicing of their own transcripts so as to alter the proportion of unproductive isoforms.[49,60] This creates a negative feedback loop, stabilizing the concentration of the splicing factor over time. Autoregulated splicing factors are generally not specific to their own transcripts, since they also affect the splicing of many other pre-mRNAs.

splicing of nPTB RNA, which is unsurprising given that PTB regulates the splicing of its own transcript, and nPTB is a close homolog of PTB. Furthermore, nPTB transcripts are alternatively spliced in a tissue-specific manner: in nonneuronal cells, PTC$^+$ isoforms are preferentially produced but, in neurons, productive isoforms are made and translated. The resulting nPTB protein may compete with PTB for binding polypyrimidine tracts on heterologous pre-mRNAs, and may inhibit splicing more strongly or weakly than does PTB, resulting in an altered splicing pattern. Consequently, pre-mRNAs whose splicing pattern is sensitive to nPTB will be spliced differently in neurons than in other cell types.[51]

Transcripts encoding splicing factors that are autoregulated by RUST may also be subject to RUST that is triggered by heterologous factors; this is seen in the alternative splicing of PTB RNA, which can be affected by the splicing regulators raver1 and CELF4.[49]

Conservation of RUST

The coordinated use of alternative splicing and NMD is seen not only in mammals but in organisms as distant as yeast[44] and possibly plants.[55] The mechanism of PTC recognition differs in organisms other than mammals since it does not seem to depend on the location of the stop codon relative to exon junctions.[56] (see chapters by Amrani and Jacobson, Behm-Ansmant and Izaurralde, and van Hoof and Green) There have been significant advances recently in elucidating the recognition mechanism in flies and yeast,[57,58] but the rules are not clear enough to allow for computational identification of NMD targets. Nonetheless, NMD affects gene expression in a variety of different organisms.

In several of the examples discussed above, analysis of orthologous and paralogous sequences suggests that splicing to generate PTC⁺ alternative isoforms, and thus RUST regulation, is conserved across species and across protein families. For PTB transcripts, the sequence and upstream regulatory elements of alternatively included PTC-containing exon 11 are very similar in transcripts that encode the *Fugu rubripes* ortholog as well as the human neuronal-specific paralog nPTB.[49] Transcripts encoding mouse and monkey orthologs of the human multidrug resistance associated transporter ABCC4 share highly conserved PTC-containing exons that are orthologous to the alternatively included exons of the human ABCC4 transcript, another apparent RUST target.[59] Particularly strong evidence of conservation of RUST is found in the Clk transcripts. Alternative splicing to exclude exon 4, introducing a frameshift and PTC, is conserved among transcripts encoding the three human paralogs (Clk1, Clk2, Clk3), the three mouse orthologs, and even the sole ortholog in the sea squirt *Ciona intestinalis* (Fig. 6).[16]

The action of NMD on some transcripts of a gene can be conserved even when the specific alternative splicing events that elicit NMD are not. As discussed above, MID1 RNA is a human RUST target. Interestingly, while PTC⁺ MID1 isoforms are found in human, mouse, and fugu, the responsible stop codons are introduced by alternative exons that show no homology between these species.[38] Thus, in this case, it appears that the RUST mode of regulation was conserved while the specific sequence elements triggering it were not. This suggests that RUST is a generally useful mechanism that is easily applied to regulate expression of individual genes in organisms that already have both alternative splicing and NMD.

Why RUST?

A substantial portion of alternatively spliced mRNAs seem to be NMD targets. We have discussed possible explanations for the prevalence of unproductive splicing: do these splice forms represent biological noise, or are they produced to regulate protein expression? It is unlikely that all such splicing is used for regulation, but the growing body of examples presented above suggests that RUST plays a significant role in the cell.

Many truncated proteins encoded by alternative transcripts would be nonfunctional even if these transcripts were not removed by NMD. Is the combination of alternative splicing and NMD inherently different from alternative splicing that produces nonfunctional protein? Or does alternative splicing alone provide the important regulatory step, with NMD acting only as a convenient but unnecessary cleanup mechanism? Some proven cases of RUST illustrate that, in fact, the coordinated action of both pathways is required for regulation. The SR protein SC35 is autoregulated by RUST; its alternative splicing occurs in the 3' UTR to create an exon junction downstream of the original stop codon, without changing the open reading frame.[60] The alternative splicing seems to have no role other than to cause the original termination codon to elicit NMD. Without NMD, the alternative mRNA would still encode the correct protein, so the alternative splice event alone could not be used to regulate protein levels. It seems, then, that some genes have evolved to take advantage of the combination of alternative splicing and NMD in a role different from that filled by either process alone.

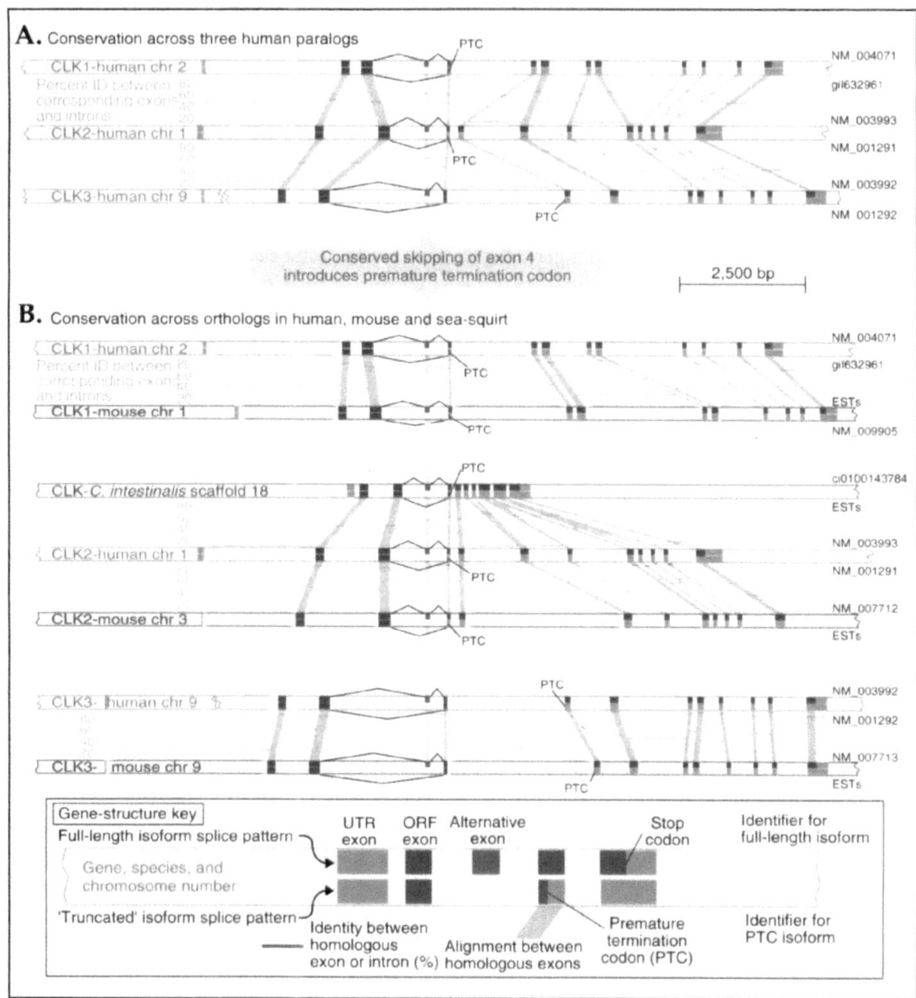

Figure 6. Splicing to generate a PTC is evolutionarily conserved in CLK transcripts. The CDC-like kinases (CLKs) are splicing regulators that affect splicing decisions through the phosphorylation of SR proteins. A) A screen of SWISS-PROT performed by Hillman et al revealed that human CLK1, CLK2 and CLK3 paralogs all generate PTC$^+$ alternative isoforms.[16] Conserved skipping of exon 4 causes a frameshift and results in a PTC. The percent identities from global alignments between corresponding exons and introns are shown in purple. B) CLK homologs were identified in mouse through existing annotation and in the predicted proteins of the sea squirt *C. intestinalis* using an hidden Markov model (HMM) constructed from sequences of annotated CLK transcripts from a variety of organisms. An EST analysis revealed that the alternative splicing pattern that generates PTC$^+$ alternative isoforms was conserved in all three sets of orthologs in human and mouse. The same splicing pattern was also found in the only identifiable *C. intestinalis* homolog. A relatively high degree of sequence similarity was found to be present in the introns flanking the alternative exon. ©2004 Hillman et al; license BioMed Central Ltd. This is an Open Access article: verbatim copying and redistribution of this article are permitted in all media for any purpose, provided this notice is preserved along with the article's original URL (http://genomebiology.com/2004/5/2/R8).[16]

RUST seems, at first, to be a wasteful process. A gene is transcribed and spliced, only to be degraded before it can produce a protein. Yet, we know that there are functional cases of RUST. The cost to the cell of transcribing apparently extraneous RNA is clearly not prohibitive. In humans, roughly 85-95% of transcribed sequence is spliced out as introns and discarded.[61] Evidently, transcription of intronic sequences is not a significant selective disadvantage, and intron splicing may even provide some general selective advantage. Similarly, the cost of transcribing a pre-mRNA only to splice it into an unproductive form must be balanced by the advantages of an additional layer of regulation of gene expression.

How is a process like RUST beneficial to the cell? Transcriptional regulation is the most-studied means of controlling gene expression, but in some cases, additional control may be beneficial. Because splicing regulation occurs after the decision to transcribe a region, RUST may provide a rapid way to change the levels of productive mRNA. In extreme cases such as the dystrophin gene, the synthesis of a single transcript can take many hours,[62] and the requirements of the cell might change after transcription begins but before a critical splicing decision that determines whether or not to introduce a PTC. Even when temporal regulation is not necessary, an extra layer of regulation can help fine-tune gene expression.

RUST is distinctive, in that it can either increase or decrease protein expression. The splicing factor PTB illustrates this point. At steady state, 20% of PTB pre-mRNA is spliced to an unproductive form.[49] In general, we expect that a RUST-regulated gene is transcribed to produce more pre-mRNA than is needed at steady state, and that under normal conditions there is a base level of downregulation by unproductive splicing. This fraction of "wasted" transcripts constitutes the headroom available to the regulatory system to increase levels of productive transcript.

The prevalence of PTC^+ alternative splice forms suggests a possible evolutionary interaction between alternative splicing and NMD.[63] The existence of NMD could have led to an increase in alternative splicing. Any splicing errors that introduced PTCs would be removed by NMD, reducing the harmful effects of missplicing. As a result, the pressure to recognize splice sites perfectly would be lowered. Functional alternative splice forms could arise through splicing errors and then become established by sequence changes that strengthen their splice sites or add regulatory elements.

In a system with prevalent alternative splicing, regulation by RUST may evolve easily. For any particular gene, there are many possible alternative splicing events that could elicit NMD, including exon skipping, splicing within the 3' UTR, or recognition of cryptic splice sites. If the sequence of a gene changes slightly to promote one of these splicing events under certain splicing environments, and the resulting downregulation of gene expression by NMD is beneficial, then a basic sort of regulation has evolved. This has clearly occurred independently many times. Without NMD, alternative splicing can still regulate gene expression by producing nonfunctional proteins. The additional advantages of coupling splicing with NMD may be that it prevents accumulation of potentially harmful truncated proteins and that it reduces wasted translation, making unproductive splicing less costly.

Splicing factors such as PTB seem to be overrepresented among the known RUST targets. Is this a coincidence, or is RUST in fact used most often to regulate a small set of proteins that are already capable of binding pre-mRNAs? A protein that has an existing role in splicing may evolve autoregulation through splicing more easily than a protein that does not bind RNA. There are only a handful of known cases in which a protein that is not a splicing factor is autoregulated by RUST, and even these are predominantly ribosomal proteins that do bind RNA in other, nonsplicing contexts. However, autoregulation is by no means the only role of RUST, and there is no reason for nonautoregulatory RUST to affect splicing factors preferentially. The examples listed in Table 1 indicate that RUST is involved in the regulation of a diverse set of proteins.

The potential for alternative splicing to regulate gene expression has been appreciated for many years. Bingham et al proposed that "on/off regulation at the level of splicing might be unexpectedly common," in a 1988 review featuring three cases of unproductive splicing in *Drosophila melanogaster*.[64] An early paper about the splicing factor ASF discussed alternative splicing as a means to control gene expression.[65] NMD adds an additional layer to the story;[66] many of the unproductive splice forms identified years ago are now known to be degraded rather than translated. The prevalence of NMD-targeted splice forms has only recently become clear. Alternative splicing and NMD are often combined in an elegant way to regulate the expression of a wide range of genes. RUST seems to be a generally applicable, widespread, and readily evolved regulatory mechanism.

Acknowledgements

We would like to thank Richard E. Green and Rajiv Bhatnagar for helpful discussions and assistance with a few sections of the text. This work was supported by National Institutes of Health grants K22 HG00056 and R01 GM071655, and an IBM SUR grant. DAWS is supported by a predoctoral fellowship from the Howard Hughes Medical Institute. LFL is supported by NIH training grant T32 GM07127l. SEB is a Searle Scholar (I-L-110).

References

1. Boue S, Letunic I, Bork P. Alternative splicing and evolution. Bioessays 2003; 25:1031-4.
2. Modrek B, Lee C. A genomic view of alternative splicing. Nat Genet 2002; 30:13-9.
3. Maniatis T, Tasic B. Alternative pre-mRNA splicing and proteome expansion in metazoans. Nature 2002; 418:236-43.
4. Garcia J, Gerber SH, Sugita S et al. A conformational switch in the Piccolo C2A domain regulated by alternative splicing. Nat Struct Mol Biol 2004; 11:45-53.
5. Resch A, Xing Y, Modrek B et al. Assessing the impact of alternative splicing on domain interactions in the human proteome. J Proteome Res 2004; 3:76-83.
6. Xing Y, Xu Q, Lee C. Widespread production of novel soluble protein isoforms by alternative splicing removal of transmembrane anchoring domains. FEBS Lett 2003; 555:572-8.
7. Nagy E, Maquat LE. A rule for termination-codon position within intron-containing genes: When nonsense affects RNA abundance Trends Biochem Sci 1998; 23:198-9.
8. Lejeune F, Maquat LE. Mechanistic links between nonsense-mediated mRNA decay and pre-mRNA splicing in mammalian cells. Curr Opin Cell Biol 2005; 17:309-15.
9. Cali BM, Anderson P. mRNA surveillance mitigates genetic dominance in Caenorhabditis elegans. Mol Gen Genet 1998; 260:176-84.
10. Maquat LE, Carmichael GG. Quality control of mRNA function. Cell 2001; 104:173-6.
11. Mendell JT, Sharifi NA, Meyers JL et al. Nonsense surveillance regulates expression of diverse classes of mammalian transcripts and mutes genomic noise. Nat Genet 2004; 36:1073-8.
12. Lewis BP, Green RE, Brenner SE. Evidence for the widespread coupling of alternative splicing and nonsense-mediated mRNA decay in humans. Proc Natl Acad Sci USA 2003; 100:189-92.
13. Pruitt KD, Maglott DR. RefSeq and LocusLink: NCBI gene-centered resources. Nucleic Acids Res 2001; 29:137-40.
14. Boguski MS, Lowe TM, Tolstoshev CM. dbEST—database for "expressed sequence tags". Nat Genet 1993; 4:332-3.
15. Boeckmann B, Bairoch A, Apweiler R et al. The SWISS-PROT protein knowledgebase and its supplement TrEMBL in 2003. Nucleic Acids Res 2003; 31:365-70.
16. Hillman RT, Green RE, Brenner SE. An unappreciated role for RNA surveillance. Genome Biol 2004; 5:R8.
17. Baek D, Green P. Nonsense-mediated decay, sequence conservation, and relative isoform frequencies in evolutionarily conserved alternative splicing. Proc Natl Acad Sci USA 2005.
18. Ishigaki Y, Li X, Serin G et al. Evidence for a pioneer round of mRNA translation: mRNAs subject to nonsense-mediated decay in mammalian cells are bound by CBP80 and CBP20. Cell 2001; 106:607-17.
19. Maquat LE. Nonsense-mediated mRNA decay: Splicing, translation and mRNP dynamics. Nat Rev Mol Cell Biol 2004; 5:89-99.
20. Cao D, Parker R. Computational modeling and experimental analysis of nonsense-mediated decay in yeast. Cell 2003; 113:533-45.
21. Chang AC, Sohlberg B, Trinkle-Mulcahy L et al. Alternative splicing regulates the production of ARD-1 endoribonuclease and NIPP-1, an inhibitor of protein phosphatase-1, as isoforms encoded by the same gene. Gene 1999; 240:45-55.
22. Veldhoen N, Metcalfe S, Milner J. A novel exon within the mdm2 gene modulates translation initiation in vitro and disrupts the p53-binding domain of mdm2 protein. Oncogene 1999; 18:7026-33.
23. Zhang J, Maquat LE. Evidence that translation reinitiation abrogates nonsense-mediated mRNA decay in mammalian cells. EMBO J 1997; 16:826-33.
24. Chester A, Somasekaram A, Tzimina M et al. The apolipoprotein B mRNA editing complex performs a multifunctional cycle and suppresses nonsense-mediated decay. EMBO J 2003; 22:3971-82.

25. Inácio A, Silva AL, Pinto J et al. Nonsense mutations in close proximity to the initiation codon fail to trigger full nonsense-mediated mRNA decay. J Biol Chem 2004; 279:32170-80.
26. Danckwardt S, Neu-Yilik G, Thermann R et al. Abnormally spliced beta-globin mRNAs: A single point mutation generates transcripts sensitive and insensitive to nonsense-mediated mRNA decay. Blood 2002; 99:1811-6.
27. Dreumont N, Maresca A, Boisclair-Lachance JF et al. A minor alternative transcript of the fumarylacetoacetate hydrolase gene produces a protein despite being likely subjected to nonsense-mediated mRNA decay. BMC Mol Biol 2005; 6:1.
28. McAdams HH, Arkin A. It's a noisy business! Genetic regulation at the nanomolar scale. Trends Genet 1999; 15:65-9.
29. Lareau LF, Green RE, Bhatnagar RS et al. The evolving roles of alternative splicing. Curr Opin Struct Biol 2004; 14:273-82.
30. Modrek B, Lee CJ. Alternative splicing in the human, mouse and rat genomes is associated with an increased frequency of exon creation and/or loss. Nat Genet 2003; 34:177-80.
31. Kan Z, States D, Gish W. Selecting for functional alternative splices in ESTs. Genome Res 2002; 12:1837-45.
32. Letunic I, Copley RR, Bork P. Common exon duplication in animals and its role in alternative splicing. Hum Mol Genet 2002; 11:1561-7.
33. Jones RB, Wang F, Luo Y et al. The nonsense-mediated decay pathway and mutually exclusive expression of alternatively spliced FGFR2IIIb and -IIIc mRNAs. J Biol Chem 2001; 276:4158-67.
34. Horikawa Y, Oda N, Cox NJ et al. Genetic variation in the gene encoding calpain-10 is associated with type 2 diabetes mellitus. Nat Genet 2000; 26:163-75.
35. Green RE, Lewis BP, Hillman RT et al. Widespread predicted nonsense-mediated mRNA decay of alternatively-spliced transcripts of human normal and disease genes. Bioinformatics 2003; 19(Suppl 1):i118-21.
36. Black DL. Mechanisms of alternative premessenger RNA splicing. Annu Rev Biochem 2003; 72:291-336.
37. Wagner EJ, Baraniak AP, Sessions OM et al. Characterization of the intronic splicing silencers flanking FGFR2 exon IIIb. J Biol Chem 2005.
38. Winter J, Lehmann T, Krauss S et al. Regulation of the MID1 protein function is fine-tuned by a complex pattern of alternative splicing. Hum Genet 2004; 114:541-52.
39. Philips AV, Timchenko LT, Cooper TA. Disruption of splicing regulated by a CUG-binding protein in myotonic dystrophy. Science 1998; 280:737-41.
40. Brook JD, McCurrach ME, Harley HG et al. Molecular basis of myotonic dystrophy: Expansion of a trinucleotide (CTG) repeat at the 3' end of a transcript encoding a protein kinase family member. Cell 1992; 68:799-808.
41. Liquori CL, Ricker K, Moseley ML et al. Myotonic dystrophy type 2 caused by a CCTG expansion in intron 1 of ZNF9. Science 2001; 293:864-7.
42. Charlet-B N, Savkur RS, Singh G et al. Loss of the muscle-specific chloride channel in type 1 myotonic dystrophy due to misregulated alternative splicing. Mol Cell 2002; 10:45-53.
43. Mankodi A, Takahashi MP, Jiang H et al. Expanded CUG repeats trigger aberrant splicing of ClC-1 chloride channel pre-mRNA and hyperexcitability of skeletal muscle in myotonic dystrophy. Mol Cell 2002; 10:35-44.
44. Vilardell J, Chartrand P, Singer RH et al. The odyssey of a regulated transcript. RNA 2000; 6:1773-80.
45. He F, Peltz SW, Donahue JL et al. Stabilization and ribosome association of unspliced pre-mRNAs in a yeast upf1- mutant. Proc Natl Acad Sci USA 1993; 90:7034-8.
46. Mitrovich QM, Anderson P. Unproductively spliced ribosomal protein mRNAs are natural targets of mRNA surveillance in C. elegans. Genes Dev 2000; 14:2173-84.
47. Valcárcel J, Gebauer F. Post-transcriptional regulation: The dawn of PTB. Curr Biol 1997; 7:R705-8.
48. Ghetti A, Piñol-Roma S, Michael WM et al. hnRNP I, the polypyrimidine tract-binding protein: Distinct nuclear localization and association with hnRNAs. Nucleic Acids Res 1992; 20:3671-8.
49. Wollerton MC, Gooding C, Wagner EJ et al. Autoregulation of polypyrimidine tract binding protein by alternative splicing leading to nonsense-mediated decay. Mol Cell 2004; 13:91-100.
50. Hamilton BJ, Genin A, Cron RQ et al. Delineation of a novel pathway that regulates CD154 (CD40 ligand) expression. Mol Cell Biol 2003; 23:510-25.
51. Rahman L, Bliskovski V, Reinhold W et al. Alternative splicing of brain-specific PTB defines a tissue-specific isoform pattern that predicts distinct functional roles. Genomics 2002; 80:245-9.
52. Spellman R, Rideau A, Matlin A et al. Regulation of alternative splicing by PTB and associated factors. Biochem Soc Trans 2005; 33:457-60.
53. Duncan PI, Stojdl DF, Marius RM et al. In vivo regulation of alternative pre-mRNA splicing by the Clk1 protein kinase. Mol Cell Biol 1997; 17:5996-6001.
54. Menegay HJ, Myers MP, Moeslein FM et al. Biochemical characterization and localization of the dual specificity kinase CLK1. J Cell Sci 2000; 113(Pt 18):3241-53.

55. Staiger D, Zecca L, Wieczorek Kirk DA et al. The circadian clock regulated RNA-binding protein AtGRP7 autoregulates its expression by influencing alternative splicing of its own pre-mRNA. Plant J 2003; 33:361-71.
56. Conti E, Izaurralde E. Nonsense-mediated mRNA decay: Molecular insights and mechanistic variations across species. Curr Opin Cell Biol 2005; 17:316-25.
57. Amrani N, Ganesan R, Kervestin S et al. A faux 3'-UTR promotes aberrant termination and triggers nonsense-mediated mRNA decay. Nature 2004; 432:112-8.
58. Gatfield D, Unterholzner L, Ciccarelli FD et al. Nonsense-mediated mRNA decay in Drosophila: At the intersection of the yeast and mammalian pathways. EMBO J 2003; 22:3960-70.
59. Lamba JK, Adachi M, Sun D et al. Nonsense mediated decay downregulates conserved alternatively spliced ABCC4 transcripts bearing nonsense codons. Hum Mol Genet 2003; 12:99-109.
60. Sureau A, Gattoni R, Dooghe Y et al. SC35 autoregulates its expression by promoting splicing events that destabilize its mRNAs. EMBO J 2001; 20:1785-96.
61. Lander ES, Linton LM, Birren B et al. Initial sequencing and analysis of the human genome. Nature 2001; 409:860-921.
62. Tennyson CN, Klamut HJ, Worton RG. The human dystrophin gene requires 16 hours to be transcribed and is cotranscriptionally spliced. Nat Genet 1995; 9:184-90.
63. Xing Y, Lee CJ. Negative selection pressure against premature protein truncation is reduced by alternative splicing and diploidy. Trends Genet 2004; 20:472-5.
64. Bingham PM, Chou TB, Mims I et al. On/off regulation of gene expression at the level of splicing. Trends Genet 1988; 4:134-8.
65. Ge H, Zuo P, Manley JL. Primary structure of the human splicing factor ASF reveals similarities with Drosophila regulators. Cell 1991; 66:373-82.
66. Hilleren P, Parker R. Mechanisms of mRNA surveillance in eukaryotes. Annu Rev Genet 1999; 33:229-60.
67. Hovhannisyan RH, Carstens RP. A novel intronic cis element, ISE/ISS-3, regulates rat fibroblast growth factor receptor 2 splicing through activation of an upstream exon and repression of a downstream exon containing a noncanonical branch point sequence. Mol Cell Biol 2005; 25:250-63.
68. Carter MS, Li S, Wilkinson MF. A splicing-dependent regulatory mechanism that detects translation signals. EMBO J 1996; 15:5965-75.
69. Gudikote JP, Wilkinson MF. T-cell receptor sequences that elicit strong down-regulation of premature termination codon-bearing transcripts. EMBO J 2002; 21:125-34.
70. Wang J, Vock VM, Li S et al. A quality control pathway that down-regulates aberrant T-cell receptor (TCR) transcripts by a mechanism requiring UPF2 and translation. J Biol Chem 2002; 277:18489-93.
71. Skandalis A, Uribe E. A survey of splice variants of the human hypoxanthine phosphoribosyl transferase and DNA polymerase beta genes: Products of alternative or aberrant splicing? Nucleic Acids Res 2004; 32:6557-64.
72. Confaloni A, Crestini A, Albani D et al. Rat nicastrin gene: cDNA isolation, mRNA variants and expression pattern analysis. Brain Res Mol Brain Res 2005; 136:12-22.
73. Blanchette M, Labourier E, Green RE et al. Genome-wide analysis reveals an unexpected function for the drosophila splicing factor U2AF(50) in the nuclear export of intronless mRNAs. Mol Cell 2004; 14:775-86.
74. Cazalla D, Newton K, Cáceres JF. A novel SR-related protein is required for the second step of pre-mRNA splicing. Mol Cell Biol 2005; 25:2969-80.
75. Henscheid KL, Shin DS, Cary SC et al. The splicing factor U2AF65 is functionally conserved in the thermotolerant deep-sea worm Alvinella pompejana. Biochim Biophys Acta 2005.
76. Lallena MJ, Chalmers KJ, Llamazares S et al. Splicing regulation at the second catalytic step by Sex-lethal involves 3'splice site recognition by SPF45. Cell 2002; 109:285-96.
77. Pacheco TR, Gomes AQ, Barbosa-Morais NL et al. Diversity of vertebrate splicing factor U2AF35: Identification of alternatively spliced U2AF1 mRNAS. J Biol Chem 2004; 279:27039-49.
78. Engebrecht JA, Voelkel-Meiman K, Roeder GS. Meiosis-specific RNA splicing in yeast. Cell 1991; 66:1257-68.
79. Petruzzella V, Panelli D, Torraco A et al. Mutations in the NDUFS4 gene of mitochondrial complex I alter stability of the splice variants. FEBS Lett 2005.
80. Tsyba L, Skrypkina I, Rynditch A et al. Alternative splicing of mammalian Intersectin 1: Domain associations and tissue specificities. Genomics 2004; 84:106-13.
81. Screaton GR, Xu XN, Olsen AL et al. LARD: A new lymphoid-specific death domain containing receptor regulated by alternative pre-mRNA splicing. Proc Natl Acad Sci USA 1997; 94:4615-9.
82. Colwill K, Pawson T, Andrews B et al. The Clk/Sty protein kinase phosphorylates SR splicing factors and regulates their intranuclear distribution. EMBO J 1996; 15:265-75.
83. Duncan PI, Stojdl DF, Marius RM et al. The Clk2 and Clk3 dual-specificity protein kinases regulate the intranuclear distribution of SR proteins and influence pre-mRNA splicing. Exp Cell Res 1998; 241:300-8.

84. Le Guiner C, Gesnel MC, Breathnach R. TIA-1 or TIAR is required for DT40 cell viability. J Biol Chem 2003; 278:10465-76.
85. Morrison M, Harris KS, Roth MB. Smg mutants affect the expression of alternatively spliced SR protein mRNAs in Caenorhabditis elegans. Proc Natl Acad Sci USA 1997; 94:9782-5.
86. Wilson GM, Sun Y, Sellers J et al. Regulation of AUF1 expression via conserved alternatively spliced elements in the 3' untranslated region. Mol Cell Biol 1999; 19:4056-64.
87. Wilson GM, Brewer G. The search for trans-acting factors controlling messenger RNA decay. Prog Nucleic Acid Res Mol Biol 1999; 62:257-91.
88. Stoilov P, Daoud R, Nayler O et al. Human tra2-beta1 autoregulates its protein concentration by influencing alternative splicing of its pre-mRNA. Hum Mol Genet 2004; 13:509-24.
89. Condorelli DF, Nicoletti VG, Barresi V et al. Structural features of the rat GFAP gene and identification of a novel alternative transcript. J Neurosci Res 1999; 56:219-28.
90. Yoshikawa T, Makino S, Gao XM et al. Splice variants of rat TR4 orphan receptor: Differential expression of novel sequences in the 5'-untranslated region and C-terminal domain. Endocrinology 1996; 137:1562-71.
91. Moreau P, Carosella E, Teyssier M et al. Soluble HLA-G molecule. An alternatively spliced HLA-G mRNA form candidate to encode it in peripheral blood mononuclear cells and human trophoblasts. Hum Immunol 1995; 43:231-6.
92. Lui VC, Ng LJ, Sat EW et al. Extensive alternative splicing within the amino-propeptide coding domain of alpha2(XI) procollagen mRNAs. Expression of transcripts encoding truncated pro-alpha chains. J Biol Chem 1996; 271:16945-51.
93. Kolpakova E, Rusten TE, Olsnes S. Characterization and tissue expression of acidic fibroblast growth factor binding protein homologue in Drosophila melanogaster. Gene 2003; 310:185-91.
94. Holder E, Maeda M, Bies RD. Expression and regulation of the dystrophin Purkinje promoter in human skeletal muscle, heart, and brain. Hum Genet 1996; 97:232-9.
95. Jumaa H, Nielsen PJ. The splicing factor SRp20 modifies splicing of its own mRNA and ASF/SF2 antagonizes this regulation. EMBO J 1997; 16:5077-85.
96. Ram D, Ziv E, Lantner F et al. Stage-specific alternative splicing of the heat-shock transcription factor during the life-cycle of Schistosoma mansoni. Parasitology 2004; 129:587-96.
97. Wollerton MC, Gooding C, Robinson F et al. Differential alternative splicing activity of isoforms of polypyrimidine tract binding protein (PTB). RNA 2001; 7:819-32.
98. Besançon R, Prost AL, Konecny L et al. Alternative exon 3 splicing of the human major protein zero gene in white blood cells and peripheral nerve tissue. FEBS Lett 1999; 457:339-42.
99. Cellier M, Govoni G, Vidal S et al. Human natural resistance-associated macrophage protein: cDNA cloning, chromosomal mapping, genomic organization, and tissue-specific expression. J Exp Med 1994; 180:1741-52.
100. Dür S, Krause K, Pluntke N et al. Gene structure and expression of the mouse APOBEC-1 complementation factor: Multiple transcriptional initiation sites and a spliced variant with a premature stop translation codon. Biochim Biophys Acta 2004; 1680:11-23.
101. Niimi T, Copeland NG, Gilbert DJ et al. Cloning, expression, and chromosomal localization of the mouse gene (Scgb3a1, alias Ugrp2) that encodes a member of the novel uteroglobin-related protein gene family. Cytogenet Genome Res 2002; 97:120-7.
102. Dear TN, Kefford RF. The WDNM1 gene product is a novel member of the 'four-disulphide core' family of proteins. Biochem Biophys Res Commun 1991; 176:247-54.
103. Golovkin M, Reddy AS. Structure and expression of a plant U1 snRNP 70K gene: Alternative splicing of U1 snRNP 70K pre-mRNAs produces two different transcripts. Plant Cell 1996; 8:1421-35.
104. Hornig H, Fischer U, Costas M et al. Analysis of genomic clones of the murine U1RNA-associated 70-kDa protein reveals a high evolutionary conservation of the protein between human and mouse. Eur J Biochem 1989; 182:45-50.
105. Spritz RA, Strunk K, Surowy CS et al. Human U1-70K ribonucleoprotein antigen gene: Organization, nucleotide sequence, and mapping to locus 19q13.3. Genomics 1990; 8:371-9.
106. Zhu S, Li W, Cao Z. A naturally occurring noncoding fusion transcript derived from scorpion venom gland: Implication for the regulation of scorpion toxin gene expression FEBS Lett 2001; 508:241-4.
107. Lemaire M, Prime J, Ducommun B et al. Evolutionary conservation of a novel splice variant of the Cds1/CHK2 checkpoint kinase restricted to its regulatory domain. Cell Cycle 2004; 3:1267-70.
108. Marshall B, Isidro G, Boavida MG. Naturally occurring splicing variants of the hMSH2 gene containing nonsense codons identify possible mRNA instability motifs within the gene coding region. Biochim Biophys Acta 1996; 1308:88-92.
109. Barros SA, Tennant RW, Cannon RE. Molecular structure and characterization of a novel murine ABC transporter, Abca13. Gene 2003; 307:191-200.
110. van Dam A, Winkel I, Zijlstra-Baalbergen J et al. Cloned human snRNP proteins B and B' differ only in their carboxy-terminal part. EMBO J 1989; 8:3853-60.

Chapter 15

NMD and the Evolution of Eukaryotic Gene Structure

Michael Lynch,* Xin Hong, and Douglas G. Scofield

Abstract

All cells are confronted with undesirable transcripts derived from mutant alleles, but the production of aberrant transcripts from otherwise normal DNA may be an even greater challenge. The substantial fraction of problematical transcripts containing premature termination codons (PTCs) are subject to elimination by the nonsense-mediated mRNA decay (NMD) pathway. Phylogenetic analysis suggests that NMD and the exon junction complex (EJC) upon which it depends in mammals (see chapter by Maquat) are ancient, raising the possibility of an early association of NMD with introns. This may help explain why introns were able to proliferate to an apparently considerable degree in the stem eukaryote, despite the mutational burden that introns impose upon their host genes. A long-term evolutionary association between NMD and introns also provides a possible explanation for the nonrandom spatial distribution of introns in the genes of multicellular species and for an apparent slow-down (and possible stabilization) of intron colonization in modern species. Several lineages, all of which are nearly devoid of ancestral introns, appear to have lost NMD and the EJC, and these taxa have exceptionally simple genomic features that minimize the chances of producing erroneous transcripts. Validation of these ideas will require empirical work on the degree of coordination between NMD, the EJC, and the locations of introns in a wide array of genes distributed across diverse phylogenetic lineages.

Introduction

The manufacture of gene products that promote survival and reproduction is confronted by challenges at three levels—replication, transcription, and translation, the first two being the focus of this paper. Mutations at the DNA level introduce deleterious alleles into populations at a level that induces an average fitness loss of ~0.1 to 1.0% per generation in eukaryotes.[1] Although natural selection can eventually purge most such mutations from populations of moderate to large size, the residence time of such alleles can still be on the order of dozens to hundreds of generations depending on the selective disadvantage of heterozygous carriers, and as a consequence, virtually every organism is a carrier of multiple defective genes. Transcriptional errors compound the problem by resulting in inaccurate messenger RNAs (mRNAs).[2] Although the residence times of erroneous transcripts will generally be no longer than one or two life spans (the latter limit possibly arising in some cases of maternal inheritance), most genes produce multiple transcripts during the life of a cell, so the cumulative effects of such errors could be considerable. Erroneous transcripts of DNA-processing genes can also leave permanent genomic effects by altering the mutation rate.[3]

*Corresponding Author: Michael Lynch—Department of Biology, Indiana University, Bloomington, Indiana 47405, U.S.A. Email: milynch@indiana.edu

Nonsense-Mediated mRNA Decay, edited by Lynne E. Maquat. ©2006 Eurekah.com.

These simple observations make it clear that all organisms could profit from surveillance mechanisms to eliminate unproductive or harmful transcripts. The challenges are particularly great in eukaryotes, for which the greater complexities of gene structure provide more opportunities for error propagation than experienced by prokaryotes. However, because information virtually always flows from mRNA to protein, and not vice versa, any mechanism to detect inappropriate transcripts must operate at the mRNA level prior to (or coincident with) translation, and this limits the options for mRNA surveillance. The only erroneous transcripts that eukaryotes appear to be capable of eliminating on the basis of protein-coding potential are those containing PTCs. Of course, PTC-containing transcripts are not a trivial subset of aberrant mRNAs, as they can arise by a variety of mechanisms, can serve as flags for other transcript problems, and in many cases can deleteriously affect fitness via their production of truncated proteins.

The purpose of this chapter is to outline the selective context in which the main mechanism for dealing with PTC-containing mRNAs, the NMD pathway, is likely to have evolved. Because NMD relies on indicator sequences encoded within genes, consideration will also be given to the reciprocal pressures that NMD imposes on genome evolution. Although our understanding of the molecular mechanisms underlying NMD is still at a rudimentary stage and quite restricted phylogenetically, enough comparative information exists to allow some reasonable speculation as to when and why NMD evolved and how it has subsequently fared in various phylogenetic lineages. These observations may provide a useful framework for the future exploration of the molecular biology and organismal consequences of NMD.

Sources of Inappropriate Transcripts

As noted above, PTC-bearing transcripts result from two sources: accurate transcription from defective alleles, and inaccurate transcription from appropriate alleles. Loss-of-function mutations can be broadly classified into amino-acid replacement (missense) base substitutions, chain-terminating (nonsense) base substitutions, insertions and deletions, and splice-site altering base substitutions. With few exceptions, the last two classes will result in downstream frameshifts, so nearly all except missense mutations will generally give rise to a PTC-bearing allele. As can be seen in Table 1, an average 81% of newly arisen defective alleles are expected to harbor a PTC in eukaryotes, whereas the average fraction is ~31% in prokaryotes. Thus, although NMD is incapable of silencing all deleterious alleles, it does address a substantial fraction of those that are expected to produce an entirely defective protein. Less clear is whether the main function of NMD is to silence defective genes or to screen for recurrent transcriptional errors.

Numerous lines of evidence suggest that eukaryotic cells are constantly confronted with aberrant transcripts derived from otherwise good genes. First, although only limited information is available on the fidelity of mRNA chain elongation, the rate of base misincorporation appears to be on the order of 10^{-5} per nucleotide.[2-4] With ~3/64ths of random codons denoting stop and 10^3 to 10^4 coding nucleotides in a typical gene, 0.05% to 0.5% of transcripts are expected to acquire a PTC by this route, and slippage at mono- or dinucleotide runs could elevate these levels substantially by inducing frameshifts.

Second, because transcription-factor binding sites and core promoters are generally very simple, extending over just five to ten base pairs each, the transcriptional apparatus can often be recruited to false-positive sites. The situation is most dramatic in humans, where ~25% of observed cDNAs contain no obvious open reading frames, many of which are derived from AT-rich genomic regions harboring spurious TATA sequences,[5] and numerous mRNAs are from antisense strands of DNA.[6] In principle, such transcripts might have a regulatory function, but the fact that nearly ten times more human genomic DNA is transcribed than can be accounted for by known exons suggests otherwise.[7] Despite their accumulation of nonsense codons, pseudogenes can also continue to be expressed as long as their regulatory sequences remain intact, as can ancient defective mobile elements.[8,9]

Table 1. Spectrum of mutations causing defective alleles

Species	Locus	Frequency				n	Reference
		ID	CT	MS	SL		
Escherichia coli	LacI	0.885	0.034	0.081		174	96
	TonB	0.682	0.133	0.185		255	103,104
Salmonella typhimurium	His	0.327	0.135	0.538		226	100
Sulfolobus acidocaldarius	PyrE	0.882	0.000	0.118		101	98
Saccharomyces cerevisiae	URA3	0.160	0.368	0.472		106	96
Homo sapiens	Factor IX	0.078	0.120	0.657	0.145	399	97
	HPRT	0.370	0.092	0.424	0.114	271	101
Mesocricetus auratus	APRT	0.067	0.033	0.833	0.067	30	95
	HPRT	0.333	0.045	0.424	0.197	132	102,105
Mus domesticus	CII	0.142	0.109	0.749		183	99
	LacI	0.149	0.159	0.692		348	94

Notes: (1) Results are obtained from studies designed to isolate defective alleles in an essentially nonselective manner, followed by sequencing to identify the mutation. (2) ID = insertions/deletions of all sizes; CT = chain-terminating base substitutions; MS = missense (amino-acid replacement) substitutions; SL = splice-site altering base substitution (either exonic or intronic); n = total number of mutations observed at the locus. (3) Results for *Mus* involve intron-free reporter genes inserted into mouse autosomes.

Third, the point of transcription initiation for functional genes appears to be variable.[10-15] For example, in humans, where the data are most extensive, many genes have a continuum of transcription-initiation sites over an average range of 62 bp.[16] This means that some transcripts are likely to initiate downstream of the correct translation-initiation codon. Two thirds of such transcripts are expected to be out of frame and hence to encode one or more PTCs. Abnormally early transcription starts can also result in the incorporation of premature initiation codons, which because of the scanning mechanism of eukaryotic translation initiation can also induce downstream frameshifts.[17]

Fourth, genes containing introns are often alternatively spliced to produce different mRNAs. At least 75% of human genes express splice variants,[18,19] and similar estimates have been made for other species.[20,21] Because such estimates are downwardly biased by the restricted size and coverage of cDNA databases, it is conceivable that essentially every intron-containing gene experiences alternative splicing at least occasionally. In principle, alternative splicing can expand the range of variation in functional protein diversity, and it is often suggested that natural selection favors the adaptive exploitation of alternatively spliced alleles.[22-25] However, the frequent production of PTC-containing mRNAs by alternative splicing (~50% in the case of humans[26]) suggests that many such transcripts are nothing more than by-products of an imperfect splicing process. Indeed, ~5% of alternatively spliced exons in humans are a consequence of intronic insertions of *Alu* retrotransposons harboring latent splice sites, many of which cause frameshifts that result in genetic disorders.[27-29] The much lower degree of conservation of minor rather than major forms of alternatively spliced exons between humans and mouse[30,31] also suggests the opposition of alternatively spliced variants by natural selection.

These observations indicate that cells bearing genomes with substantial amounts of intergenic and intronic DNA, most notably those of animals and land plants,[32] produce a large fraction of primary transcripts with the potential to produce deleterious truncated proteins. To put this problem into perspective relative to the load from mutations at the DNA level, a lethal recessive mutation will have an allele frequency equal to the square root of the mutation rate, or ~0.001, at selection-mutation balance.[33,34] Because the cumulative transcription error rate from the four sources noted above is almost certainly >0.1%, the average cell must encounter at least as many problematical mRNAs from transcription errors as from mutant alleles.

Although it has been argued that the construction of PTC-containing transcripts can provide an advantageous means for regulating gene expression,[35,36] (see chapter by Soergel et al) there is little direct evidence for such adaptation, and a less costly (but less intricate) regulatory mechanism is the simple control of expression at the site of transcript initiation. In any event, there is little question that the NMD pathway relieves most species from a significant burden of inappropriate mRNAs. Knockouts of genes in the NMD pathway have small phenotypic effects in *Saccharomyces cerevisiae* and *Schizosaccharomyces pombe*,[37-39] moderate fitness effects in *Caenorhabditis elegans*,[40] and lethal effects in mice.[41] Many human genetic disorders associated with PTC-containing alleles are rendered recessive in the heterozygous state.[42]

Phylogenetic Roots of NMD

An effective NMD pathway requires a mechanism for reliably distinguishing PTCs from correct termination codons. In mammals, such guidance is often provided by a series of proteins that form an exon junction complex (EJC), thought to be deposited ~20-24 nucleotides 5' to every exon-exon junction at the time of splicing.[43] During the first round of translation, these ornamented junctions allow the cell to infer the position of the true termination codon, which generally lies in the final exon. If a termination codon is detected ~50 or more nucleotides upstream of an EJC, the mRNA is targeted for selective degradation.[44] If a PTC is not encountered, the EJC is stripped off, and the mRNA is free to engage in further rounds of translation. A central role of introns in this process is confirmed by the observation that intron-free mammalian genes are generally NMD insensitive.[45,46] However, introns are not always required for NMD. For example, the yeast *Saccharomyces cerevisiae*, whose nuclear genome contains only about 250 introns, is still able to deploy NMD, with PTC recognition in some genes relying on downstream sequence elements (DSEs) embedded within coding DNA[47] and in others on the distance of the PTC from some aspect of the 3' untranslated region (UTR) (see chapter by Amrani et al).[48,49] The latter mechanism has also been suggested for *Drosophila*, as intron-free genes from other species imported into fly cells are subject to NMD.[50]

It is unclear whether the mammalian EJC-based or one of the *S. cerevisiae* PTC-recognition pathways more closely represents the ancestral mode of NMD. However, since empirical work demonstrates that NMD can operate on intron-free genes in *S. pombe*,[39] at least some intron-free genes in nematodes (see chapter by Anderson),[51] dipterans (see chapter by Behm-Ansmant and Izaurralde),[50] and plants (see chapter by van Hoof and Green),[52,53] and in at least some genes in mammals in which a PTC is not followed by an EJC (see chapters by Maquat, and Gudikote and Wilkinson),[54-57] it appears that multiple NMD mechanisms may exist within some eukaryotic lineages.

To put an upper limit on the age of the EJC-based NMD pathway, we have queried a wide phylogenetic array of fully sequenced genomes for putative orthologs of proteins known to be involved in the EJC (Table 2). To understand the implication of the results, a brief overview of the EJC is necessary. Almost all of what is known about the EJC derives from human cell lines (see chapter by Maquat).[58-60] Five proteins appear to form the core of the EJC: a heterodimer comprised of Y14 and Mago; RNPS1; Barentz (Btz); and a helicase, eIF4AIII, which anchors the former four to the mRNA.[61] It is thought that Y14, Mago, RNPS1, and Btz remain on mRNAs until removed during the first round of translation.[62] Furthermore, all four seem to play a central role in NMD in humans, as each elicits mRNA decay if tethered to the 3' untranslated region of a human mRNA sufficiently downstream of the normal termination codon.[58,63,64] The extent to which random proteins would have a similar effect is unknown, but eIF4AIII appears to be incapable of recruiting the NMD pathway components.[58] Several other proteins appear to be peripherally involved in the EJC in humans, among them SRm160, REF/Aly, UAP56, TAP, and PYM. Most of these appear to join the mRNA at sites of transcription and seem to have a primary role as nuclear export factors,[65-67] although SRm160 and UAP56 are also involved in splicing.[68,69] PYM is a cytoplasmic RNA-binding protein that interacts with Y14:Mago and, like the latter, elicits NMD when tethered to 3' UTRs of human mRNAs.[70]

Inferences about orthology and shared functions based on sequence-homology searches have limitations, as the acceptance criteria are necessarily arbitrary (we have used 15% shared amino-acid identity for this purpose), but a stronger case can be made when all components of a pathway are present or absent. Although we are unable to make a firm statement of this nature for every species examined, our observations on these genes encourage several preliminary conclusions. First, although the EJC has only been formally demonstrated in mammals, it is clear that similar proteins exist in a wide array of eukaryotic lineages. For example, putative orthologs to all of the human EJC proteins are also found in the dipteran *Drosophila* and the nematode *Caenorhabditis*, and so they must have been present in the basal bilaterian. This need not imply that such orthologs always share identical functions. In *C. elegans*, for example, REF/Aly and RNPS1 appear to be nonessential for nuclear export,[71] and in *D. melanogaster*, Mago, RNSP1, and Y14 appear not to play a role in NMD.[50] However, the latter observations have been restricted to a small number of mRNAs, and it is perhaps premature to entirely rule out shared functions of the EJC between mammals and invertebrates. Given that putative orthologs to all of the EJC core proteins as well as the splicing cofactor UAP56 are broadly distributed throughout the eukaryotic tree (Fig. 1), and that similar sorts of observations have been made for the components of the spliceosome,[72,73] an ancient origin for the EJC, and by extension the EJC-based NMD pathway, cannot be ruled out.

Second, although REF/Aly appears to be almost universally present throughout eukaryotes, other core components of the EJC seem to have been lost from several lineages. Numerous fungi appear not to harbor Mago while retaining Y14 (see footnotes to Table 2), an unexpected

Table 2. **Phylogenetic distribution of various proteins known to be involved in the exon junction complex in humans**

		EJC Core Components					Nuclear Export Factors			
Group	**Introns/Gene**	**Mago**	**Y14**	**REF/Aly**	**Btz**	**eIF4AIII**	**SRm160**	**RNPS1**	**UAP56**	**PYM**
Vertebrates	6.6(5.2 – 7.2)	+	+	+	+	+	+	+	+	+
Urochordate	5.8	+	+	+	+	+	–	+	+	+
Insects	4.0 (3.1 – 5.5)	+	+	+	+	+	–	+	+	+
Nematodes	5.1 (4.9 – 5.3)	+	+	+	+	+	–	+	+	+
Fungi	1.2 (0.0 – 5.3)	±	+	+	–	+	±	±	+	–
Slime mold	1.3	+	+	+	–	+	+	–	+	+
Entamoeboid	0.3	+	+	+	–	+	+	–	+	–
Chlorophytes	4.2 (3.4 – 6.2)	+	+	+	–	+	±	–	+	–
Rhodophyte	0.0	–	–	+	–	+	–	–	+	–
Apicomplexans	0.8 (0.1 – 1.4)	+	+	+	–	+	–	+	+	–
Oomycetes	2.8	+	+	+	±	–	+	–	+	–
Ciliate	2.3	+	+	–	–	+	–	–	+	–
Kinetoplastids	0.0 (0.0 – 0.0)	±	±	+	–	±	–	–	±	±
Diatom	1.4	+	+	+	–	+	–	–	+	–
Trichomonad	2.6	+	+	+	–	+	–	+	+	–
Diplomonad	0.0	–	–	+	–	+	–	–	+	–

continued on next page

Table 2. Continued

Notes: (1) Queried species: Vertebrates—*Danio rerio, Fugu rubripes, Gallus gallus, Homo sapiens, Mus musculus, Pan troglodytes, Rattus norvegicus, Tetraodon nigroviridis, Xenopus tropicalis*; Urochordate—*Ciona intestinalis*; Insects—*Anopheles gambiae, Apis mellifera, Drosophila melanogaster*; Nematode—*Caenorhabditis elegans*; Fungi—*Ashbya gossypii, Aspergillus nidulans, Candida albicans, Candida glabrata, Chaetomium globosum, Cryptococcus neoformans, Debaryomyces hansenii, Encephalitozoon cuniculi, Fusarium graminearum, Kluyveromyces lactis, Magnaporthe grisea, Neurospora crassa, Saccharomyces cerevisiae, Schizosaccharomyces pombe, Stagonospora nodorum, Trichoderma ressei, Ustilago maydis, Yarrowia lipolytica*; Slime mold—*Dictyostelium discoideum*; Entamoeboid—*Entamoeba histolytica*; Chlorophytes—*Arabidopsis thaliana, Chlamydomonas reinhardtii, Oryza sativa*; Rhodophyte — *Cyanidioschyzon merolae*; Apicomplexans—*Cryptosporidium hominis, Cryptosporidium parvum, Plasmodium falciparum, Plasmodium yoelii*; Oomycetes—*Phytophthora ramorum, Phytophthora sojae*; Ciliate—*Tetrahymena themophila*; Kinetoplastids—*Leishmania major, Trypanosoma brucei*; Diatom—*Thalassiosira pseudonana*; Trichomonad—*Trichomonas vaginalis*; Diplomonad—*Giardia lamblia*. (2) Of the unicellular groups for which a range in intron number is not given, for most only a single genome has been analyzed, so the group-wide implications are uncertain. (3) For species with 0.0 reported as the average intron number per protein-coding gene, the actual number is in the range of 0.00 to 0.01, except in the case of kinetoplastids for which no spliceosomal intron has yet been found. (4) Proteins were located by applying Blastp[93] to the databases of fully sequenced genomes, using functionally defined genes from *Homo sapiens* and *Saccharomyces cerevisiae* as query sequences. Sequence identities <0.15 were taken to infer the absence of an ortholog. A ± indicates that not all members of the group contain the protein of interest by our criterion of homology. (5) Species for which Mago is missing for groups with mixed results (with the average number of introns per protein-coding gene in parentheses): Fungi—*Ashbya gossypii* (0.05), *Debaryomyces hansenii* (0.01), *Encephalitozoon cuniculi* (0.01), *Kluyveromyces lactis* (0.00), *Saccharomyces cerevisiae* (0.05); Apicomplexans—absent from *Trypanosoma* (0.00), but present in *Leishmania* (0.00). (Here and below, a—implies that all queried members of the group were missing the gene, so the species identities can be found in the above list of queried sequences). (6) Species for which Y14 is missing: Apicomplexans—absent from *Trypanosoma*, but present in *Leishmania*. (7) Species for which eIF4AIII is missing: Kinetoplastids—*Trypanosoma brucei* (0.00). (8) Species for which SRm160 is missing: Fungi—all species except *Saccharomyces pombe* (0.98), *Stagonospora nodorum* (1.65), *Trichoderma ressei* (2.63), *Ustilago maydis* (0.75), *Yarrowia lipolytica* (0.02); Green plants—*Chlamydomonas reinhardtii* (4.55). (9) Species for which RNPS1 is missing: Fungi—all species, with the possible exception of *Candida albicans* (unknown) and *Schizosaccharomyces pombe* (0.98). (10) Species for which UAP56 is missing: Apicomplexans—absent from *Trypanosoma* (0.00), but present in *Leishmania* (0.00).

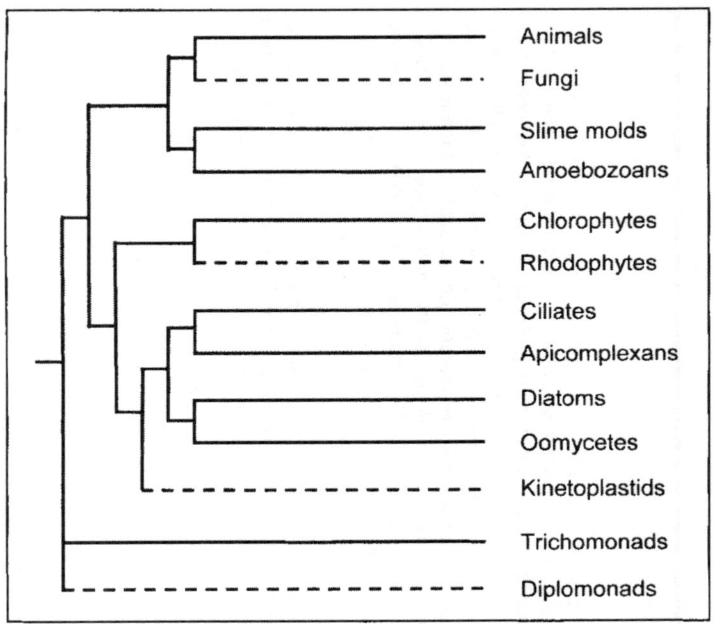

Figure 1. A current view of the phylogenetic relationships of the major eukaryotic groups (ref. 92 and additional sources). Dashed lines denote lineages that contain one or more species that appear to have lost NMD, the EJC, and substantial numbers of spliceosomal introns.

result given that these two proteins form heterodimers in mammals, and the diminutive red alga *Cyanidioschyzon*, the kinetoplastid *Trypanosoma*, and the diplomonad *Giardia* have lost both Mago and Y14. Notably, each species that has lost Mago and/or Y14 is nearly devoid of spliceosomal introns (Table 2), and the phylogenetic evidence suggests that each species reached this point independently (Fig. 1). No introns have been found in kinetoplastids, and only one has been located in *Giardia*.[74] Surprisingly, *Leishmania*, another kinetoplastid with no known introns, does have readily identifiable copies of Mago and Y14. Because the 5' ends of all kinetoplastid mRNAs are trans-spliced to a short leader sequence,[75] some species in this group might recruit an EJC to the resultant splice junctions.

We now turn to a similar analysis of the NMD pathway (Table 3). The working hypothesis for NMD in mammalian cells is that Upf3 becomes associated with the EJC within the nucleus, with Upf2 and then Upf1 joining during or after export of the mRNA to the cytoplasm, where the complex then promotes either the termination of translation of a proper mRNA or the elimination of an aberrant transcript.[76] Less well understood are the roles of SMG1, SMG5, SMG6, and SMG7, all of which were discovered in *C. elegans*, where they are implicated in either the phosphorylation or dephosphorylation of Upf1 (see chapter by Yamashita et al).[77,78]

Our results for core NMD proteins yield a phylogenetic perspective similar to that provided by proteins of the EJC. First, the core components of the NMD pathway (Upf1, Upf2, and Upf3; also known as SMG2, SMG3, and SMG4 in *C. elegans*) are broadly distributed across the eukaryotic phylogeny, implying that, like the EJC, NMD was present prior to the divergence of most of the major eukaryotic lineages (Fig. 1). Second, the kinetoplastids *Trypanosoma* and *Leishmania*, the rhodophyte *Cyanidioschyzon*, the microsporidian (fungus) *Encephalitozoon*, and the diplomonad *Giardia* appear to have lost the capacity for NMD. Again, all of these groups are nearly devoid of spliceosomal introns and apparently have lost the EJC. Third, evidence for the presence of the accessory NMD proteins (SMG1, SMG5, SMG6, and SMG7)

Table 3. Phylogenetic distribution of various proteins known to be involved in the NMD pathway

Group	SMG1	Upf1	Upf2	Upf3	SMG5	SMG6	SMG7
Vertebrates	+	+	+	+	–	±	+
Urochordate	+	+	+	+	–	+	–
Insects	+	+	+	+	+	+	–
Nematodes	+	+	+	+	+	+	+
Fungi	–	±	±	±	–	–	–
Slime mold	+	+	+	+	–	+	+
Entamoebid	+	+	–	+	–	–	–
Chlorophytes	–	+	+	+	–	–	–
Rhodophyte	–	–	–	–	–	–	–
Apicomplexans	–	+	+	+	–	±	–
Oomycetes	–	+	+	+	–	–	–
Ciliate	–	+	+	+	–	–	–
Kinetoplastids	–	–	–	+	–	–	–
Diatom	–	+	+	+	–	+	–
Trichomonad	+	+	+	–	–	–	–
Diplomonad	–	+	–	–	–	–	–

Notes: (1) See Table 1 for queried taxa and general methodology. (2) SMG2 = Upf1; SMG3 = Upf2; SMG4 = Upf3, where the first notation is that originally used for *Caenorhabditis elegans*. (3) The situation with respect to orthology to SMG5 and SMG7 in other species is unclear. In humans, SMG6 (initially called hSmg5/7a) appears to be homologous to both SMG5 and SMG7 of *C. elegans*.[106] (4) Among the fungi, *Encephalitozoon cuniculi* is missing Upf1, Upf2, and Upf3, whereas *Candida albicans* and *Trichoderma ressei* are missing Upf1.

is weak outside of animals, although the slime mold *Dictyostelium* appears to contain three of the four (with the SMG5 falling just below our homology cutoff at 13%).

These and other observations suggest that the stem eukaryote was quite complex from the standpoint of transcript processing. It had at least one spliceosome (the phylogenetic evidence suggests two),[72] and it most likely contained large numbers of introns,[79,83] so the evolutionary context necessary for the origin of the EJC was certainly present. There is no evidence that any of this machinery ever existed in a prokaryote. However, despite the strong statistical association among the presence of NMD, the EJC and intron abundance, the causal connections among the three is less certain. Because experimental results obtained using a broad range of species clearly indicate that introns (and therefore EJCs) are not a physical necessity for the operation of NMD, an alternative explanation for the parallel loss of NMD and introns from specific lineages is required. One possibility is that splicing errors are the primary selective forces maintaining the NMD pathway. It should be noted, however, that most of the species that have apparently lost the NMD pathway have exceptionally compact genomes in all respects.[32] In addition to having lost most introns, they are nearly devoid of mobile elements and have greatly reduced levels of intergenic DNA. Because all of these features can reduce the production of erroneous transcripts (as noted above), the parallel loss of NMD and introns may reflect a broader set of conditions that minimize the need for mRNA surveillance.

Less clear are the conditions that led to the origin and maintenance of the EJC. Tange et al[59] suggested that the EJC may have evolved initially as a mechanism for labeling mRNAs for export, only taking on an NMD-associated function after errors from alternative splicing became common in vertebrates. However, as noted above, alternative splicing is ubiquitous in all organisms with introns, and this generality, combined with the consistent phylogenetic cooccurrence of NMD and the EJC, argues for the coordination of NMD and the EJC since the very dawn of eukaryotic evolution.

Recent studies with mammalian cells indicate that the presence of the EJC-NMD complex on normal mRNAs enhances the efficiency of their translation.[59,80,81] This observation raises the important caveat that the key to understanding the evolution of NMD may not be found in mRNA surveillance alone. If, however, the enhancement of translation efficiency was the ancestral function of the NMD-EJC complex, then an explanation is needed for how such an essential feature was either eliminated or replaced in lineages that have lost most of their introns. To resolve these issues, further work is required on the functions of the EJC and NMD machinery in diverse lineages.

NMD and the Proliferation of Introns

One of the great unsolved mysteries of eukaryotic genome evolution is the origin of spliceosomal introns. Aside from the physical challenges that they present for the production of proper mRNAs, introns are inherently disadvantageous to host genes. This is because the need to conserve nucleotide signatures associated with proper splicing increases the genic mutation rate to defective alleles relative to the situation for intron-free genes.[82] The fact that 7 to 20% of nonfunctionalizing mutations in mammalian genes alter splicing makes this case quite strongly (Table 1). The selective disadvantage of an individual intron is small—on the order of the product of the number of nucleotide sites required to guide the spliceosome times the per-nucleotide mutation rate. Thus, unless the population size is sufficiently large that the power of random genetic drift is weaker than the excess mutation rate to defective alleles, introns are able to colonize a host genome in an effectively neutral fashion. The phylogenetic distribution of introns is consistent with this hypothesis—multicellular animals and plants have small effective population sizes and, on average, harbor four to seven introns per protein-coding gene, whereas unicellular lineages tend to have much larger effective population sizes and many fewer introns.[32] Given that a number of unicellular species average two to three introns per protein-coding gene (Table 2) and that many of the introns of animals and land plants are conserved in location,[83] it is plausible that an initial proliferation of introns occurred during a period in which the stem eukaryote experienced relatively small population sizes for an extended period of time. This initial phase of intron proliferation was then followed by substantial losses in numerous unicellular lineages (as noted above) and secondary gains in the animal and land-plant lineages.

A significant challenge for understanding the origin of spliceosomal introns is the need for a mechanism that is quantitatively sufficient to explain the early phase of relatively rapid intron proliferation (within a window <1 billion years (BY), the approximate period between the origin of prokaryotes and eukaryotes) while also explaining why runaway intron proliferation has not occurred (no known eukaryote contains > 7.5 introns per average protein-coding gene). Using birth-death process theory[82] and estimated rates of intron turnover between pairs of invertebrate species,[72] one can roughly infer the rate of intron birth within the past ~100 million years (MY) to be ~0.001/nucleotide site/BY. Using different methods, Roy and Gilbert[84] arrived at a similar estimate. Thus, assuming an average coding-region length of 1000 bp, the accumulation of six introns per protein-coding gene (the average situation in today's animals and land plants) by neutral processes would require about six billion years. It is possible that the earliest introns had more efficient means of colonization, but a simpler explanation for a more rapid accumulation of introns is an elevation of the fixation probability above the neutral expectation by some form of positive selection. For such a scenario to be plausible, any selective advantage of introns would have to more than offset the mutational disadvantage. Once established, an EJC-based mechanism for NMD orientation may have had this effect.[85] At this point in eukaryotic evolution, introns that increased the net capacity of their host gene to guard against transcriptional errors (including and beyond those associated with introns) may have been promoted by positive selection. However, once sufficient spatial coverage for efficient PTC detection was achieved, the inherent disadvantages of introns would have begun to outweigh their NMD-associated advantages, providing a natural barrier to runaway intron colonization.

Under this model, the selective consequences of a new intron are expected to depend on the point of insertion and on the geometric constraints associated with the surveillance process.[85] If the NMD machinery were capable of scanning the entire length of a transcript (and if there were no other selective factors associated with introns), the optimal configuration would be a single intron less than 50- to 55-nt downstream of the normal termination codon, so that almost all PTCs but not the normal termination codon would elicit NMD[44] (see chapter by Maquat) 3' end of every gene. However, if NMD is not fully capable of scanning hundreds to thousands of bases, a more uniform scattering of introns would be necessary for complete coverage. Unfortunately, empirical information on the spatial requirements of NMD is quite limited. In humans, NMD is elicited by PTCs as far as 550-bp upstream of an intron in the triosephosphate isomerase gene,[55] 700-bp upstream of an intron in the β-globin gene,[53] and 1200-bp upstream in the HSP70 gene.[45] However, the **quantitative** relationship between NMD efficiency and the PTC-EJC distance may be more gradual than a threshold function, multiple introns may facilitate the process,[56,86] and studies of human genetic disorders have revealed cases in which PTCs do not elicit much NMD despite the consistency of their location with the 50- to 55-nt criteria (e.g., see ref. 87). A much stronger spatial dependency is found in *S. cerevisiae*, where sensitivity to NMD is completely eliminated if the PTC is more than 300-bp upstream of a DSE.[47] In contrast, some nonmammalian studies have suggested that more 5'-located PTCs elicit a higher level of NMD than more 3'-located PTCs.[51,52,88] As the topological features of mRNAs might vary dramatically from gene to gene in ways that influence the efficiency of NMD, it is clear that a fuller understanding of the spatial constraints on the NMD process will ultimately require studies of multiple genes in multiple species.

Provided there are spatial limitations to the NMD scanning mechanism, two simple predictions can be made under this model: (1) exon sizes should be more uniform than under a model of random intron insertion; and (2) the average number of introns per gene should increase linearly with the length of the gene, i.e., with the amount of territory that needs to be scanned. Genome-wide surveys of multicellular species have revealed that although the quantitative relationship differs among species, intron number is consistently linearly related with gene size, and the average positions of consecutive introns tend to evenly partition the coding sequence (see also refs. 84,88,89). Although these patterns could arise with complete random colonization of introns, the variance of exon size is too low to be compatible with such a model, suggesting the presence of stabilizing selection on exon size.[85] Additional support for the idea that selection favors architectural features of genes that influence the properties of their mRNAs derives from the observation that although average intron lengths decline dramatically with increasing levels of gene expression in both nematode and mammalian genes, intron number remains constant.[91]

Acknowledgements

This work has been supported by NIH grant GM36827 and NSF grant MCB-0342431 to M. Lynch, and NSF Postdoctoral Fellowship in Biological Informatics DBI-0434671 to D.G. Scofield.

References

1. Lynch M, Blanchard J, Houle D et al. Spontaneous deleterious mutation. Evolution 1999; 53:645-663.
2. Ninio J. Connections between translation, transcription and replication error rates. Biochimie 1991a; 73:1517-1523.
3. Ninio J. Transient mutators: A semiquantitative analysis of the influence of translation and transcription errors on mutation rates. Genetics 1991b; 129:957-962.
4. Shaw RJ, Bonawitz ND, Reines D. Use of an in vivo reporter assay to test for transcriptional and translational fidelity in yeast. J Biol Chem 2002; 277:24420-24426.

5. Ota T et al. Complete sequencing and characterization of 21,243 full-length human cDNAs. Nature Genet 2004; 36:40-45.
6. Cawley S et al. Unbiased mapping of transcription factor binding sites along human chromosomes 21 and 22 points to widespread regulation of noncoding RNAs. Cell 2004; 116:499-509.
7. Kapranov P, Cawley SE, Drenkow J et al. Large-scale transcriptional activity in chromosomes 21 and 22. Science 2002; 296:916-919.
8. Mendell JT, Sharifi NA, Meyers JL et al. Nonsense surveillance regulates expression of diverse classes of mammalian transcripts and mutes genomic noise. Nature Genet 2004; 36:1073-1078.
9. Mitrovich QM, Anderson P. mRNA surveillance of expressed pseudogenes in C. elegans. Curr. Biol 2005; 15:963-967.
10. Hahn SE, Hoar T, Guarente L. Each of three "TATA elements" specifies a subset of the transcription initiation sites at the CYC-1 promoter of Saccharomyces cerevisiae. Proc Natl Acad Sci USA 1985; 82:8562-8566.
11. Nagawa F, Fink GR. The relationship between the "TATA" sequence and transcription initiation sites at the HIS4 gene of Saccharomyces cerevisiae. Proc Natl Acad Sci USA 1985; 82:8557-8561.
12. Bergsma DJ, Ai Y, Skach WR et al. Fine structure of the human galactokinase GALK1 gene. Genome Res 1996; 6:980-985.
13. Yu YS, Suzuki Y, Yoshitomo K et al. The promoter structure of TGF-β type II receptor revealed by "oligo-capping" method and deletion analysis. Biochem Biophys Res Comm 1996; 225:302-306.
14. Kaji H, Tai S, Okimura Y et al. Cloning and characterization of the 5'-flanking region of the human growth hormone secretagogue receptor gene. J Biol Chem 1998; 273:33885-33888.
15. Watanabe J, Sasaki M, Suzuki Y et al. Analysis of transcriptomes of human malaria parasite Plasmodium falciparum using full-length enriched library: Identification of novel genes and diverse transcription start sites of messenger RNAs. Gene 2002; 291:105-113.
16. Suzuki Y, Tsunoda T, Sese J et al. Identification and characterization of the potential promoter regions of 1031 kinds of human genes. Genome Res 2001; 11:677-684.
17. Lynch M, Scofield D, Hong X. The evolution of transcription-initiation sites. Mol Biol Evol 2005; 22:1137-1146.
18. Boue S, Letunic I, Bork P. Alternative splicing and evolution. Bioessays 2003; 25:1031-1034.
19. Johnson J, Castle MJ, Garrett-Engele P et al. Genome-wide survey of human alternative pre-mRNA splicing with exon junction microarrays. Science 2003; 302:2141-2144.
20. Brett D, Pospisil H, Valcárcel J et al. Alternative splicing and genome complexity. Nature Genet 2002; 30:29-30.
21. Okazaki Y et al. Analysis of the mouse transcriptome based on functional annotation of 60,770 full-length cDNAs. Nature 2002; 420:563-573.
22. Kondrashov FA, Koonin EV. Origin of alternative splicing by tandem exon duplication. Hum Mol Genet 2001; 10:2661-2669.
23. Letunic I, Copley RR, Bork P. Common exon duplication in animals and its role in alternative splicing. Hum Mol Genet 2002; 11:1561-1567.
24. Xu Q, Modrek B, Lee C. Genome-wide detection of tissue-specific alternative splicing in the human transcriptome. Nucleic Acids Res 2002; 30:3754-3766.
25. Sorek R, Ast G. Intronic sequences flanking alternatively spliced exons are conserved between human and mouse. Genome Res 2003; 13:1631-1637.
26. Sorek R, Shamir R, Ast G. How prevalent is functional alternative splicing in the human genome? Trends Genet 2004; 20:68-71.
27. Sorek R, Ast G, Graur D. Alu-containing exons are alternatively spliced. Genome Res 2002; 12:1060-1067.
28. Lev-Maor G, Sorek R, Shomron N et al. The birth of an alternatively spliced exon: 3' splice-site selection in Alu exons. Science 2003; 300:1288-1291.
29. Kreahling J, Graveley BR. The origins and implications of alternative splicing. Trends Genet 2004; 20:1-4.
30. Modrek B, Lee CJ. Alternative splicing in the human, mouse and rat genomes is associated with an increased frequency of exon creation and/or loss. Nature Genet 2003; 34:177-180.
31. Thanaraj TA, Clark F, Muilu J. Conservation of human alternative splice events in mouse. Nucleic Acids Res 2003;. 31:2544-2552.
32. Lynch M, Conery JS. The origins of genome complexity. Science 2003; 302:1401-1404.
33. Crow JF, Kimura M. An Introduction to population genetics theory. New York, NY: Harper and Row Publ, 1970.
34. Simmons MJ, Crow JF. Mutations affecting fitness in Drosophila populations. Annu Rev Genet 1977; 11:49-78.

35. Lewis BP, Green GE, Brenner SE. Evidence for the widespread coupling of alternative splicing and nonsense-mediated mRNA decay in humans. Proc Natl Acad Sci USA 2003; 100:189-192.
36. Mitrovich QM, Anderson P. Unproductively spliced ribosomal protein mRNAs are natural targets of mRNA surveillance in C. elegans. Genes Dev 2000; 14:2173-2184.
37. Leeds P, Wood JM, Lee BS et al. Gene products that promote mRNA turnover in Saccharomyces cerevisiae. Mol Cell Biol 1992; 12:2165-2177.
38. Dahlseid JN, Puziss J, Shirley RL et al. Accumulation of mRNA coding for the ctf13p kinetochore subunit of Saccharomyces cerevisiae depends on the same factors that promote rapid decay of nonsense mRNAs. Genetics 1998; 150:1019-1035.
39. Mendell JT, Medghalchi SM, Lake RG et al. Novel Upf2p orthologues suggest a functional link between translation initiation and nonsense surveillance complexes. Mol Cell Biol 2000; 20:8944-8957.
40. Hodgkin J, Papp A, Pulak R. A new kind of informational suppression in the nematode Caenorhabditis elegans. Genetics 1989; 123:301-313.
41. Medghalchi SM, Frischmeyer PA, Mendell JT et al. Rent1, a trans-effector of nonsense-mediated mRNA decay, is essential for mammalian embryonic viability. Hum Mol Genet 2001; 10:99-105.
42. Frischmeyer PA, Dietz HC. Nonsense-mediated mRNA decay in health and disease. Hum Mol Genet 1999; 8:1893-1900.
43. Le Hir H, Izaurralde E, Maquat LE, Moore MJ. The spliceosome deposits multiple proteins 20-24 nucleotides upstream of mRNA exon-exon junctions. EMBO J 2000; 19:6860-6869.
44. Nagy E, Maquat LE. A rule for termination-codon position within intron-containing genes: When nonsense affects RNA abundance. Trends Biochem Sci 1998; 6:198-199.
45. Maquat LE, Li X. Mammalian heat shock p70 and histone H4 transcripts, which derive from naturally intronless genes, are immune to nonsense-mediated decay. RNA 2001; 7:445-456.
46. Brocke KS, Neu-Yilik G, Gehring NH et al. The human intronless melanocortin 4-receptor gene is NMD insensitive. Hum Mol Genet 2002; 11:331-335.
47. Ruiz-Echevarria MJ, González CI, Peltz SW. Identifying the right stop: Determining how the surveillance complex recognizes and degrades an aberrant mRNA. EMBO J 1998; 17:575-589.
48. Hilleren P, Parker R. mRNA surveillance in eukaryotes: Kinetic proofreading of proper translation termination as assessed by mRNP domain organization? RNA 1999; 5:71171-71179.
49. Amrani N, Ganesan R, Kervestin S et al. A faux 3'-UTR promotes aberrant termination and triggers nonsense-mediated mRNA decay. Nature 2004; 432:112-118.
50. Gatfield D, Unterholzner L, Ciccarelli FD et al. Nonsense-mediated mRNA decay in Drosophila: At the intersection of the yeast and mammalian pathways. EMBO J 2003; 22:3960-3970.
51. Pulak R, Anderson P. mRNA surveillance by the Caenorhabditis elegans smg genes. Genes Dev 1993; 7:1885-1897.
52. van Hoof A, Green PJ. Premature nonsense codons decrease the stability of phytohemagglutinin mRNA in a position-dependent manner. Plant J 1996; 10:415-424.
53. Neu-Yilik G, Gehring NH, Thermann R et al. Splicing and 3' end formation in the definition of nonsense-mediated decay-competent human β-globin mRNPs. EMBO J 2001; 20:532-540.
54. Cheng J, Belgrader P, Zhou X et al. Introns are cis-effectors of the nonsense-codon-mediated reduction in nuclear mRNA abundance. Mol Cell Biol 1994; 14:6317-6325.
55. Zhang J, Sun X, Qian Y et al. At least one intron is required for the nonsense-mediated decay of triosephosphate isomerase mRNA: A possible link between nuclear splicing and cytoplasmic translation. Mol Cell Biol 1998; 18:5272-5283.
56. Rajavel KS, Neufeld EF. Nonsense-mediated decay of human HEXA mRNA. Mol Cell Biol 2001; 21:5512-5519.
57. Wang J, Gudikote JP, Olivas OR et al. Boundary-independent polar nonsense-mediated decay. EMBO Rep 2002; 3:274-279.
58. Palacios IM, Gatfield D, St Johnston D et al. An eIF4AIII-containing complex required for mRNA localization and nonsense-mediated mRNA decay. Nature 2004; 427:753-757.
59. Tange TO, Nott A, Moore MJ. The ever-increasing complexities of the exon junction complex. Curr Opin Cell Biol 2004; 16:279-284.
60. Lejeune F, Maquat LE. Mechanistic links between nonsense-mediated mRNA decay and pre-mRNA splicing in mammalian cells. Curr Opin Cell Biol 2005; 17:309-315.
61. Shibuya T, Tange TO, Sonenberg N et al. eIF4AIII binds spliced mRNA in the exon junction complex and is essential for nonsense-mediated decay. Nat Struct Mol Biol 2004; 11:346-351.
62. Dostie J, Dreyfuss G. Translation is required to remove Y14 from mRNAs in the cytoplasm. Curr Biol 2002; 12:1060-1067.
63. Lykke-Andersen J, Shu MD, Steitz JA. Human Upf proteins target an mRNA for nonsense-mediated decay when bound downstream of a termination codon. Cell 2000; 103:1121-1131.

64. Lykke-Andersen J, Shu MD, Steitz JA. Communication of the position of exon-exon junctions to the mRNA surveillance machinery by the protein RNPS1. Science 2001; 293:1836-1839.
65. Kim VN, Kataoka N, Dreyfuss G. Role of the nonsense-mediated decay factor hUpf3 in the splicing-dependent exon-exon junction complex. Science 2001; 293:1832-1836.
66. Le Hir H, Gatfield D, Izaurralde E et al. The exon-exon junction complex provides a binding platform for factors involved in mRNA export and nonsense-mediated mRNA decay. EMBO J 2001; 20:4987-4997.
67. Custodio NC, Carvalho I, Condado M et al. In vivo recruitment of exon junction complex proteins to transcription sites in mammalian cell nuclei. RNA 2004; 10:622-633.
68. Blencowe BJ, Issner R, Nickerson JA Et al. A coactivator of pre-mRNA splicing. Genes Dev 1998.; 12:996-1009.
69. Luo ML, Zhou Z, Magni K et al. Pre-mRNA splicing and mRNA export linked by direct interactions between UAP56 and Aly. Nature 2001; 413:644-647.
70. Bono F, Ebert J, Unterholzner L et al. Molecular insights into the interaction of PYM with the Mago-Y14 core of the exon junction complex. EMBO Rep 2004; 5:304-310.
71. Longman D, Johnstone IL, Caceres JF. The Ref/Aly proteins are dispensable for mRNA export and development in Caenorhabditis elegans. RNA 2003; 9:881-891.
72. Lynch M, Richardson A. The evolution of spliceosomal introns. Curr Opin Gen Devel 2002; 12:701-710.
73. Collins L, Penny D. Complex spliceosomal organization ancestral to extant eukaryotes. Mol Biol Evol 2005; 22:1053-1066.
74. Nixon JE, Wang A, Morrison HG et al. A spliceosomal intron in Giardia lamblia. Proc Natl Acad Sci USA 2002; 99:3701-3705.
75. Perry K, Agabian N. mRNA processing in the Trypanosomatidae. Experientia 1991; 47:118-128.
76. Maquat LE. Nonsense-mediated mRNA decay: A comparative analysis of different species. Curr Genomics 2004; 5:175-190.
77. Ohnishi T, Yamashita A, Kashima I et al. Phosphorylation of hUPF1 induces formation of mRNA surveillance complexes containing hSMG-5 and hSMG-7. Mol Cell 2003; 12:1187-1200.
78. Grimson AS, O'Connor C, Newman L et al. SMG-1 is a phosphatidylinositol kinase-related protein kinase required for nonsense-mediated mRNA decay in Caenorhabditis elegans. Mol Cell Biol 2004; 24:7483-7490.
79. Roy SW, Gilbert W. Complex early genes. Proc Natl Acad Sci USA 2005a; 102:1986-1991.
80. Wiegand HL, Lu S, Cullen BR. Exon junction complexes mediate the enhancing effect of splicing on mRNA expression. Proc Natl Acad Sci USA 2003; 100:11327-11332.
81. Wilkinson MF. A new function for nonsense-mediated mRNA-decay factors. Trends Genet 2005; 21:143-148.
82. Lynch M. Intron evolution as a population-genetic process. Proc Natl Acad Sci USA 2002; 99:6118-6123.
83. Rogozin IB, Wolf YI, Sorokin AV et al. Remarkable interkingdom conservation of intron positions and massive, lineage-specific intron loss and gain in eukaryotic evolution. Curr Biol 2003; 13:1512-1517.
84. Roy SW, Gilbert W. Rates of intron loss and gain: Implications for early eukaryotic evolution. Proc Natl Acad Sci USA 2005b; 102:5773-5778.
85. Lynch M, Kewalramani A. Messenger RNA surveillance and the evolutionary proliferation of introns. Mol Biol Evol 2003; 20:563-571.
86. Mango SE. Stop making nonSense: The C. elegans smg genes. Trends Genet 2001; 17:646-653.
87. Boyer J, Crosnier C, Driancourt C et al. Expression of mutant JAGGED1 alleles in patients with Alagille syndrome. Hum Genet 2005; 116:445-453.
88. Isshiki M, Yamamoto Y, Satoh H et al. Nonsense-mediated decay of mutant waxy mRNA in rice. Plant Physiol 2001; 125:1388-1395.
89. Sakurai A, Fujimori S, Kochiwa H et al. On biased distribution of introns in various eukaryotes. Gene 2002; 300:89-95.
90. Mourier T, Jeffares DC. Eukaryotic intron loss. Science 2003; 300:1393.
91. Castillo-Davis CI, Mekhedov SL, Hartl DL et al. Selection for short introns in highly expressed genes. Nature Genet 2002; 31:415-418.
92. Baldauf SL, Roger AJ, Wenk-Siefert I et al. A kingdom-level phylogeny of eukaryotes based on combined protein data. Science 2000; 290:972-977.
93. Altschul SF, Madden TL, Schaffer AA et al. Gapped BLAST and PSI-BLAST: A new generation of protein database search programs. Nucleic Acids Res 1997; 25:3389-3402.

94. de Boer JG, Erfle H, Walsh D et al. Glickman. Spectrum of spontaneous mutations in liver tissue of lacI transgenic mice. Environ Mol Mutagen 1997; 30:273-286.
95. de Jong PJ, Grosovsky AJ, Glickman BW. Spectrum of spontaneous mutation at the APRT locus of Chinese hamster ovary cells: An analysis at the DNA sequence level. Proc Natl Acad Sci USA 1988; 85:3499-3503.
96. Drake JW. A constant rate of spontaneous mutation in DNA-based microbes. Proc Natl Acad Sci USA 1991; 88:7160-7164.
97. Green PM, Saad S, Lewis CM et al. Mutation rates in humans. I. Overall and sex-specific rates obtained from a population study of hemophilia B. Amer J Hum Genet 1999; 65:1572-1579.
98. Grogan DW, Carver GT, Drake JW. Genetic fidelity under harsh conditions: Analysis of spontaneous mutation in the thermoacidophilic archaeon Sulfolobus acidocaldarius. Proc Natl Acad Sci USA 2001; 98:7928-7933.
99. Harbach PR, Zimmer DM, Filipunas AL et al. Spontaneous mutation spectrum at the lambda cII locus in liver, lung, and spleen tissue of Big Blue transgenic mice. Environ Mol Mutagen 1999; 33:132-143.
100. Hartman PE, Hartman Z, Stahl RC. Classification and mapping of spontaneous and induced mutations in the histidine operon of Salmonella. Adv Genet 1971; 16:1-34.
101. Jinnah HA, De Gregorio L, Harris JC et al. The spectrum of inherited mutations causing HPRT deficiency: 75 new cases and a review of 196 previously reported cases. Mutat Res 2000; 463:309-326.
102. Xu Z, Yu Y, Schwartz JL et al. Molecular nature of spontaneous mutations at the hypoxanthine-guanine phosphoribosyltransferase (hprt) locus in Chinese hamster ovary cells. Environ Mol Mutagen 1995; 26:127-138.
103. Yamamura E, Lee EH, Kuzumaki A et al. Characterization of spontaneous mutation in the delta soxR and SoxS overproducing strains of Escherichia coli. J Radiat Res 2002; 43:195-203.
104. Yamamura E, Nunoshiba T, Kawata M et al. Characterization of spontaneous mutation in the oxyR strain of Escherichia coli. Biochem Biophys Res Comm 2000; 279:427-432.
105. Zhang LH, Vrieling H, van Zeeland AA et al. Spectrum of spontaneously occurring mutations in the hprt gene of V79 Chinese hamster cells. J Mol Biol 1992; 223:627-635.
106. Chiu SY, Serin G, Ohara O et al. Characterization of human Smg5/7a: A protein with similarities to Caenorhabditis elegans SMG5 and SMG7 that functions in the dephosphorylation of Upf1. RNA 2003; 9:77-87.

Section VII

Roles for NMD Factors Other Than in NMD

Chapter 16

The SMG-1 Kinase:

At the Crossroads of mRNA Surveillance and Stress Response Pathways

Robert T. Abraham* and Vasco Oliveira

Abstract

The SMG-1 kinase is the newest member of an unusual family of protein serine-threonine kinases that have catalytic domains bearing significant similarity to those of the phosphoinositide 3-kinases. Studies in worms and mammalian cells have established that SMG-1 is a key component of the nonsense-mediated mRNA decay (NMD) machinery in metazoan organisms. However, accumulating evidence suggests that the signaling functions of human (h)SMG-1 are not restricted to the regulation of NMD. In this chapter, we provide an overview of the structure, biochemical properties, and functions of hSMG-1. We highlight recent findings indicating that hSMG-1 plays dual roles in mRNA surveillance and stress-induced signaling pathways in human cells.

Introduction

The NMD pathway is operational in all eukaryotic organisms, ranging from single-cell fungi to metazoans, including humans. As reviewed elsewhere in this volume (see chapters by Baker and Parker, Maquat, Behm-Ansmant and Izaurralde, Anderson, and van Hoof and Green), the core function of the NMD pathway is to degrade mRNAs containing premature termination codons (PTCs). PTCs can arise in mRNAs due to gene mutations, errors in transcription, or errors in RNA processing. NMD preserves the quality of the transcriptome and promotes organismal survival through the elimination of aberrant mRNAs, which have the potential to encode truncated proteins that could poison intracellular regulatory pathways (see chapters by Holbrook et al and Sharifi and Dietz). It is tempting to draw some conceptual parallels between the NMD machinery and the genome surveillance network, which detects and repairs damaged DNA before mutations become fixed in the genome. Indeed, the DNA repair apparatus complements the NMD system, in that both surveillance mechanisms play crucial roles in the maintenance of transcriptome fidelity. However, NMD is much more than simply a cytoprotective mechanism against abnormal mRNAs. Eukaryotic cells rely heavily on alternative slicing of pre-mRNAs and NMD to regulate gene expression during development or in response to changing environmental conditions[1,2] (see chapters by Shafari and Dietz and Soergel et al). Recent evidence also indicates that the tandem processes of alternative mRNA splicing and NMD are centrally involved in the determination of cell fate—survival or death by apoptosis.[3]

The proteins that comprise the NMD pathway were initially defined through genetic studies in tractable model systems, mainly the budding yeast *Saccharomyces cerevisiae* and the worm

*Corresponding Author: Robert T. Abraham—Oncology Research, Wyeth Pharmaceutical Research, 401 N. Middletown Road, Pearl River, New York 10965, U.S.A.
Email: abrahar@wyeth.com

Nonsense-Mediated mRNA Decay, edited by Lynne E. Maquat. ©2006 Eurekah.com.

Caenorhabditis elegans.[4-6] A somewhat surprising observation, given the evolutionary conservation of NMD, was that several essential components of the NMD machinery in metazoan organisms were not expressed in budding yeast.[2,5] A noteworthy example is SMG-1, which was genetically defined as an essential component of the NMD machinery in *C. elegans*.[4,6] Orthologs of *C. elegans* (Ce)SMG-1 have been described in mammalian cells, and are likely present in most if not all metazoan cells.[7-10] While the yeast genome encodes a number of SMG-1-related proteins, yeast cells do not express a true SMG-1 ortholog. We now know that these disparities are at least partially explained by mechanistic differences in the recognition of PTCs between these evolutionarily divergent organisms.[2]

The cloning of cDNAs encoding worm and human SMG-1 proteins revealed that SMG-1 is a member of an unusual family of protein kinases termed the phosphoinositide 3-kinase related kinases (PIKKs).[7,8,10] This observation implied that phosphorylation of one or more substrates was required for NMD in metazoan cells. Human (h)UPF1 has emerged as a major recipient of hSMG-1-mediated regulatory signals (see chapter by Yamashita et al), just as CeSMG-2, which is orthologous to hUPF1, is the major recipient of CeSMG-1-mediated regulatory signals (see chapter by Anderson). As discussed in greater detail below, the PIKK family members are centrally involved in cell signaling pathways triggered by metabolic or genetic stress.[11-13] This chapter will focus on the hSMG-1 kinase, beginning with an overview of its structural features and biochemical properties, many but not all of which are shared with other PIKK family members. With this information as a backdrop, we will then briefly discuss the role of hSMG-1 in the regulation of hUPF1, a pivotal effector in the NMD pathway. Finally, we will present some provocative new findings that suggest hSMG-1 participates in stress-response pathways that appear unrelated to its evolutionarily conserved role in NMD.

Structural Features of SMG-1 and Related Kinases

The mid-1990's witnessed the emergence of a novel family of kinases that were first predicted through genetic studies in yeast. Independent research teams led by Michael Hall[14] and George Livi[15] were studying the mechanism of action of rapamycin, a bacterial product that was known at the time to bear potent antifungal and immunosuppressive activities.[16] These studies led to the identification of two novel and highly related genes, termed TOR (Target Of Rapamycin) 1 and TOR2, respectively, that appeared to encode novel members of the phosphoinositide (PI) kinase superfamily. cDNA encoding a mammalian ortholog to yeast TOR was isolated shortly thereafter, followed closely by the isolation of related cDNAs encoding the mammalian proteins ATM, ATR, and DNA-PK, as well as their orthologs in budding and fission yeast.[17,18,19] Like the TOR proteins, these novel kinases uniformly expressed, at their carboxyl-termini, catalytic domains bearing significant sequence homology to the catalytic domains of lipid PI 3- and 4-kinases. The evolutionary conservation of this lipid kinase-like domain led to the name 'PI 3-kinase related kinases' (PIKKs), and triggered speculation that the substrates for the PIKKs were membrane-derived inositol phospholipids—an assumption that was eventually proven incorrect (see below). The second common feature of the newly cloned PIKK cDNAs was that their open reading frames encoded unusually large polypeptides (≥ ~290 kilodaltons), particularly when compared to other members of the extended kinase superfamily.

Efforts from many different laboratories led to the conclusion that the PIKKs represent a nonconventional subfamily of protein serine-threonine kinases. Human cells express a total of six PIKKs, including hSMG-1, which is the most recent addition to this unusual group of kinases.[20] As mentioned above, a ubiquitously expressed PIKK in eukaryotic cells is TOR, a key regulator of mRNA translation and cell growth, and a specific target of the clinically approved anti-proliferative drug, rapamycin.[13,21] Two members of this kinase family, ATM and ATR, serve as apical signal transducers in cell-cycle checkpoint pathways initiated by DNA double-strand breaks and DNA replication stress, respectively.[12,22] The ATM kinase is noteworthy because loss of ATM function in humans gives rise to the heritable disorder ataxia-telangiectasia (A-T). A-T patients exhibit a devastating array of pathologies, including neurodegeneration, immunodeficiency, and a heightened risk of developing cancer. Parenthetically, the majority of *ATM* gene mutations that underlie A-T yield PTC-containing *ATM* mRNA that is eliminated from the cell via NMD,[23,24] underscoring the central role of NMD in the etiology of human genetic diseases (see chapters by Holbrook et al and Keeling et al). The fifth mammalian PIKK, DNA-PK, is required for a specialized form of DNA double-strand break repair termed nonhomologous end joining. The five PIKKs described above have been biochemically characterized as protein serine-threonine kinases; indeed, with the exception of mammalian (m)TOR, these PIKKs preferentially phosphorylate protein substrates bearing a glutamine (Q) residue immediately carboxyl-terminal to the targeted serine (S) or threonine (T) residue. The real outlier in this protein kinase subfamily is the TRRAP protein, which lacks protein kinase activity due to nucleotide sequence alterations in the catalytic domain. TRRAP serves as a scaffold for Myc-dependent transcriptional complexes, and may participate in cell-cycle related changes in chromatin structure.[25,26]

From a structural perspective, CeSMG-1 and hSMG-1 manifest several motifs that are shared with the remaining PIKK family members (Fig. 1). These proteins contain extended amino-terminal regions (>1500 amino acids) of poorly defined function. A computational analysis predicted that the amino termini of hSMG-1, ATM, ATR, mTOR, and DNA-PK consist of several tandem arrays of α-helical subdomains termed HEAT (Huntington, Elongation factor 3, A subunit of protein phosphatase 2A, TOR1) repeats.[27] A single HEAT repeat comprises a pair of interacting, anti-parallel alpha helices linked by a flexible loop consisting of 30-50 amino acids. Remarkably, amino-terminal regions of the PIKKs contain 40-54 such HEAT repeats, arranged in four or more clustered arrays. The HEAT repeat regions appear to form large superhelical domains that offer numerous, potentially redundant sites for interactions with other proteins. The presence of such large HEAT arrays in PIKK family members

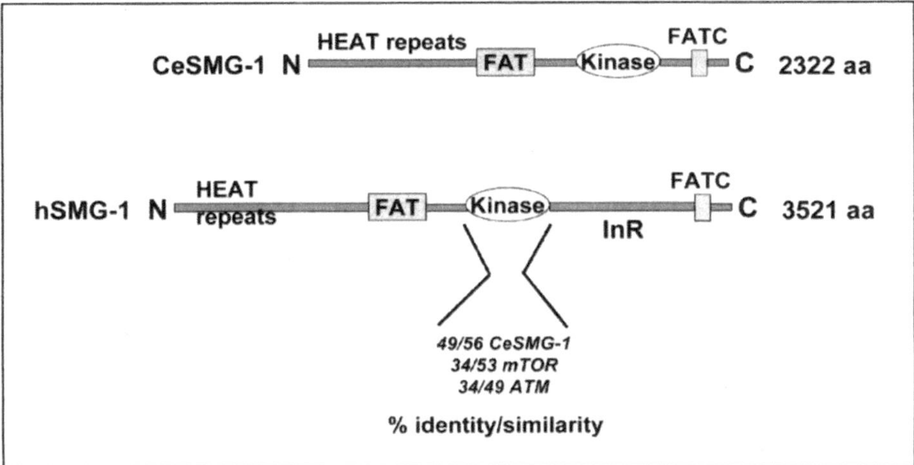

Figure 1. Domain structures of *C. elegans* and human SMG-1 proteins. The amino-terminal regions of both proteins contain an array of HEAT repeat domains that represent potential sites for protein-protein interactions. The conserved FAT and FATC domains are found in all PIKK family members, and play important roles in maintaining the intervening kinase domain in an enzymatically active conformation. Numbers below the hSMG-1 kinase domain indicate the percentage amino acid identity and similarity in the indicated PIKK catalytic regions. Human (h)SMG-1 contains an extended insert region (InR) not found in the *C. elegans* (Ce)SMG-1 ortholog. N, amino terminal; C, carboxyl terminal.

hints that these kinases form dynamic complexes with their respective partner proteins, some of which regulate PIKK function, while others likely mediate the presentation of specific substrates to the downstream catalytic domain. More detailed structural analyses of the amino-terminal regions of the various PIKK family members are clearly needed to understand the mechanisms through which these protein kinases communicate with both upstream regulators and downstream target proteins in eukaryotic cells.

As stated previously, the defining sequence motif of the PIKKs is the catalytic domain, which comprises only 5-10% of the total amino acids present in these very large polypeptides. A closer inspection of this domain revealed that the sequence similarity to the catalytic domains of the PI 3- and 4-kinases is confined largely to those residues involved in utilizing ATP, particularly with respect to those residues involved in the ATP-binding and phosphotransferase activities.[18] The catalytic domains of all PIKK family members are flanked by FAT (FRAP, ATM, TRRAP) and FATC (FAT C-terminal), domains at their amino and carboxyl ends, respectively.[28] The FAT domain encompasses ~500 amino acids, while the FATC domain spans only ~35 amino acids. Although the functions of FAT and FATC are poorly understood, mutational analyses suggest that both domains are critical for the activity of the intervening kinase domain (R.T. Abraham, unpublished observations). In general, the PIKK catalytic domain is separated from the FATC domain and the adjacent carboxyl terminus by ≤ 200 amino acids. The outstanding exception to this rule is hSMG-1, which contains an extended 'insert region' (InR) of >1000 amino acids between the kinase domain and the FATC domain.

The unique InR of hSMG-1 bears special mention, because it likely harbors some of the keys to understanding the specialized functions of this kinase in mammalian cells (Fig. 1). This prediction is underscored by the finding that CeSMG-1 lacks an extended InR.[7] A speculative proposal is that the InR expanded during metazoan evolution to allow SMG-1 access to a broadened spectrum of regulatory inputs and/or downstream substrates. Recent studies in our laboratory indicate that deletion of the entire InR generates a catalytically inactive hSMG-1 protein, suggesting that, in addition to its proposed roles in protein-protein interactions, the InR provides structural support for the catalytic domain (R.T. Abraham, unpublished observations). The creation of a series of mutated hSMG-1 proteins bearing partial deletions in the InR will yield important insights into the critical sub-regions that are needed for protein kinase activity as well as hSMG-1 function in intact cells. Furthermore, studies aimed toward the identification of putative binding partners for the InR of hSMG-1 are clearly warranted. A yeast two-hybrid screen using the InR as bait identified the protein serine-threonine phosphatase, PP5, as one such interacting protein (V. Oliveira and R.T. Abraham, unpublished observations). Interestingly, PP5 was previously shown to regulate signaling through at least two other PIKKs, ATM and DNA-PK.[29,30] While PP5 is certainly an intriguing partner for hSMG-1, more work is needed to clarify the biological significance of this interaction.

Characterizing hSMG-1 Kinase Activity

Human cDNA clones encoding hSMG-1 were isolated independently by three research groups.[8-10] Each group identified multiple *hSMG-1* mRNAs in various cell and tissue types, with the principal variation being the length of the 5'-termini. Our *hSMG-1* cDNA encodes a 3,521-amino acid polypeptide with a deduced molecular mass of 395 kDa, and it is eight amino acids shorter at the amino-terminus than the cDNA studied by Yamashita et al.[8] Whether multiple forms of hSMG-1 protein with different amino-termini are actually expressed in mammalian cells remains unclear. The subcellular localization of hSMG-1 has not been precisely defined, but the bulk of this protein is found in the cytoplasmic fraction of human cells.[10] However, a minor subpopulation of hSMG-1 is also observed in the nucleus and/or perinuclear compartment[10,31] (V. Oliveira and R.T. Abraham, unpublished observations). The subcellular distribution of hSMG-1 is clearly distinct from that of ATM and ATR, which are nuclear proteins, but is quite similar to that of mTOR, another PIKK family member with close ties to the translational machinery.[32,33]

To date, the major focus of biochemical studies has been on hSMG-1 kinase activity. Immunoprecipitates containing either native or recombinant hSMG-1 were subjected to immune complex kinase assays under reaction conditions previously shown to support the phosphotransferase activities of other PIKK family members, particularly ATM and mTOR. These experiments revealed that the protein kinase activity of hSMG-1 is more closely aligned with those of ATM/ATR/DNA-PK than with that of mTOR.[8,10] The immunoprecipitated protein did not recognize the translational repressor protein, 4E-BP1 (eukaryotic translation initiation factor 4E-Binding Protein), as a substrate in immune complex kinase reactions (R.T. Abraham, unpublished observations). 4E-BP1 contains several S/T-proline (P) motifs that are phosphorylated by mTOR, both in kinase reactions and in intact cells.[34,35] Unlike mTOR, immunoprecipitated hSMG-1 phosphorylates substrates bearing the S/T-Q motif favored by ATM, ATR, and DNA-PK.[11,36] The physiological substrates for ATM/ATR/DNA-PK are typically proteins involved in cell-cycle checkpoint signaling and DNA repair, and these substrates often express multiple phosphorylation sites clustered into S/T-Q-rich domains.[37] Notably, the hUPF1 protein, which is a well-established intracellular substrate for hSMG-1, contains a carboxyl-terminal S/T-Q-rich domain with fourteen candidate sites for phosphorylation by hSMG-1 (and possibly other PIKK family members; see below for further discussion). At least four of these S/T-Q motifs are modified by hSMG-1 in vitro[8] (see chapter by Yamashita et al). In immune complex kinase assays, hSMG-1 phosphorylated, at multiple S-Q motifs, a GST (Glutathione S-Transferase) fusion protein containing the carboxyl-terminal region (amino acid residues 1019-1118) of hUPF1.[8,10] This GST-hUPF1$^{1019-1118}$ fusion protein was also a good substrate for ATM, but not mTOR, under identical assay conditions.[10] The biochemical parallels between hSMG-1 and ATM were further strengthened by the finding that hSMG-1 avidly phosphorylated a GST-p53$^{1-70}$ fusion protein at Ser-15, which resides in an optimal peptide sequence (L^{15}SQE) for phosphorylation by the ATM kinase domain.[38,39] The phosphorylation of p53 at Ser-15 occurs after cellular exposure to various genotoxic agents, and has been implicated in both the stabilization and transcriptional activation of p53 in damaged cells.[40] Quantitative assays suggested that recombinant hSMG-1 was considerably more active as a GST-p53$^{1-70}$ kinase than was recombinant ATM, and recent reports indicate that hSMG-1 contributes to the phosphorylation of the Ser-15 site in cells exposed to ionizing radiation (IR).[10,41] An intriguing but as yet unproven model is that hSMG-1 and ATM phosphorylate partially overlapping sets of substrates in intact cells, particularly under stressful environmental conditions.

Pharmacological inhibitors of PIKK catalytic activities have played prominent roles in the functional characterization of these protein kinases. An extraordinary example is provided by the natural product, rapamycin, which is a highly potent and specific inhibitor of mTOR.[16] Rapamycin is actually a pro-drug that first binds to a ubiquitously expressed cytoplasmic receptor, FKBP12, to generate the proximate inhibitor of mTOR. The FKBP12•rapamycin complex interacts with a ~110-amino acid region of mTOR, termed the FRB (FKBP12•Rapamycin-Binding) domain, which lies upstream of the kinase domain and overlaps with the FAT domain.[42] Interestingly, a FRB-like domain was found in hSMG-1.[8,10] However, FKBP12•rapamycin neither binds to hSMG-1 nor inhibits hSMG-1 kinase activity (R.T. Abraham, unpublished observations). Two other well-studied PIKK inhibitors are wortmannin and caffeine, which, unlike rapamycin, exhibit broad inhibitory activities against this class of protein kinases.[43,44] Both wortmannin and caffeine are competitive inhibitors of ATP binding to PIKK catalytic domains. While caffeine reversibly inhibits PIKK kinase activities at millimolar concentrations, wortmannin is a far more potent (nanomolar to micromolar range), irreversible inhibitor of these enzymes.

Wortmannin is also a useful inhibitor because different PIKK family members display variable sensitivities to this drug. In immune complex kinase assays, ATM activity is inhibited by 50% (IC_{50}) with 80 nM wortmannin, whereas ATR kinase activity is far less sensitive to this drug (IC_{50}, 1800 nM).[44] In our hands, hSMG-1 and ATM kinase activities exhibited virtually identical sensitivity profiles to wortmannin in immune complex kinase assays[10] (R.T. Abraham, unpublished observations). Thus, studies with wortmannin as a kinase-domain probe further suggest that the ATM and hSMG-1 kinase domains are conformationally more similar to each other than to the other PIKK catalytic domains. For unknown reasons, wortmannin is a slightly more potent inhibitor of hSMG-1 than ATM in intact cells. This potency difference can be used advantageously, because treating human cells with 5 μM wortmannin virtually abolishes hSMG-1 activity, but spares >50% of the total ATM kinase activity (R.T. Abraham, unpublished observations). Thus, wortmannin (at the 5 μM concentration) can be deployed as a relatively selective probe of hSMG-1 functions in intact cells.

Roles of hSMG-1 in NMD

The only known substrate for hSMG-1 function in the NMD pathway is hUPF1, which is arguably the pivotal molecule that marks PTC-bearing mRNAs for rapid degradation. The phosphorylation of hUPF1, and the regulation of hUPF1 function by cycles of phosphorylation and dephosphorylation, are critical for NMD. As stated above, the carboxyl-terminal region of hUPF1 is termed an S/T-Q-rich domain by virtue of its complement of 14 candidate phosphorylation sites for PIKK family kinases. Phosphorylation of the S/T-Q-rich region (and perhaps other sites) in hUPF1 drives critical protein-protein interactions, including but by no means restricted to that involving the hSMG-7 protein.[31] Remarkably, the interaction between phosphorylated hUPF-1 and hSMG-7 ultimately recruits a PP2A-like protein phosphatase.[31,45] This elegant mechanism insures that phosphate turnover on hUPF1 is tightly coupled to its modification by hSMG-1. Exactly how the cyclic phosphorylation - dephosphorylation of hUPF1 promotes the degradation of PTC-bearing mRNAs is an area of active investigation.

While considerable progress has been made toward understanding the mechanism of hUPF1 dephosphorylation, the events that govern phosphorylation remain elusive. One scenario is that the SMG-1 kinases simply rephosphorylate CeSMG-2/hUPF1 after the substrate is dephosphorylated and released from mRNA surveillance complexes. The alternative model, which we favor, is that the phosphorylation of CeSMG-2/hUPF1 by the respective SMG-1 (and other protein kinases) is highly regulated to insure that CeSMG-2/hUPF1 is modified with the proper timing and in the appropriate location. In the absence of definitive data, we can look to other members of the PIKK family for clues regarding the regulation of hSMG-1 activity toward hUPF1. For conventional protein serine-threonine or tyrosine kinases, substrate phosphorylation is often preceded by phosphorylation or other covalent modifications of the protein kinase itself that lead to a measurable increase in the activity of the catalytic domain. Within the PIKK family, efforts to observe stimulus-dependent increases in phosphotransferase activity have yielded mixed results. DNA-PK participates in the repair of DNA double-strand breaks, and its catalytic activity in vitro is increased by binding to double-stranded DNA.[46] More recent studies have uncovered an elegant model for ATM activation by genotoxic agents.[47,48] In its inactive state, ATM resides in the nucleus in homo-dimeric or homo-oligomeric complexes. Exposure of cells to DNA-damaging agents triggers auto-and/or trans-phosphorylation of the ATM polypeptide at a single Ser residue, which in turn causes dissociation of the ATM complexes into active ATM monomers. In contrast, it has been difficult to document a reliable increase in the protein kinase activity of ATR during genotoxic stress, or of mTOR in cells exposed to mitogens or nutrients.

A recent report from our laboratory indicated that IR or ultraviolet (UV) light triggers a relatively rapid increase in hSMG-1 kinase activity.[10] Although the exact mechanism is unknown, these results suggest that hSMG-1 is activated by agents that induce 'macromolecular stress', involving widespread damage to DNA, RNA, proteins, and lipids. A key unresolved question is whether the presence of a PTC-containing mRNA delivers a stimulatory or inhibitory signal to the hSMG-1 kinase domain. An alternative paradigm for the regulation of hUPF1 phosphorylation by hSMG-1 is an induced-proximity model, in which hUPF1 is presented to the hSMG-1 kinase domain at the appropriate time and/or in the proper conformation for phosphorylation. An instructive precedent for this model is provided by mTOR, which shares with hSMG-1 a tight linkage to the translational machinery. Genetic and proteomic analyses revealed that mTOR resides in at least two multi-protein complexes in eukaryotic cells.[49] These complexes, termed TORC1 and TORC2, phosphorylate distinct sets of substrates, and they exhibit strikingly different sensitivities to the mTOR kinase inhibitor, rapamycin.[50-52] A key determinant of substrate specificity for each of the complexes is not the mTOR-kinase domain itself, but rather the set of auxiliary proteins that surround the mTOR polypeptide in these complexes. For the TORC1 complex, the mTOR-associated Raptor protein plays a pivotal role in the selection of the major TORC1 substrates, 4E-BP1 and p70S6 kinase, for presentation to the mTOR catalytic domain. Similar substrate selection and presentation functions may be carried out by an analogous protein, named Rictor, in the TORC2 complex.[50,51] The TORC2 complex appears to target AKT as one of its key substrates.[53] The exclusive presence of Raptor in TORC1 and Rictor in TORC2 therefore dictates the distinct substrate specificities of each complex. By analogy, the kinase activity of hSMG-1 may also be regulated by an induced-proximity mechanism involving the expression of several multi-protein complexes with differential phosphotransferase activities toward hUPF1 versus other (yet to be defined) hSMG-1 substrates. With the availability of recombinant hSMG-1, together with higher-quality antibodies directed against the endogenous protein, the identification of hSMG-1-binding proteins will no doubt be pursued in earnest. We expect that these studies will yield crucial insights into the mechanism of substrate phosphorylation by hSMG-1, as well as the regulation of this PIKK by signals from the external environment and intracellular milieu.

Cytoprotective Functions of hSMG-1

Although some major conceptual gaps remain to be filled, the central role of hSMG-1 in NMD in human cells is now firmly established. A considerably more speculative area of research focuses on the participation of hSMG-1 in stress-response pathways in these cells. As discussed above, the related PIKKs, ATM, ATR, and DNA-PK, are crucial players in the signaling network that promotes cellular survival after potentially catastrophic damage to genomic DNA.[12] Likewise, mTOR is a key regulator of the switch from anabolic metabolism in the presence of adequate nutrient supplies to an energy-conserving starvation program that allows cells to cope with inadequate supplies of bioenergetic precursors.[54] A recent study by our group offered suggestive evidence that hSMG-1 responds to cellular stress induced by agents that damage DNA and other intracellular macromolecules.[10,55] We will discuss these provocative findings, together with some more recent data suggesting that hSMG-1 regulates cellular sensitivity to apoptosis induced by stimulation of members of the TNF-R (Tumor Necrosis Factor-Receptor) family. Collectively, these studies lead us to propose that hSMG-1 carries out additional 'moonlighting' functions, unrelated to NMD, that protect human cells against DNA damage and apoptotic death.

As described above, the biochemical and pharmacologic characteristics of hSMG-1 most closely resemble those of a subset of the PIKK family members that are exemplified by ATM and, to a lesser extent, ATR. Indeed, the similarity was sufficiently striking that we originally designated our *hSMG-1* cDNA as '*ATX*' cDNA, a name that acquired a temporary foothold among investigators in the DNA damage-response field.[12] The first supporting evidence for the idea that ATM, ATR, and ATX/hSMG-1 form a triad of stress-responsive, S/T-Q-directed kinases in human cells came from the above-cited studies showing that hSMG-1 phosphorylated a peptide derived from the p53 tumor suppressor protein at the same site (Ser-15) known to be targeted by the immunoprecipitated ATM or ATR kinases. The latter two PIKK family members are thought to be major effectors of p53 (Ser-15) phosphorylation in cells exposed to IR or other DNA-damaging agents.[56-58] Indeed, we found that hSMG-1 was the most avid p53 (Ser-15) kinase among all of the PIKK family members tested in these in vitro kinase assays—approximately 4-fold more active on a protein basis than was the 'gold standard' p53 (Ser-15) kinase, ATM. Obviously, however, the results of such in vitro kinase assays must be extrapolated to the in vivo situation with considerable caution.

The phosphorylation of p53 by hSMG-1 suggested that this protein kinase might share with ATM and ATR a cytoprotective function during genotoxic stress. Indeed, cells depleted of hSMG-1 using small-interfering (si)RNA exhibit a modest but consistent increase in radiosensitivity, suggesting that, like ATM,[12] hSMG-1 transmits pro-survival signals that contribute to cellular recovery from radiation-induced macromolecular damage.[10] IR-damaged cells typically accumulate in the G_2-phase of the cell cycle, a response that is mediated through the 'G_2 DNA-damage checkpoint'.[59] Once again, both ATM and ATR have been implicated as important mediators of G_2 checkpoint signaling induced by genotoxic stress. We observed that hSMG-1-depleted cells also exhibited a partial defect in G_2 damage checkpoint function, suggesting once again that three PIKKs—ATM, ATR, and hSMG-1—contribute to optimal activation of the DNA damage response network in human cells. A key question is whether hSMG-1 responds directly to damaged DNA or to other sequelae, such as protein damage or oxidative stress, which inevitably result from cellular exposure to radiation and many other genotoxic agents.

Perhaps the most compelling support for the hypothesis that hSMG-1 functions as a stress-responsive kinase is that exposure of human cells to IR or UV light leads to a rapid increase in hSMG-1 phosphotransferase activity toward p53 in immune complex kinase assays.[10] In this regard, hSMG-1 resembles ATM, and it will be interesting to learn whether, like ATM, hSMG-1 undergoes transitions from dimeric/oligomeric complexes to a monomeric polypeptide[47] during the activation process. In addition, studies with human cells that over-express wild-type or catalytically inactive hSMG-1 protein, or the same cells transfected with hSMG-1-specific siRNA, displayed phenotypes consistent with the idea that hSMG-1 function is required for optimal p53 stabilization and Ser-15 phosphorylation in IR-damaged cells.[10] More recent, unpublished experiments performed in our laboratory demonstrated that hSMG-1 coimmunoprecipitates with p53 in IR-treated cells, suggesting that these two proteins are partners in an enzyme-substrate relationship. Interestingly, an independent group has recently suggested that the chemopreventative antioxidant, tempol, triggers p53 (Ser-15) phosphorylation in a hSMG-1-dependent fashion.[41] The possibility that changes in the intracellular redox status influence hSMG-1, as well as NMD activity, is intriguing and warrants further investigation.

Consistent with a role for hSMG-1 in genome surveillance and/or repair, we have found that hSMG-1 depletion using siRNA causes the accumulation of DNA double-strand breaks in human cancer cells.[10] These results suggest that hSMG-1 function is required for the DNA damage response, even in the absence of extrinsic genotoxic agents. This finding brings to mind the related PIKK family member, ATR, which is essential for chromosome maintenance and viability in mammalian cells.[12,22] The critical contribution of ATR is to insure that DNA replication forks generate a complete and accurate copy of the genome during S phase. Whether hSMG-1 plays a cell-cycle, phase-specific function in DNA damage detection and repair, or more general roles in these critical processes, remains to be established. Furthermore, the possibility that the spontaneous DNA damage observed in hSMG-1-depleted cells is an indirect consequence of the suppression of NMD activity has not been formally ruled out. If loss of hSMG-1 function increases genetic instability, we might predict that this protein kinase would behave as a tumor suppressor in human cancer. In fact, a recent report documents the opposite outcome; i.e., the human *SMG1* gene is overexpressed in a subset of patients with acute myeloid leukemia (AML).[60] The functional significance of this finding is unclear, and, once again, further studies are needed to determine whether loss or gain of hSMG-1 function contributes in any way to cancer development in humans.

Anti-Apoptotic Functions of hSMG-1

The predominant localization of hSMG-1 in the cytoplasmic compartment positions this protein to interact with the signaling machineries engaged by plasma membrane receptors for extracellular ligands. In follow-up studies to our earlier subcellular localization assessments, we used higher-quality anti-SMG-1 antibodies to examine in greater detail the compartmentalization of hSMG-1 in human cell lines, including the U2-OS osteosarcoma line. A particularly noteworthy outcome of these experiments is that a subpopulation of hSMG-1 decorated the

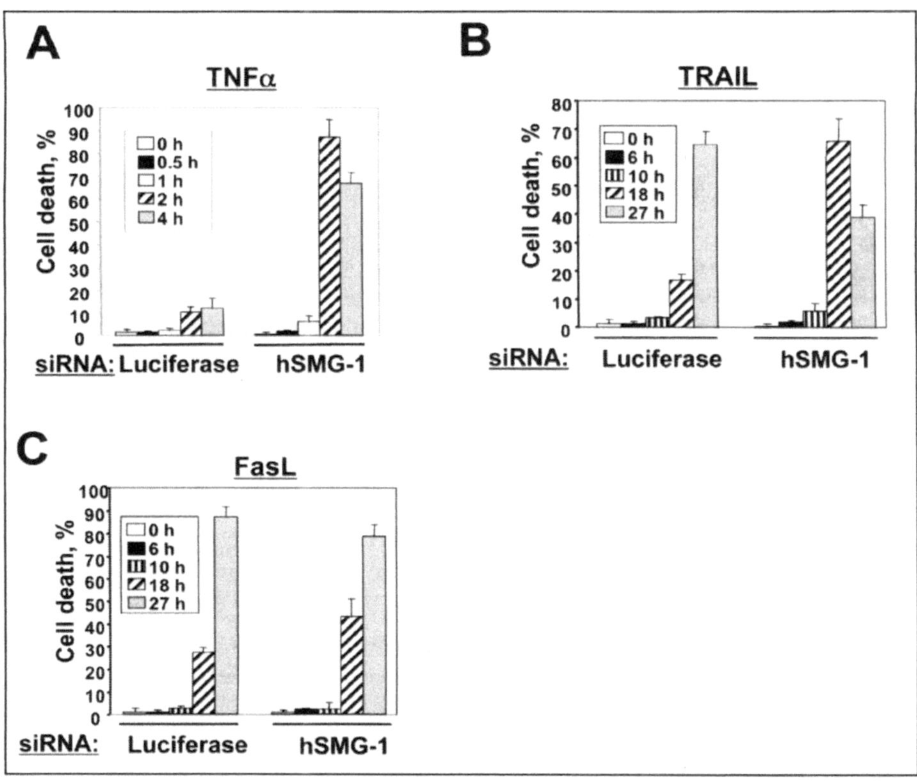

Figure 2. Loss of hSMG-1 sensitizes cells to TNFα- or TRAIL-induced apoptosis. U2-OS cells were transfected with control (Luciferase)-siRNA or hSMG-1-siRNA duplexes, and, after 48 h, were treated for the indicated times with 20 ng per ml of TNFα (A), 200 ng per ml of TRAIL (B), or 200 ng per ml of FasL (C). The percentage of cell death was determined using a Trypan blue dye-exclusion assay and scanning electron microscopy. Results are presented as the mean from three independent trials.

outer surface of mitochondria in these cells. As mitochondrial damage is a well-established initiating event in the "intrinsic" pathway of cellular apoptosis,[61] we hypothesized that hSMG-1 might regulate mitochondrial integrity, thereby modulating cellular sensitivity to apoptosis. Consequently, we examined the effects of several stress-inducing stimuli on human cells depleted of hSMG-1 using siRNA. These studies revealed that loss of hSMG-1 function led to a dramatic increase in cellular sensitivity to apoptosis induced by TNFα or TRAIL (TNF-Related Apoptosis-Inducing Ligand) (Fig. 2). The sensitizing effects of hSMG-1 siRNA were relatively specific, in that cell killing by Fas L (Fas Ligand) was not increased in hSMG-1-depleted cells. TNFα, TRAIL, and FasL are evolutionarily related cytokines that bind to members of the TNFR (Tumor Necrosis Factor-Receptor) family.[62] Our studies with these TNFR ligands were initially performed using fully transformed human cancer cell lines, such as U2-OS osteosarcoma cells. We therefore repeated these experiments with a noncancerous human mammary epithelial cell line, MCF-10A, which expresses receptors for both TNFα and TRAIL. In sharp contrast to the human cancer cell lines, MCF-10A cells displayed no increase in TNFα- or TRAIL-induced killing after depletion of hSMG-1 with siRNA. The selective effect of human *SMG-1* gene silencing on transformed cells is obviously intriguing, and warrants a more comprehensive investigation.

Under normal circumstances, many cells survive exposure to TNF due to the activation of counterbalancing pro-apoptotic and pro-survival pathways.[62] Engagement of trimeric TNFR-1 clusters by the cytokine elicits the assembly of the DISC (Death-Inducing Signaling Complex) on the receptor cytoplasmic domains. The DISC ultimately mediates the auto-catalytic activation of either caspase-8 or caspase-10, which are DISC components. These activated apoptotic proteases in turn cleave specific substrates, including additional caspase isoforms, initiating a cascade of events culminating in apoptotic death. Depending on the cell type and circumstances, the apoptotic cascade may kill the cell via the extrinsic (mitochondrial damage-independent) pathway or the intrinsic pathway, which involves the release of additional apoptotic mediators from damaged mitochondria. This TNFR-1 triggered march toward cell death is held in check by the activation of parallel signaling pathways that favor cell survival, most notably a protein kinase-dependent cascade that results in the nuclear translocation of the heterodimeric transcription factor, NFκB. This transcription factor controls a genotypic response program that leads to a general increase in cellular resistance to stress. TNFα-induced NFκB activation blocks apoptosis, in part via NFκB-induced expression of FLIP (Flice-Inhibitory Protein), which is a potent inhibitor of DISC-mediated caspase-8/10 activation.[63] Consequently, manipulations that interfere with the activation or trans-activating functions of NFκB dramatically increase the level of apoptotic cell death provoked by TNFα. We therefore tested the hypothesis that the loss of hSMG-1 suppresses the activation and/or function of NFκB in TNFα-stimulated cells, and we found that NFκB function is not impaired in hSMG-1-depleted cells. Moreover, the striking effects of hSMG-1 depletion on TNFα/TRAIL sensitivity appear unrelated to the inhibition of NMD, because siRNA-mediated *UPF1* gene silencing, which leads to a profound inhibition of NMD, fails to sensitize cancer cells to TNFα/TRAIL-induced apoptosis. Ongoing experiments are testing the hypothesis that the localization of hSMG-1 to the mitochondrial outer membrane prevents damage to this key organelle, which suppresses activation of the intrinsic pathway of apoptosis in TNFα-stimulated cells. An equally plausible model posits that hSMG-1 suppresses TNFα/TRAIL-induced apoptosis by direct phosphorylation of DISC components. The outcomes of these studies could have major implications for our understanding of the signaling machinery engaged by TNFR family members, and may lead to efforts to modulate hSMG-1 function in patients with cancer or chronic inflammatory diseases.

Conclusions and Perspectives

The first report describing the molecular cloning and functional characterization of full-length hSMG-1 appeared in 2001,[8] making this protein kinase the most recent addition to the sub-family of PIKKs expressed in mammalian cells. In contrast, research efforts surrounding the other mammalian PIKKs—ATM, ATR, DNA-PK, and mTOR—have been intensively pursued for more than a decade. Studies of the PIKKs have had a disproportionately broad impact on our understanding of the signaling networks that govern the growth and stress responsiveness of mammalian cells. Furthermore, this body of work has established strong ties between aberrant signaling through the PIKKs and major human diseases, including cancer, immunodeficiency and neurodegenerative disorders. Indeed, one member of the PIKK family, mTOR, is a validated drug target for cancer therapy and immunosuppression.[16,21] We expect that ongoing studies of the hSMG-1 kinase will yield equally profound insights into both cell-signaling mechanisms and disease pathogenesis in humans. Our current view of the regulation and functions of hSMG-1 is, at this stage, more speculative than factual (Fig. 3). On the regulatory side, we know that hSMG-1 catalytic activity is stimulated by external stress, such as that provoked by IR, but the mechanism of kinase activation is unknown. A related and an equally important area of research concerns the identification of hSMG-1-interacting proteins, as well as the full range of downstream target proteins that are influenced by hSMG-1. Finally, the significance of hSMG-1 localization on mitochondria is a critical issue that warrants detailed investigation, particularly with respect to the anti-apoptotic activity of this protein kinase.

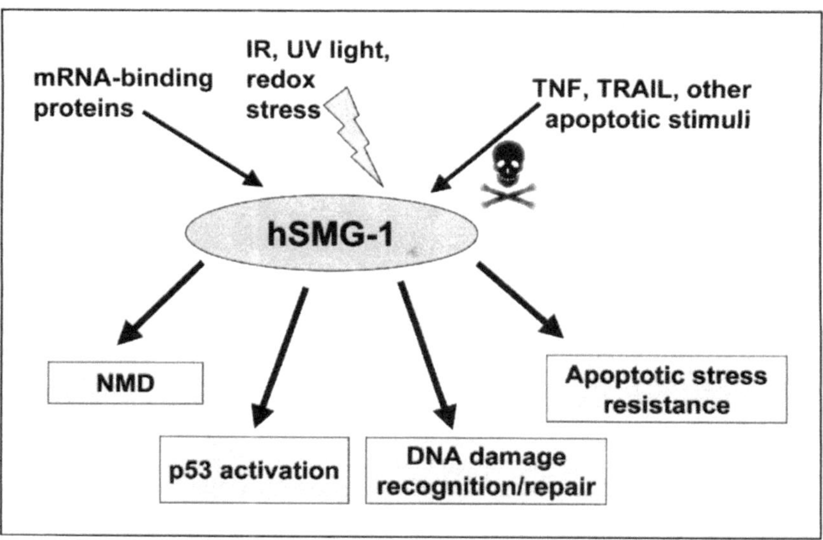

Figure 3. A working model for the regulation and functions of hSMG-1 in human cells. The most well-documented function of hSMG-1 is its involvement in the NMD pathway. We speculate that hSMG-1 is recruited to and/or activated by PTC-bearing mRNAs based on the binding of specific hSMG-1-interacting proteins to the targeted mRNA transcripts. Exposure of cells to ionizing radiation (IR) or ultraviolet (UV) light activates the kinase domain, and stimulates the phosphorylation of proteins, such as p53, which in turn orchestrate cellular responses to DNA damage and other forms of macromolecular stress. Finally, hSMG-1 protects cells against TNFα- and TRAIL-induced apoptosis, possibly by suppressing damage to mitochondria.

A major objective of this chapter was to highlight evidence that hSMG-1 carries out pleiotropic functions, not all of which are directly related to its evolutionarily conserved role in NMD. This discussion would be incomplete without a brief overview of the findings that the major hSMG-1 substrate, hUPF1, is not only a central player in NMD, but also a protein that participates in other intracellular signaling pathways. The NMD-related functions of hUPF1 are described in detail in chapters by Kim and Maquat and Kaygun and Marzluff, who present provocative evidence that hUPF1 steps outside of its role in NMD and participates in, respectively, Staufen1-mediated mRNA decay and the degradation of histone mRNA in mammalian cells. Our attention was drawn to the carboxyl-terminal, S/T-Q-rich domain of hSMG-1, which we predicted might target this protein for phosphorylation by PIKK family members other than hSMG-1. Based on our earlier biochemical and pharmacological studies,[10] we considered ATM to be a strong candidate for the proposed 'alternate' effector of hUPF1 phosphorylation. The fact that ATM, unlike hSMG-1, is primarily localized in the nucleus does not present a major conceptual roadblock to this hypothesis since hUPF1 is known to cycle between cytoplasmic and nuclear compartments.[64] Preliminary studies from our laboratory indicate that cellular exposure to IR provokes a rapid increase in hUPF1 phosphorylation at S/T-Q motifs distinct from the two S-Q sites (Ser-1078 and Ser-1096) identified as in vivo targets for hSMG-1 by Ohno and coworkers.[8] Furthermore, siRNA-mediated depletion of hUPF1, but not the distinct NMD pathway component, hUPF2, causes a dramatic increase in DNA strand breaks within 48 h (C. Geisen, V. Oliveira and R.T. Abraham, unpublished observations). Because both *UPF1* and *UPF2* gene knockdowns inhibit NMD in human cells, the DNA damage induced by hUPF1 deficiency appears unrelated to suppression of the NMD pathway. Along these lines, it is intriguing that phosphorylated S/T-Q motifs in certain DNA damage response proteins mediate the recruitment of partner proteins containing BRCT (BRCA1 Carboxyl-Terminal) domains.[65,66] Perhaps phosphorylation of hUPF1 by hSMG-1 and/or ATM triggers interactions with BRCT domain-containing proteins, which are common components of the cell-cycle checkpoint signaling and DNA repair machinery. Additional studies of hUPF1 function, and the role of the S/T-Q-rich domain in the modulation of those functions, will undoubtedly provide some critical information regarding the mechanism of NMD, and may well yield some surprising insights into the DNA damage response network in human cells.

As a potential drug target, hSMG-1 presents an attractive point of intervention in the NMD pathway. In principle, inhibitors of hSMG-1, and in turn NMD, could restore the expression of at least partially functional proteins in cells crippled by the complete absence of key structural proteins or enzymatic activities due to mutations or aberrant mRNA splicing.[67] Obviously, the risk of generally inhibiting NMD could outweigh the benefits of enhancing the expression of specific, disease-related mRNAs. In the near term, cancer is one disease in which a hSMG-1 inhibitor might display a favorable risk-benefit ratio. In a broad sense, cancer cells might be sensitized to drug-mediated suppression of NMD, based on the hypothesis that these cells are overloaded with potentially toxic, PTC-bearing mRNAs due to both gene mutations and aberrant splicing. An analogous mechanism may underlie the clinically useful antitumor activities of proteasome inhibitors, such as bortezomib.[68] The latter agents may derive their favorable therapeutic indices from the presence of an overload of misfolded proteins in certain malignant cells, which renders them more sensitive to agents that interfere with the major mechanism whereby these potentially toxic polypeptides are removed from the cell. In certain cancer cells, suppressing NMD might have the added benefit of restoring the expression of tumor suppressor genes, such as p53 or retinoblastoma, in situations where mutations introduce a PTC into the mRNA transcript. If the truncated gene product retains some of its normal functions, treatment with a hSMG-1 inhibitor could trigger senescence or death in the malignant host cell. Finally, our own recent observation that depletion of hSMG-1 from certain malignant cell lines sensitizes these cells to TNFα- and TRAIL-induced apoptosis raises some obvious possibilities for combination therapy with hSMG-1 inhibitors in cancer patients. If any of these predictions regarding hSMG-1 inhibitors are substantiated in preclinical investigations, hSMG-1 could become an exciting target for the discovery of novel anticancer agents.

References

1. Lejeune F, Maquat LE. Mechanistic links between nonsense-mediated mRNA decay and pre-mRNA splicing in mammalian cells. Curr Opin Cell Biol 2005; 17:309-315.
2. Conti E, Izaurralde E. Nonsense-mediated mRNA decay: Molecular insights and mechanistic variations across species. Curr Opin Cell Biol 2005; 17:316-325.
3. Schwerk C, Schulze-Osthoff K. Regulation of apoptosis by alternative pre-mRNA splicing. Mol Cell 2005; 19:1-13.
4. Hodgkin J, Papp A, Pulak R et al. A new kind of informational suppression in the nematode Caenorhabditis elegans. Genetics 1989; 123:301-313.
5. Culbertson MR, Leeds PF. Looking at mRNA decay pathways through the window of molecular evolution. Curr Opin Genet Dev 2003; 13:207-214.
6. Pulak R, Anderson P. mRNA surveillance by the Caenorhabditis elegans smg genes. Genes Dev 1993; 7:1885-1897.
7. Grimson A, O'Connor S, Newman CL et al. SMG-1 is a phosphatidylinositol kinase-related protein kinase required for nonsense-mediated mRNA Decay in Caenorhabditis elegans. Mol Cell Biol 2004; 24:7483-7490.
8. Yamashita A, Ohnishi T, Kashima I et al. Human SMG-1, a novel phosphatidylinositol 3-kinase-related protein kinase, associates with components of the mRNA surveillance complex and is involved in the regulation of nonsense-mediated mRNA decay. Genes Dev 2001; 15:2215-2228.
9. Denning G, Jamieson L, Maquat LE et al. Cloning of a novel phosphatidylinositol kinase-related kinase: Characterization of the human SMG-1 RNA surveillance protein. J Biol Chem 2001; 276:22709-22714.
10. Brumbaugh KM, Otterness DM, Geisen C et al. The mRNA surveillance protein hSMG-1 functions in genotoxic stress response pathways in mammalian cells. Mol Cell 2004; 14:585-598.

11. Abraham RT. PI 3-kinase related kinases: 'Big' players in stress-induced signaling pathways. DNA Repair 2004; 3:883-887.
12. Shiloh Y. ATM and related protein kinases: Safeguarding genome integrity. Nat Rev Cancer 2003; 3:155-168.
13. Fingar DC, Blenis J. Target of rapamycin (TOR): An integrator of nutrient and growth factor signals and coordinator of cell growth and cell cycle progression. Oncogene 2004; 23:3151-3171.
14. Kunz J, Henriquez R, Schneider U et al. Target of rapamycin in yeast, TOR2, is an essential phosphatidylinositol kinase homolog required for G1 progression. Cell 1993; 73:585-596.
15. Cafferkey R, McLaughlin MM, Young PR et al. Yeast TOR (DRR) proteins: Amino-acid sequence alignment and identification of structural motifs. Gene 1994; 141:133-136.
16. Abraham RT, Wiederrecht GJ. Immunopharmacology of rapamycin. Ann Rev Immunol 1996; 14:483-510.
17. Keith CT, Schreiber SL. PIK-related kinases: DNA repair, recombination, and cell cycle checkpoints. Science 1995; 270:50-51.
18. Hunter T. When is a lipid kinase not a lipid kinase? When it is a protein kinase. Cell 1995; 83:1-4.
19. Zakian VA. ATM-related genes: What do they tell us about functions of the human gene? Cell 1995; 82:685-687.
20. Tibbetts RS, Abraham RT. PI3K-related kinases - Roles in cell-cycle regulation and DNA damage responses. Signaling Networks and Cell Cycle Control: The Molecular Basis of Cancer and Other Diseases. Bethesda: Humana Press, 2000:267-301.
21. Bjornsti MA, Houghton PJ. The tor pathway: A target for cancer therapy. Nat Rev Cancer 2004; 4:335-348.
22. Abraham RT. Cell cycle checkpoint signaling through the ATM and ATR kinases. Genes Dev 2001; 15:2177-2196.
23. Teraoka SN, Telatar M, Becker-Catania S et al. Splicing defects in the ataxia-telangiectasia gene, ATM: Underlying mutations and consequences. Am J Human Genetics 1999; 64:1617-1631.
24. Lai CH, Chun HH, Nahas SA et al. Correction of ATM gene function by aminoglycoside-induced read-through of premature termination codons. Proc Natl Acad Sci USA 2004; 101:15676-15681.
25. McMahon SB, Van Buskirk HA, Dugan KA et al. The novel ATM-related protein TRRAP is an essential cofactor for the c-Myc and E2F oncoproteins. Cell 1998; 94:363-374.
26. Herceg Z, Wang ZQ. Rendez-vous at mitosis: TRRAPed in the chromatin. Cell Cycle 2005; 4:383-387.
27. Perry J, Kleckner N. The ATRs, ATMs, and TORs are giant HEAT repeat proteins. Cell 2003; 112:151-155.
28. Bosotti R, Isacchi A, Sonnhammer EL. FAT: A novel domain in PIK-related kinases. Trends Biochem Sci 2000; 25:225-227.
29. Ali A, Zhang J, Bao S et al. Requirement of protein phosphatase 5 in DNA-damage-induced ATM activation. Genes Dev 2004; 18:249-254.
30. Wechsler T, Chen BP, Harper R et al. DNA-PKcs function regulated specifically by protein phosphatase 5. Proc Natl Acad Sci USA 2004; 101:1247-1252.
31. Ohnishi T, Yamashita A, Kashima I et al. Phosphorylation of hUPF1 induces formation of mRNA surveillance complexes containing hSMG-5 and hSMG-7. Mol Cell 2003; 12:1187-1200.
32. Gingras AC, Raught B, Sonenberg N. Regulation of translation initiation by FRAP/mTOR. Genes Dev 2001; 15:807-826.
33. Kim JE, Chen J. Cytoplasmic-nuclear shuttling of FKBP12-rapamycin-associated protein is involved in rapamycin-sensitive signaling and translation initiation. Proc Natl Acad Sci USA 2000; 97:14340-14345.
34. Brunn GJ, Hudson CC, Sekulic A et al. Phosphorylation of the translational repressor PHAS-I by the mammalian target of rapamycin. Science 1997; 277:99-101.
35. Brunn GJ, Fadden P, Haystead TAJ et al. The mammalian target of rapamycin phosphorylates sites having a (Ser/Thr)-Pro motif and is activated by antibodies to a region near its COOH terminus. J Biol Chem 1997; 272:32547-32550.
36. Chiang GG, Abraham RT. Determination of the catalytic activity of mTOR and other members of the phosphoinositide 3-kinase related kinase family. Meth Mol Biol 2004; 281:125-141.
37. Traven A, Heierhorst J. SQ/TQ cluster domains: Concentrated ATM/ATR kinase phosphorylation site regions in DNA-damage-response proteins. Bioessays 2005; 27:397-407.
38. Kim ST, Lim DS, Canman CE et al. Substrate specificities and identification of putative substrates of ATM kinase family members. J Biol Chem 1999; 274:37538-37543.
39. O'Neill T, Dwyer AJ, Ziv Y et al. Utilization of oriented peptide libraries to identify substrate motifs selected by ATM. J Biol Chem 2000; 275:22719-22727.

40. Giaccia AJ, Kastan MB. The complexity of p53 modulation: Emerging patterns from divergent signals. Genes Dev 1998; 12:2973-2983.
41. Erker L, Schubert R, Yakushiji H et al. Cancer chemoprevention by the antioxidant tempol acts partially via the p53 tumor suppressor. Hum Mol Genet 2005; 14:1699-1708.
42. Vilella-Bach M, Nuzzi P, Fang Y et al. The FKBP12-rapamycin-binding domain is required for FKBP12-rapamycin-associated protein kinase activity and G1 progression. J Biol Chem 1999; 274:4266-4272.
43. Sarkaria JN, Busby EC, Tibbetts RS et al. Inhibition of ATM and ATR kinase activities by the radiosensitizing agent, caffeine. Cancer Res 1999; 59:4375-4382.
44. Sarkaria JN, Tibbetts RS, Busby EC et al. Inhibition of phosphoinositide 3-kinase related kinases by the radiosensitizing agent wortmannin. Cancer Res 1998; 58:4375-4382.
45. Anders KR, Grimson A, Anderson P. SMG-5, required for C. elegans nonsense-mediated mRNA decay, associates with SMG-2 and protein phosphatase 2A. EMBO J 2003; 22:641-650.
46. Smith GCM, Jackson SP. The DNA-dependent protein kinase. Genes Dev 1999; 13:916-934.
47. Bakkenist CJ, Kastan MB. DNA damage activates ATM through intermolecular autophosphorylation and dimer dissociation. Nature 2003; 421:499-506.
48. Bakkenist CJ, Kastan MB. Initiating cellular stress responses. Cell 2004; 118:9-17.
49. Abraham RT. Identification of TOR signaling complexes: More TORC for the cell growth engine. Cell 2002; 111:9-12.
50. Sarbassov dos D, Ali SM, Kim DH et al. Rictor, a novel binding partner of mTOR, defines a rapamycin-insensitive and raptor-independent pathway that regulates the cytoskeleton. Curr Biol 2004; 14:1296-1302.
51. Jacinto E, Loewith R, Schmidt A et al. Mammalian TOR complex 2 controls the actin cytoskeleton and is rapamycin insensitive. Nat Cell Biol 2004; 6:1122-1128.
52. Jacinto E, Hall MN. Tor signalling in bugs, brain and brawn. Nat Rev Mol Cell Biol 2003; 4:117-126.
53. Sarbassov DD, Guertin DA, Ali SM et al. Phosphorylation and regulation of Akt/PKB by the rictor-mTOR complex. Science 2005; 307:1098-1101.
54. Lum JJ, DeBerardinis RJ, Thompson CB. Autophagy in metazoans: Cell survival in the land of plenty. Nat Rev Mol Cell Biol 2005; 6:439-448.
55. Abraham RT. The ATM-related kinase, hSMG-1, bridges genome and RNA surveillance pathways. DNA Repair 2004; 3:919-925.
56. Canman CE, Lim DS, Cimprich KA et al. Activation of the ATM kinase by ionizing radiation and phosphorylation of p53. Science 1998; 281:1677-1679.
57. Banin S, Moyal L, Shieh S et al. Enhanced phosphorylation of p53 by ATM in response to DNA damage. Science 1998; 281:1674-1677.
58. Tibbetts RS, Brumbaugh KM, Williams JM et al. A role for ATR in the DNA damage-induced phosphorylation of p53. Genes Dev 1999; 13:152-157.
59. O'Connell MJ, Walworth NC, Carr AM. The G2-phase DNA-damage checkpoint. Trends Cell Biol 2000; 10:296-303.
60. Neben K, Schnittger S, Brors B et al. Distinct gene expression patterns associated with FLT3- and NRAS-activating mutations in acute myeloid leukemia with normal karyotype. Oncogene 2005; 24:1580-1588.
61. Zhivotovsky B, Kroemer G. Apoptosis and genomic instability. Nat Rev Mol Cell Biol 2004; 5:752-762.
62. Aggarwal BB. Signalling pathways of the TNF superfamily: A double-edged sword. Nat Rev Immunol 2003; 3:745-756.
63. Kreuz S, Siegmund D, Scheurich P et al. NF-kappaB inducers upregulate cFLIP, a cycloheximide-sensitive inhibitor of death receptor signaling. Mol Cell Biol 2001; 21:3964-3973.
64. Mendell JT, Rhys C, Dietz HC. Separable roles for rent1/hUpf1 in altered splicing and decay of nonsense transcripts. Science 2002; 298:419-422.
65. Manke IA, Lowery DM, Nguyen A et al. BRCT repeats as phosphopeptide-binding modules involved in protein targeting. Science 2003; 302:636-639.
66. Yu X, Chini CC, He M et al. The BRCT domain is a phospho-protein binding domain. Science 2003; 302:639-642.
67. Holbrook JA, Neu-Yilik G, Hentze MW et al. Nonsense-mediated decay approaches the clinic. Nat Genet 2004; 36:801-808.
68. Cardoso F, Ross JS, Picart MJ et al. Targeting the ubiquitin-proteasome pathway in breast cancer. Clin Breast Cancer 2004; 5:148-157.

CHAPTER 17

Staufen1-Mediated mRNA Decay: A Upf1-Dependent Pathway

Yoon Ki Kim and Lynne E. Maquat*

Abstract

In mammalian cells, nonsense-mediated mRNA decay (NMD) degrades mRNAs that terminate translation sufficiently upstream of a post-splicing exon junction complex (EJC) of proteins. Components of the EJC include the NMD factor Upf1. We have found that Upf1 also interacts with the RNA binding protein Staufen (Stau)1. To understand the significance of this finding, microarray analyses were used to identify mRNAs that bind Stau1. For those mRNAs tested, Stau1 interacts with the 3' untranslated region, and down-regulating the cellular level of Stau1 or Upf1 increases mRNA half-life. These and other results indicate that Stau1 mediates the decay of specific cellular mRNAs when translation terminates normally. This Stau1-mediated mRNA decay (SMD) contributes significantly to the network of post-transcriptional regulatory pathways in mammalian cells. SMD differs from NMD because it occurs independently of splicing and after down-regulating Upf NMD factors other than Upf1. However, SMD is similar to NMD because it requires translation and the recruitment of Upf1 sufficiently downstream of a termination codon.

Human Upf1 Interacts with Human Staufen1

The role of Staufen (Stau) is best characterized in *Drosophila*, where it functions in the transport, localization and translational control of particular mRNAs during oogenesis and neuronal development (reviewed in ref. 1). Mammalian Stau1, which is homologous to *Drosophila* Stau, is ubiquitously expressed as isoforms having apparent molecular weights of 55 and 63 kDa.[2-5] The 55-kDa isoform associates with 40S and 60S ribosomal subunits as well as the rough endoplasmic reticulum.[3,5,6] Stau1 is also a component of mRNA-containing particles in dendrites of hippocampal neurons.[2,7-11] Furthermore, Stau1 is encapsidated together with HIV-1 RNA in virus particles.[12] These findings suggest that Stau1 in mammalian cells, like Stau in *Drosophila*, functions in mRNA transport and translational control.

Consistent with a role in RNA metabolism, Stau1 binds to single-stranded and double-stranded (ds) RNAs with extensive secondary structure.[3,5] Stau1 contains four dsRNA-binding domains (dsRBDs), which have been named dsRBD2 to dsRBD5 for consistency with *Drosophila* Stau[13] (Fig. 1). dsRBD3 efficiently binds to dsRNA, whereas dsRBD4 has only weak RNA-binding activity.[5] Stau1 also contains a tubulin-binding domain (TBD) that has been shown to bind tubulin in vitro[5] (Fig. 1). The TBD is thought to mediate, either directly or indirectly, the association of Stau1 with microtubules in hippocampal neurons.[2,4]

Upf1 is an RNA-dependent ATPase and 5'-to-3' helicase that is required for NMD.[14-17] Remarkably, we identified Stau1 in a yeast two-hybrid screen for protein products of a human

*Corresponding Author: Lynne E. Maquat—Department of Biochemistry and Biophysics, School of Medicine and Dentistry, 601 Elmwood Avenue, Box 712, University of Rochester, Rochester, New York 14642, U.S.A. Email: lynne_maquat@urmc.rochester.edu

Nonsense-Mediated mRNA Decay, edited by Lynne E. Maquat. ©2006 Eurekah.com.

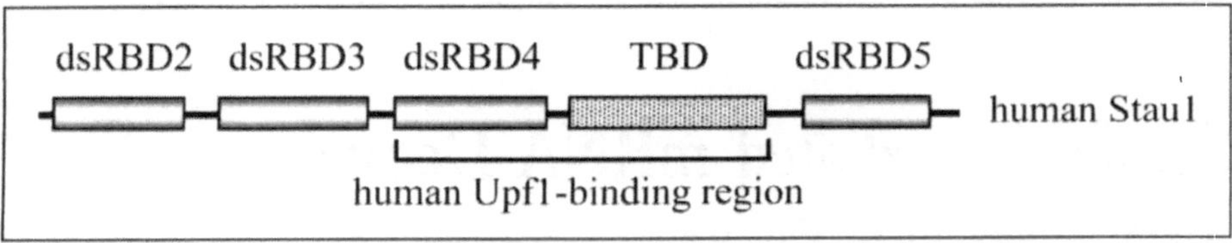

Figure 1. Region of human Staufen (Stau)1 that interacts with human Upf1. Human Upf1, when used as bait in a yeast two-hybrid screen, interacts with human Stau1 encoded by the pMyr-cDNA library.[14] The region of Stau1 that interacts with Upf1 was mapped to within the double-stranded RNA binding domain (dsRNA)4 and tubulin binding domain (TBD).

HeLa-cell cDNA expression library that interact with human Upf1.[14] Results from sequencing partial cDNAs that were obtained in the screen demonstrated that Upf1 interacts with Stau1 within the dsRBD4, TBD or both (Fig. 1). Studies that followed revealed that Stau1 is involved in an mRNA decay pathway that is related to NMD.

Initially, the interaction between Stau1 and Upf1 was corroborated using a number of methods.[14] Purified GST-tagged human Upf1 that had been produced in *E. coli* was shown to bind purified 6xHis-tagged human Stau1 that had been separately produced in *E. coli*. This indicated that the two proteins interact independently of RNA. Additionally, HA-tagged Stau1 that had been immunopurified from monkey Cos cells using anti-HA antibody was found to copurify with cellular Upf1. Copurification was resistant to the addition of RNase A, indicating that the interaction is stable in the absence of RNA. Furthermore, FLAG-tagged Upf1 that had been isolated from HeLa cells using anti-FLAG antibody was found to copurify with both the 55-kDa and 63-kDa isoforms of cellular Stau1. Taken together, these results indicate that Stau1 and Upf1 are components of the same complex in mammalian cells.

Stau1 Binding to the 3' Untranslated Region of an mRNA Reduces mRNA Abundance Independently of an EJC

The finding that Stau1 interacts with the NMD factor Upf1 raised the possibility that Stau1 functions in NMD. NMD generally occurs when translation terminates more than ~50-55 nucleotides upstream of a splicing-generated exon-exon junction (see chapter by Maquat). The dependence of NMD on splicing reflects the deposition of an exon junction complex (EJC) of proteins ~20-24 nucleotides upstream of exon-exon junctions (see chapter by Maquat). On average, EJCs consist of at least 13 proteins that include the Upf NMD factors (see chapter by Singh and Lykke-Andersen). The order in which the Upf factors function in NMD appears to be first Upf3 or Upf3X, then Upf2, and finally Upf1.[14,17] Notably, Upf3 and Upf3X (which are also called Upf3a and Upf3b, respectively) seem to function comparably except that Upf3, relative to Upf3X, preferentially associates with the EJC component Y14.[18]

Even though Stau1 interacts with Upf1, it did not coimmunopurify with other components of the EJC.[14] Therefore, Stau1 does not appear to be a stable component of the EJC. Furthermore, down-regulating the cellular level of Stau1 using small interfering (si)RNA did not inhibit the EJC-dependent NMD of β-globin or glutathione peroxidase 1 mRNAs.[14] Therefore, Stau1 does not appear to function in EJC-dependent NMD.

However, three lines of evidence raised the possibility that Stau1 could conceivably elicit mRNA decay if it binds mRNA sufficiently downstream of a termination codon. First, Stau1 interacts with Upf1. Second, Upf1 is the last of the Upf factors to function in NMD. Third, Upf1 binding to mRNA sufficiently downstream of a termination codon elicits NMD. In support of this possibility, tethering a fusion of HA-tagged Stau1 and the bacteriophage MS2 coat protein to MS2 coat protein binding sites that are situated more than 50 nucleotides downstream of an mRNA translation termination codon reduced mRNA abundance (Fig. 2A).[14] Furthermore, moving the termination codon to a position that resides downstream of

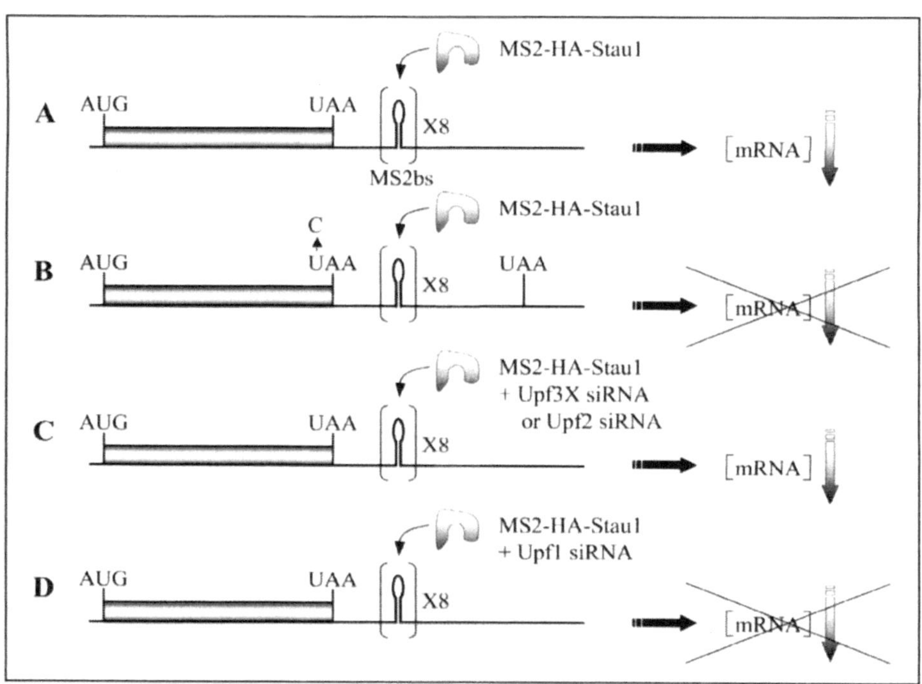

Figure 2. Schematic representations of mRNAs harboring MS2 coat protein binding sites, and the effects on mRNA abundance when MS2-HA-Stau1 is tethered under different conditions. The open box represents the open translational reading frame (ORF). The 3' untranslated region begins immediately downstream of the ORF with the normal termination codon (UAA). X8 specifies the eight tandem repeats of the MS2 coat protein-binding site (bs), to which the fusion protein MS2-HA-Stau1 binds (i.e., can be tethered). Where specified, conversion of the UAA termination codon to a CAA glutamine codon moves the termination codon to a site downstream of the distal-most MS2bs. [mRNA]↓ designates a decrease in the abundance of the MS2bs-containing mRNA. Also where specified, Upf3X siRNA, Upf2 siRNA or Upf1 siRNA down-regulate the level of the Upf3X, Upf2 or Upf1 NMD factor, respectively. A) Tethering MS2-HA-Stau1 downstream of a termination codon decreases mRNA abundance. B) Moving the translation termination codon from a position that resides upstream of the MS2-HA-Stau1 tethering site to a position that resides downstream of the site inhibits the decrease in mRNA abundance. It is possible that inhibition is due to MS2-HA-Stau1 removal by translating ribosomes. However, data demonstrating that Stau1 reduces mRNA abundance when a premature termination codon is introduced upstream of a natural Stau1 binding site supports the conclusion that a Stau1-mediated reduction in mRNA abundance requires an upstream termination codon.[14] C) Down-regulating the level of either Upf3X or Upf2 using, respectively, Upf3X siRNA or Upf2 siRNA is of no consequence to the reduction in mRNA abundance mediated by tethering MS2-HA-Stau1. D) Down-regulating the level of Upf1 using Upf1 siRNA inhibits the reduction in mRNA abundance mediated by tethering MS2-HA-Stau1.

the MS2 coat protein binding sites inhibited the Stau1-mediated reduction in mRNA abundance (Fig. 2B). Additionally, down-regulating the cellular level of Upf3X or Upf2 using Upf3X siRNA or Upf2 siRNA, respectively, was of no consequence to the Stau1-mediated reduction in mRNA abundance, whereas down-regulating the cellular level of Upf1 using Upf1 siRNA inhibited the Stau1-mediated reduction in mRNA abundance (Fig. 2C,D).[14] These data indicate that tethered Stau1 reduces mRNA abundance by a mechanism in which Upf1 functions after Stau1 and independently of Upf3X and Upf2.

mRNAs That Bind Stau1 within Their 3' Untranslated Regions Are Subject to SMD

The existence of natural targets of Stau1-mediated effects was verified by using microarray analyses. These analyses conservatively identified 23 human 293-cell mRNAs that were immunopurified using anti-HA antibody from cells that expressed HA-tagged Stau1.[14] One of these mRNAs, which encodes ADP-ribosylation factor (Arf)1, was chosen for detailed analysis.[14] Arf1 is a Ras-related G protein that regulates membrane traffic and organelle structure.[19]

First, RT-PCR using primers that specifically amplify Arf1 mRNA affirmed that Arf1 mRNA was among the cellular transcripts that coimmunopurified with HA-tagged Stau1. RT-PCR using the same primers demonstrated that Arf1 mRNA also coimmunopurified with cellular Stau1. Second, down-regulating the cellular level of Upf3X or Upf2 was of no consequence to the level of Arf1 mRNA. In contrast, down-regulating the cellular level of Stau1 or Upf1 increased the level of Arf1 mRNA. These results indicate that the cellular level of Arf1 mRNA is decreased by a mechanism that involves Stau1 and Upf1 but not the other Upf NMD factors.

In view of results obtained when HA-tagged Stau1 was tethered downstream of a termination codon, the simplest explanation of these findings is that Stau1 binds downstream of the normal termination codon of Arf1 mRNA. Consistent with this possibility, Arf1 mRNA abundance was no longer influenced by the cellular level of either Stau1 or Upf1 upon deletion of the part of the 3' untranslated region that resides downstream of the normal termination codon. Furthermore, RT-PCR analysis of Arf1 mRNAs that harbors a series of deletions within the 3' untranslated region demonstrated that Stau1 binds to the 231-nucleotide sequence that resides 67-nucleotides downstream of the normal termination codon. We called this sequence the minimized Stau1 binding site (SBS).

Replacing the entire 3' untranslated region of either Arf1 mRNA or a reporter mRNA with the SBS reduced the half-life of each mRNA by a mechanism that was inhibited by down-regulating the cellular level of Stau1 or Upf1 but not Upf2.[14] Therefore, Stau1 together with Upf1 mediates the decay of Arf1 mRNA by a process that we called Stau1-mediated mRNA decay (SMD; Fig. 3).

Stau1 was also found to bind the 3' untranslated region of PAICS mRNA, and down-regulating the cellular level of Stau1 increased the abundance of this mRNA.[14] Detailed

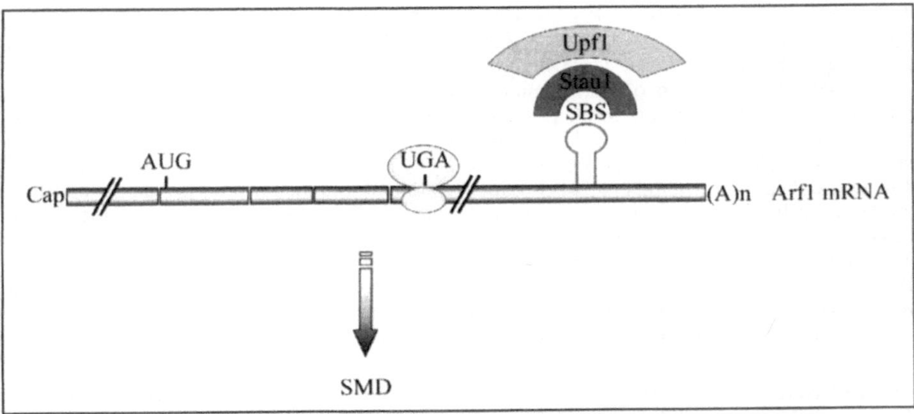

Figure 3. Model for Arf1 SMD. Arf1 mRNA consists of five exons (boxes). Translation initiates within the second exon (AUG) and terminates within the last exon (UGA). A 231-nucleotide (nt) minimized Stau1 binding domain (SBS) is located within the 3′ untranslated region, approximately 67 nucleotides downstream of the normal termination codon. SMD would occur when Stau1 together with Upf1 are bound to the SBS and translation terminates normally.

studies of the 23 Stau1-binding mRNAs that were identified using microarray analyses were limited to Arf1 and PAICS mRNAs. In fact, some of the remaining 21 mRNAs may not be targeted for SMD since Stau1 binding may not be downstream of a normal termination codon. Realistically, however, it is likely that a significant number of cellular mRNAs are targeted for SMD since efficient SMD may preclude detectable Stau1 binding. Consistent with this possibility, microarray analyses that identified mRNAs that were up-regulated when the cellular level of Stau1 was down-regulated indicate that SMD is widely used by cells as a means of post-transcriptional gene control (Y.K. Kim, L. Furic, L. DesGroseillers and L.E. Maquat, unpublished data).

The Relationship between NMD and SMD

NMD and SMD have many common features (see chapter by Maquat; Y.K. Kim and L.E. Maquat, unpublished data; Fig. 4).[14] Both are inhibited when translation is inhibited. This result, at least in part, reflects a requirement for translation termination (see below). Both NMD and SMD require Upf1. Upf1 functions in NMD after Upf3 or Upf3X and Upf2.[14,17] Based on cellular distribution, each appears to be recruited to the post-splicing EJC in order of its function in NMD.[17,20] In contrast, Upf1 functions in SMD by binding directly to the dsRNA binding protein Stau1 independently of the other Upf factors.[14]

Both NMD and SMD necessitate that Upf1 is recruited to mRNA at a position that resides sufficiently downstream of a translation termination codon. Upf1 is known to bind eukaryotic translation release factor (eRF) 1 and eRF3 (G. Singh and J. Lykke-Andersen, personal communication; F. Lejeune and L.E. Maquat, unpublished data), both of which function in translation termination. Therefore, Upf1 may interact with eRFs before, at the same time, or after it interacts with Upf2 during NMD or Stau1 during SMD. Data suggest that if translation terminates more than ~25 nucleotides upstream of an EJC or an SBS, then there will be NMD or SMD, respectively (Fig. 4).

There are an estimated 4×10^6 molecules of Upf1 per HeLa cell.[21,22] In contrast, there are an estimated 2×10^5 molecules of Upf2, and the sum of Upf3 and Upf3X molecules is estimated to be 1×10^5.[22] Notably, Upf1 appears to interact with either Stau1 or Upf2 rather than with Stau1 and Upf2 simultaneously (Y.K. Kim and L.E. Maquat, unpublished data). Therefore, Upf1 that is bound by Upf2 would not be available for binding to Stau1. Neither the cellular abundance of Stau1 nor the stability of the Upf1 interaction with either Stau1 or Upf2 is known. Nevertheless, there appear to be more Upf1 molecules per cell than are sequestered by Upf2.

It remains to be determined if NMD and SMD are competitive pathways. Upf1 is primarily hypophosphorylated.[21,23] While hyperphosphorylated Upf1 preferentially binds Upf2,[23] the phosphorylation status of Upf1 that preferentially binds Stau1 remains to be determined. Furthermore, at least three lines of evidence suggest that Stau1 function in SMD is likely to be modulated in different tissues not only by the type and level of each expressed isoform but also by phosphorylation status. First, even though Stau1 is constitutively expressed, its level of expression varies among tissues.[3-5] Second, alternative splicing generates several Stau1 isoforms that are differentially expressed among different tissue. Third, Stau1 is likely to be a phosphoprotein, considering that it contains several putative phosphorylation sites and interacts with protein phosphatase-1.[24]

Future Directions

The discovery that an RNA binding protein can recruit the Upf1 NMD factor to the 3' untranslated region of an mRNA and, by so doing, destabilize that mRNA, raises the possibility that other RNA binding proteins do likewise. In fact, there is at least one other example that is provided by the stem loop binding protein. This protein binds to the 3' untranslated region of replication-dependent histone mRNAs and elicits mRNA decay at the end of S phase of the cell cycle[25] (see chapter by Kaygun and Marzluff). Future studies are almost certain to reveal other examples.

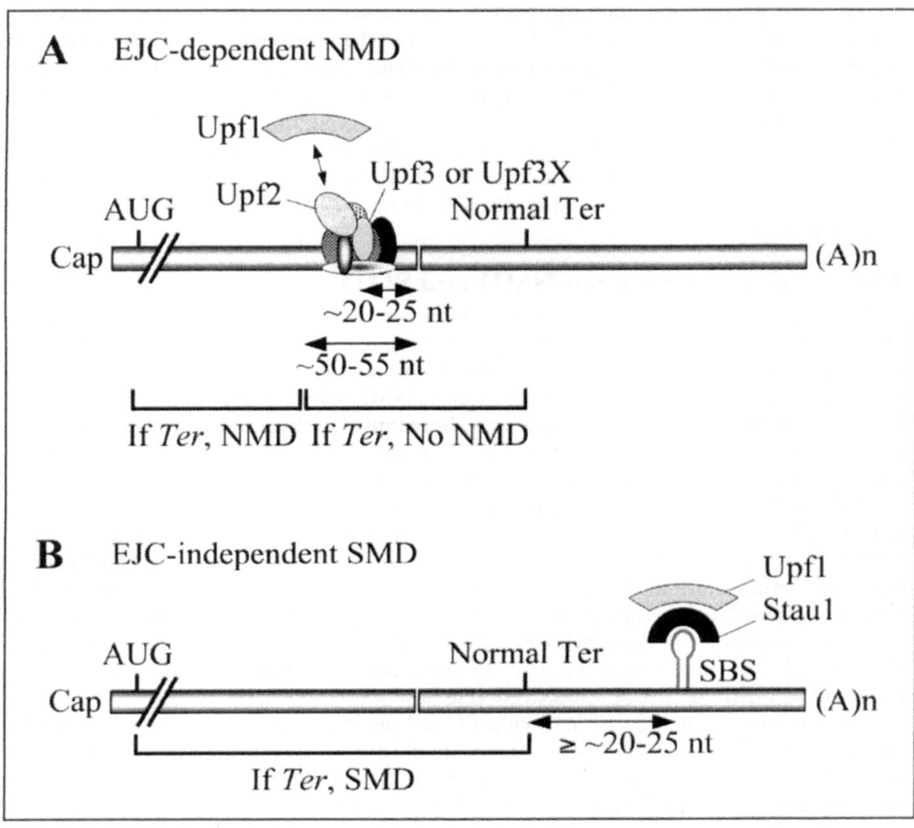

Figure 4. Models for EJC-dependent NMD and EJC-independent SMD in mammalian cells. A) NMD generally involves the recruitment of Upf1 to an exon junction complex (EJC) via Upf3 or Upf3X and Upf2. Within the nucleus, pre-mRNA splicing deposits an EJC of proteins ~20-25 nucleotides (nt) upstream of each exon-exon junction. The EJC consists of the NMD factor Upf3 or Upf3X, each of which binds Upf2. Upf2 is thought to subsequently bind Upf1. EJC-dependent NMD occurs if translation terminates (*Ter*) more than ~50-55 nucleotides upstream of any exon-exon junction, which would be more than ~25 nucleotides upstream of the corresponding EJC. Notably, only the 3'-most exon-exon junction is shown. AUG specifies the initiation codon, and Normal *Ter* denotes the normal termination codon. B) SMD involves the recruitment of Upf1 to a 3'untranslated region via Stau1. Stau1 binding to the 3'untranslated region of an mRNA recruits Upf1 independently of an EJC. Data suggest that SMD would occur when translation terminates normally. By analogy to EJC-dependent NMD, SMD would occur when translation terminates more than ~25 nucleotides upstream of the Stau1 binding site.

Interestingly, Stau1 is not the only example of a protein that functions in mammalian-cell mRNA decay and has a homolog in *Drosophila* that functions in mRNA transport and translational control. For example, Barentsz (also known as MLN51), Magoh, Y14 (also known as Tsunagi) and eIF4AIII are components of the mammalian-cell EJC that is required for NMD. The homolog of each associates with and is required for the localization of *oskar* mRNA in the *Drosophila* oocyte (reviewed in ref. 26). It will be important to determine if Stau functions in mRNA decay in *Drosophila*. It will also be important to determine if different regions of Stau1 support functionally different associations with RNA, as is the case for *Drosophila* Stau (reviewed in ref. 26). Possibly, Stau1 binding to RNA via one domain elicits SMD, and Stau1 binding to RNA via another domain functions in mRNA transport.

Questions that pertain to SMD and remain unresolved include the following. What types of transcripts are targeted for SMD? Does the specificity of Stau1 binding to RNA derive solely from Stau1, or do auxiliary proteins confer specificity? Is there a single consensus Stau1 binding site, or do classes of binding sites exist? How is SMD regulated in general and in tissue-specific ways? Are NMD, SMD and other degradative pathways in competition because each involves Upf1? We recently found that whereas EJC-dependent NMD is restricted to CBP80-bound mRNA (see chapter by Maquat,) SMD targets both CBP80-bound mRNA as well as eukaryotic translation initiation factor (eIF)4E-bound mRNA.[27]

Acknowledgements

We thank members of the Maquat lab, particularly Holly Kuzmiak, for comments on the manuscript. We also thank Luc Furic and Luc DesGroseillers for a very enjoyable and fruitful collaborative study of SMD. Y.K. Kim was supported by NIH R01 DK33938, and L.E. Maquat was supported by NIH R01 DK33938 and GM059614.

References

1. Roegiers F, Jan YN. Staufen: A common component of mRNA transport in oocytes and neurons? Trends Cell Biol 2000; 10:220-224.
2. Kiebler MA, Hemraj I, Verkade P et al. The mammalian staufen protein localizes to the somatodendritic domain of cultured hippocampal neurons: Implications for its involvement in mRNA transport. J Neurosci 1999; 19:288-297.
3. Marion RM, Fortes P, Beloso A et al. A human sequence homologue of Staufen is an RNA-binding protein that is associated with polysomes and localizes to the rough endoplasmic reticulum. Mol Cell Biol 1999; 19:2212-2219.
4. Monshausen M, Putz U, Rehbein M et al. Two rat brain staufen isoforms differentially bind RNA. J Neurochem 2001; 76:155-165.
5. Wickham L, Duchaine T, Luo M et al. Mammalian staufen is a double-stranded-RNA- and tubulin-binding protein which localizes to the rough endoplasmic reticulum. Mol Cell Biol 1999; 19:2220-2230.
6. Luo M, Duchaine TF, DesGroseillers L. Molecular mapping of the determinants involved in human Staufen-ribosome association. Biochem J 2002; 365:817-824.
7. Kanai Y, Dohmae N, Hirokawa N. Kinesin transports RNA: Isolation and characterization of an RNA-transporting granule. Neuron 2004; 43:513-525.
8. Kohrmann M, Luo M, Kaether C et al. Microtubule-dependent recruitment of Staufen-green fluorescent protein into large RNA-containing granules and subsequent dendritic transport in living hippocampal neurons. Mol Biol Cell 1999; 10:2945-2953.
9. Krichevsky AM, Kosik KS. Neuronal RNA granules: A link between RNA localization and stimulation-dependent translation. Neuron 2001; 32:683-696.
10. Mallardo M, Deitinghoff A, Muller J et al. Isolation and characterization of Staufen-containing ribonucleoprotein particles from rat brain. Proc Natl Acad Sci USA 2003; 100:2100-2105.
11. Ohashi S, Koike K, Omori A et al. Identification of mRNA/protein (mRNP) complexes containing Puralpha, mStaufen, fragile X protein, and myosin Va and their association with rough endoplasmic reticulum equipped with a kinesin motor. J Biol Chem 2002; 277:37804-37810.
12. Mouland AJ, Mercier J, Luo M et al. The double-stranded RNA-binding protein Staufen is incorporated in human immunodeficiency virus type 1: Evidence for a role in genomic RNA encapsidation. J Virol 2000; 74:5441-5451.
13. St Johnston D. The intracellular localization of messenger RNAs. Cell 1995; 81:161-170.
14. Kim YK, Furic L, Desgroseillers L et al. Mammalian Staufen1 recruits Upf1 to specific mRNA 3'UTRs so as to elicit mRNA decay. Cell 2005; 120:195-208.
15. Sun X, Perlick HA, Dietz HC et al. A mutated human homologue to yeast Upf1 protein has a dominant-negative effect on the decay of nonsense-containing mRNAs in mammalian cells. Proc Natl Acad Sci USA 1998; 95:10009-10014.
16. Bhattacharya A, Czaplinski K, Trifillis P et al. Characterization of the biochemical properties of the human Upf1 gene product that is involved in nonsense-mediated mRNA decay. RNA 2000; 6:1226-1235.

17. Lykke-Andersen J, Shu MD, Steitz JA. Human Upf proteins target an mRNA for nonsense-mediated decay when bound downstream of a termination codon. Cell 2000; 103:1121-1131.
18. Gehring NH, Neu-Yilik G, Schell T et al. Y14 and hUpf3b form an NMD-activating complex. Mol Cell 2003; 11:939-949.
19. Donaldson JG, Jackson CL. Regulators and effectors of the ARF GTPases. Curr Opin Cell Biol 2000; 12:475-482.
20. Serin G, Gersappe A, Black JD et al. Identification and characterization of human orthologues to Saccharomyces cerevisiae Upf2 protein and Upf3 protein (Caenorhabditis elegans SMG-4). Mol Cell Biol 2001; 21:209-223.
21. Pal M, Ishigaki Y, Nagy E et al. Evidence that phosphorylation of human Upfl protein varies with intracellular location and is mediated by a wortmannin-sensitive and rapamycin-sensitive PI 3-kinase-related kinase signaling pathway. RNA 2001; 7:5-15.
22. Serin G, Maquat LE. Nonsense-mediated mRNA decay: Insights into mechanism from the cellular abundance of human Upfl, Upf2, Upf3, and Upf3X proteins. Cold Spring Harb Symp Quant Biol 2001; 66:313-320.
23. Ohnishi T, Yamashita A, Kashima I et al. Phosphorylation of hUPF1 induces formation of mRNA surveillance complexes containing hSMG-5 and hSMG-7. Mol Cell 2003; 12:1187-1200.
24. Monshausen M, Rehbein M, Richter D et al. The RNA-binding protein Staufen from rat brain interacts with protein phosphatase-1. J Neurochem 2002; 81:557-564.
25. Kaygun H, Marzluff WF. Regulated degradation of replication-dependent histone mRNAs requires both ATR and Upfl. Nat Struct Mol Biol 2005; 12:794-800.
26. Meyer EL, Gavis ER. Staufen does double duty. Nat Struct Mol Biol 2005; 12:291-292.
27. Hosada N, Kim YK, Lejeune F et al. CBP80 promotes interaction of Upfl with Upf2 during nonsense-mediated mRNA decay in mammalian cells. Nat Struct Mol Biol 2005; 12:893-901.

Chapter 18

Upf1 Regulates Mammalian Histone mRNA Decay

Handan Kaygun and William F. Marzluff*

Abstract

Metazoan replication-dependent histone mRNAs are the only eukaryotic mRNAs that do not end in a poly(A) tail. Instead, they end in a conserved stem-loop structure. The half-lives of histone mRNAs are regulated coordinately with DNA replication. Histone mRNAs are rapidly degraded both after the inhibition of DNA synthesis during S phase and at the end of S phase. The stem-loop, which is recognized by the stem-loop binding protein (SLBP), is responsible for the post-transcriptional regulation of histone mRNAs. The regulated degradation of histone mRNAs in the absence of DNA synthesis requires active translation of these mRNAs as well as the nonsense-mediated mRNA decay (NMD) factor Upf1. Upf1 is recruited to the 3' end of the histone mRNA when DNA replication is inhibited. In addition, the position of the stem-loop/SLBP complex with respect to the translation termination codon is critical for proper regulation of histone mRNA half-life. Therefore, the regulation of histone mRNA degradation is mediated by an interaction among SLBP, Upf1, and the terminating ribosome.

Introduction

Metazoan replication-dependent histone mRNAs are the only eukaryotic mRNAs that are not polyadenylated. Instead, they end in a conserved 26-nucleotide sequence that contains a stem-loop (Fig. 1). The histone pre-mRNAs do not contain any introns. Hence, the only processing step necessary for the production of the mature histone mRNA is an endonucleolytic cleavage occurring five-nucleotides downstream of the stem-loop.[1] This cleavage reaction requires a 31-kDa protein, the stem-loop binding protein (SLBP), which specifically binds the stem-loop,[2] as well as U7 snRNP (Fig. 1A). U7 snRNP binds a purine-rich region called the histone downstream element (HDE), which is located about 15-nucleotides downstream of the stem-loop[3] (Fig. 1A). After cleavage, SLBP remains associated with the mature histone mRNA and accompanies the mRNA to the cytoplasm,[4,5] where it participates in the translation of histone mRNAs.[6]

Regulation of Histone mRNA Half-Life

Histone mRNAs are present at high levels only during S phase,[7] serving as templates for the synthesis of histone proteins to package newly replicated DNA into chromatin. The synthesis of DNA and histones is tightly coupled and, in fact, the rate of histone protein synthesis is coordinated with DNA replication by continuously adjusting the levels of histone mRNAs.[8] A major component of this coordination is the regulation of the half-life of histone mRNA. The

*Corresponding Author: William F. Marzluff—Program in Molecular Biology and Biotechnology, CB #7100, University of North Carolina, Chapel Hill, North Carolina 27599, U.S.A. Email: marzluff@med.unc.edu

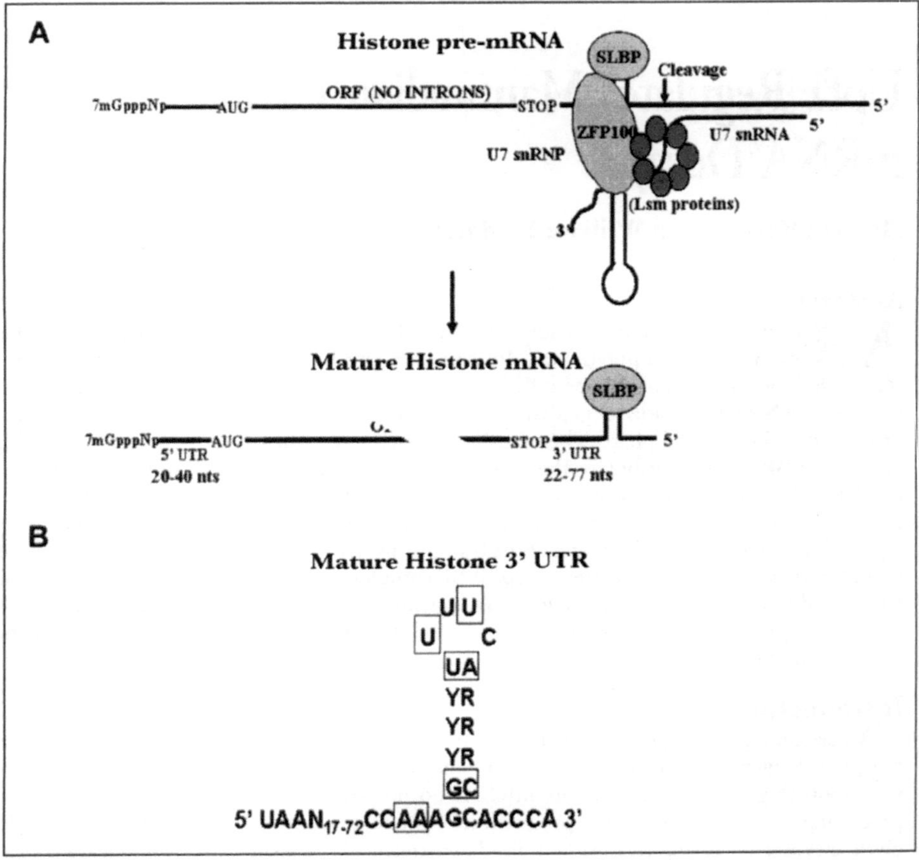

Figure 1. Histone pre-mRNA structure and processing. A) Schematic of a histone pre-mRNA and the processing reaction that forms the 3' end. Histone pre-mRNAs do not have introns. Hence, the only processing reaction they undergo in the nucleus to form the mature histone mRNA is an endonucleolytic cleavage occurring 5-nucleotides downstream of the stem-loop. SLBP binds the stem-loop, and U7 snRNP binds the histone downstream element (HDE), which base-pairs with the 5' end of U7 snRNA. U7 snRNP-specific proteins Lsm 10, Lsm 11 and ZFP100 are involved in bridging U7 snRNP to the stem-loop/SLBP complex. ORF, open translational reading frame. B) Consensus sequence of the 26-nucleotide conserved sequence that constitutes the 3' end of vertebrate histone mRNAs. Nucleotides that are particularly important for SLBP binding are boxed.[14,49] The consensus sequence starts 22-72 nucleotides downstream of the translation termination codon, depending on the histone transcript.

only cis-acting element for this regulation is the 3' end of histone mRNA, and the trans-acting factor that likely mediates this regulation is SLBP.

There are two regulatory programs that impact the levels of histone mRNAs. The first is the cell-cycle program, which allows the accumulation of large amounts of histone mRNA only during S phase. SLBP is the only known cell-cycle regulated factor in histone mRNA metabolism, and it is present at high levels only during S phase.[9,10] The activation of histone mRNA synthesis at the end of G1 requires the synthesis of SLBP, which activates histone pre-mRNA processing, and SLBP is rapidly degraded at the end of S phase, blocking further synthesis of histone mRNAs.[9]

The second regulatory program is responsible for coordinating the level of histone mRNAs with the rate of DNA replication. During S phase, inhibition of DNA synthesis results in rapid degradation of histone mRNAs. In addition, the levels of histone mRNAs are rapidly restored to their S-phase levels when DNA synthesis is resumed.[11] However, the level of SLBP protein does not change after the inhibition of DNA synthesis during S phase.[4] The finding that globin mRNA is rapidly degraded following the inhibition of DNA synthesis when its poly(A) tail is replaced with the 26-nucleotide stem-loop from the histone mRNA 3' end demonstrates that the 3' end of histone mRNA is necessary and sufficient for the coordination of histone mRNA levels with the rate of DNA synthesis.[12]

There is evidence that histone mRNA degradation is initiated by a 3'-to-5' decay pathway,[13] and we[14] and others[15] have identified 3'-to-5' exonucleases capable of degrading histone mRNAs. In addition, all histone mRNAs have very short 3'-untranslated regions (UTRs) with the stem-loop starting 22- to 77-nucleotides downstream of the translation termination codon.[16] Moreover, moving the stem-loop further 3' (to more than 300 nucleotides from the termination codon) results in stabilization of the mRNA.[17] Our recent studies have shed light on both the requirement that histone mRNAs be translated for their rapid degradation as well as the role the 3' end is likely to play in degradation.[18] Surprisingly, these studies reveal that Upf1, a key component of the NMD pathway, interacts with SLBP, and thereby the 3' end of histone mRNP in vivo, and is involved in histone mRNA degradation.[19]

NMD and Histone mRNA Degradation

Aberrant mRNAs containing a premature termination codon (PTC) are degraded by a surveillance pathway termed NMD[20](see chapter by Maquat). In mammalian cells, degradation of a PTC-containing mRNA generally requires translation and presence of an exon junction complex (EJC) located at least 20-24 nucleotides downstream of the termination codon. Since there are usually no introns in the 3'-UTR of normal mRNAs, all EJCs are normally removed during the initial, "pioneer", round of translation. The presence of a PTC prevents complete removal of all EJCs, and EJC(s) located downstream of the PTC are sensed by the Upf1-containing surveillance complex, which is recruited to the termination site, leading to the degradation of the mRNA by the NMD pathway. Upf1 can sense the presence of an EJC downstream of the termination codon because it interacts with the EJC component Upf2, which in turn interacts with the EJC component Upf3. Histone pre-mRNAs do not have any introns. Therefore, mammalian histone mRNAs are not sensitive to NMD, since mature histone mRNAs do not contain any EJCs.[21]

Histone mRNA degradation shares some properties with the NMD pathway in mammalian cells. In the mammalian NMD pathway, the EJC must be located in the 3'-UTR to elicit NMD and the recognition of a termination codon as a PTC requires translation. Similarly, we recently showed that the rapid degradation of histone mRNAs in the absence of DNA synthesis, either following the inhibition of DNA synthesis or at the end of S phase, requires translation of these mRNAs[19] and the presence of the stem-loop/SLBP complex in the 3'-UTR and at a proper distance from the translation termination codon.[18]

Translation Is Required for Histone mRNA Degradation

Treatment of cells with inhibitors of protein synthesis prevents histone mRNA degradation following the inhibition of DNA synthesis[11] and increases histone mRNA levels in S-phase cells.[22] Two possible explanations for this requirement are: first, the continuous synthesis of some proteins that play a role in histone mRNA degradation might be required for histone mRNA degradation. Alternatively, the mechanism of histone mRNA degradation may involve histone mRNA translation.

Using two different approaches, we recently showed that regulated degradation of histone mRNAs, either following the inhibition of DNA synthesis by antimetabolites or at the end of S phase, requires active histone mRNA translation.[18] First, we constructed a mouse histone H2a gene, mH2a-IRE, which contains the iron response element (IRE) in the 5'-UTR.

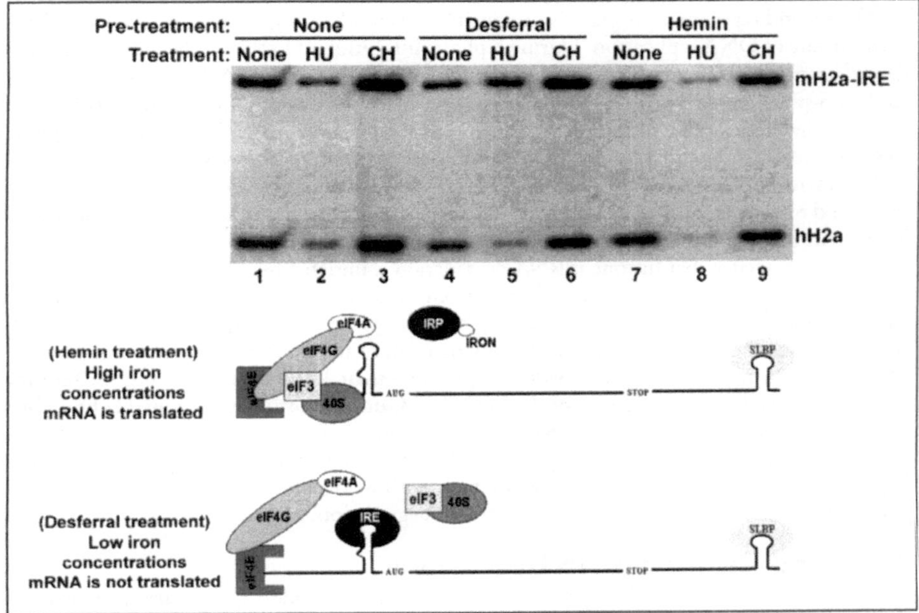

Figure 2. Degradation of mH2a-IRE mRNA requires translation. HeLa cells were stably transfected with a mouse (m) histone H2a gene containing an iron response element (IRE), diagrammed below the figure. Cells growing in normal medium (lanes 1-3) were treated for 2 hrs with desferral (lanes 4-6) or hemin (lanes 7-9). Cells were then treated for 45 min with 5 mM hydroxyurea (HU) to inhibit DNA synthesis (lanes 2,5,8), 0.1 mM cycloheximide (CH) to inhibit protein synthesis, or nothing (None, lanes 1,4,7), and the levels of mH2a-IRE mRNA as well as endogenous human histone H2a (hH2a) mRNA were monitored using an S1 nuclease protection assay. Desferral is an iron chelator and decreases available iron levels. Hemin is an iron source and increases intracellular iron levels. (Adapted from ref. 19.)

Translation of the mH2a-IRE mRNA can be regulated simply by changing the intracellular iron concentration (Fig. 2).[23] At low iron concentrations, the iron response protein (IRP) binds the IRE and prevents the recruitment of the small ribosomal subunit to the 5'-UTR, inhibiting translation of the mRNA.[24] However, at high iron concentrations, the IRP cannot bind the IRE, and the mRNA is translated (Fig. 2). When cells were treated with desferral, an iron chelator that lowers available iron concentrations, the mH2a-IRE mRNA was not degraded following the inhibition of DNA synthesis by hydroxyurea (HU). However, the mH2a-IRE mRNA was degraded after the inhibition of DNA synthesis if the iron concentration was high as a result of treatment of cells with hemin (Fig. 2).[19] Regulation of both wild-type endogenous human histone H2a mRNA (hH2a) (Fig. 2) and mH2a-IRE$_{MUT}$ mRNA, which has a mutated IRE sequence at the 3'-UTR, was not affected by either desferral or hemin pretreatment (not shown).[18] Therefore, the inhibition of mH2a-IRE mRNA degradation after the inhibition of DNA synthesis is due to altering the translation of this mRNA, clearly demonstrating that degradation of histone mRNAs following the inhibition of DNA synthesis requires active histone mRNA translation.

The second approach that we used to specifically inhibit the translation of histone mRNAs was to express a mutated SLBP, which is inactive in histone mRNA translation.[6] Over-expression of mutated SLBP reduced the rate of degradation of histone mRNA following the inhibition of DNA synthesis.[18] In addition, histone mRNA degradation at the end of S phase was delayed in these cells compared to the HeLa cells stably expressing wild-type

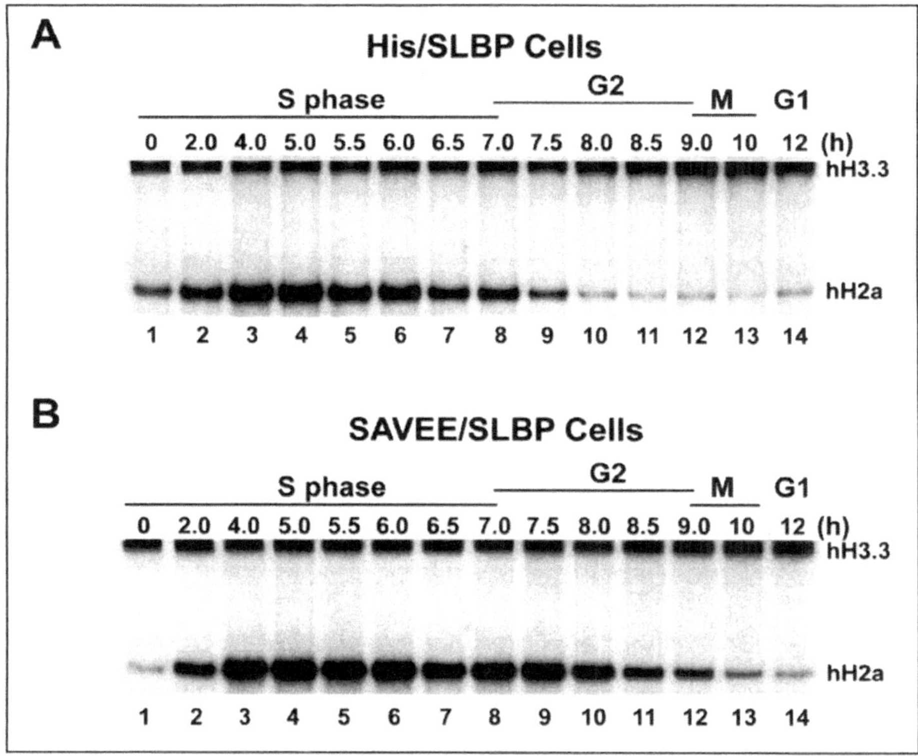

Figure 3. Stable expression of a dominant-negative SLBP, which is inactive in translating histone mRNAs, delays histone mRNA degradation as cells exit S phase. HeLa cells stably expressing either His-tagged wild-type SLBP (His/SLBP) (panel A) or His-tagged SAVEE-mutated SLBP (SAVEE/SLBP, panel B) were synchronized using a double-thymidine block and subsequently released into S phase. Total-cell RNA was prepared at the indicated time (h) after the release, and the levels of both replacement-type human histone H3 (hH3.3) mRNA and replication-dependent hH2a mRNA were determined using an S1 nuclease protection assay. hH3.3 histone mRNA has a poly(A) tail and is constitutively expressed during the cell cycle. (Adapted from ref. 19.)

SLBP (Fig. 3).[19] The finding that histone mRNA degradation requires histone mRNA translation following the inhibition of DNA synthesis and at the end of S phase suggests that the two decay pathways are similar.

The Position of the Stem-Loop Is a Critical Determinant of Histone mRNA Stability

All known metazoan replication-dependent histone mRNAs have a 3'-UTR of less than 100 nucleotides, including the last 26 nucleotides comprising the consensus SLBP binding site (Fig. 1B). Since there are over 1000 different metazoan histone mRNAs currently in the databases, there has clearly been strong selection that maintains this distance constraint. This distance places the stem-loop/SLBP complex just beyond the distance a terminating ribosome is likely to cover. Moving the stem-loop more than 300 nucleotides downstream of the termination codon prevents the rapid degradation of histone mRNAs following the inhibition of DNA synthesis,[17,18] demonstrating that the stem-loop must be located a proper distance from the termination codon in order to regulate the stability of histone mRNAs.

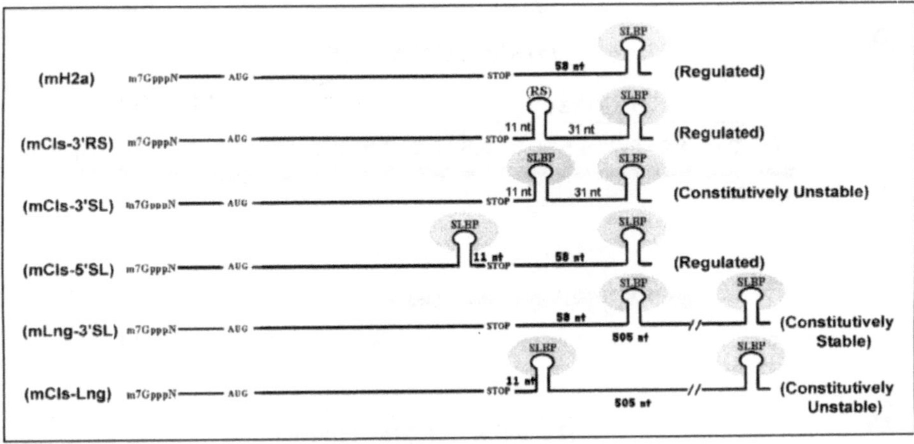

Figure 4. Effect stem-loop position on histone mRNA stability. The structure of a number of different mutated histone mRNAs containing two stem-loop sequences is shown together with the effect of the mutation on histone mRNA production. The HDE, which is absolutely required for histone pre-mRNA processing, was placed downstream of the second stem-loop to ensure that histone pre-mRNA is processed after the second stem-loop. Reverse stem (RS) is a mutated stem-loop that does not bind SLBP. Distances between the translation termination codon and the beginning of the stem-loop are specified. Regulated signifies the altered structure had no effect on either the steady-state levels or the rapid degradation of the mRNA following the inhibition of DNA synthesis. Constitutively Unstable denotes the altered structure resulted in a very low steady-state mRNA level. Constitutively Stable means the altered structure prevented the degradation of the mRNA following the inhibition of DNA synthesis. (Adapted from ref. 19.)

To further define the interaction of the stem-loop/SLBP complex with the translational machinery during histone mRNA destabilization, we systematically altered the position of the stem-loop with respect to the termination codon.[19] The schematic representation of the constructs and their stabilities are summarized in Figure 4. We placed a stem-loop 11 nucleotides either upstream or downstream of the termination codon in a construct containing the normal 3' end of the histone mRNA at the original position. This will result in the removal of the first SLBP from the histone mRNA during the initial round of translation: Placing the stem-loop 11 nucleotides upstream of the termination codon results in removal of SLBP by the elongating ribosome, while placing the stem-loop 11 nucleotides downstream of the termination codon results in removal of SLBP by the terminating ribosome.

We found that placing the stem-loop 11 nucleotides downstream of the termination codon resulted in a constitutively unstable histone mRNA, whether the second stem-loop was at the normal position or 500 nucleotides downstream (Fig. 4). However, placing the stem-loop 11 nucleotides upstream of the termination codon did not affect the regulation of the histone mRNA. Therefore, removal of SLBP by the terminating ribosome affects histone mRNA half-life, whereas removal of SLBP by the elongating ribosome has no effect on histone mRNA stability. On the other hand, increasing the distance between the stem-loop and the translation termination codon to over 300 nucleotides prevented the degradation of the histone mRNA after inhibition of DNA synthesis, even if the mRNA had an additional stem-loop at the original position. Thus, these experiments demonstrated that the SLBP/stem-loop complex has a critical impact on both the steady-state half-life and the rapid degradation of histone mRNAs following the inhibition of DNA synthesis.[18] Hence, for proper regulation of histone mRNA half-life, the stem-loop/SLBP complex must be at the 3' end and at a proper distance downstream of the translation termination codon.

Upf1 Is Required for Rapid Degradation of Histone mRNA

Given the similarities between regulated histone mRNA degradation and NMD, we investigated whether similar proteins are involved in both pathways. The key players in the NMD pathway in yeast as well as metazoans are three proteins termed Upf1, Upf2, and Upf3. Therefore, we tested whether Upf1, Upf2 or both are involved in regulated histone mRNA degradation. Upf1 is an ATP dependent, 5'-to-3' helicase. The R843C mutation abolishes its helicase activity,[25] and the K498A mutation decreases its affinity for ATP so that the protein lacks both ATPase and helicase activities.[26,27] First, we showed that expression of either of these mutated Upf1 proteins stably in HeLa cells resulted in less efficient histone mRNA degradation both following the inhibition of DNA synthesis[20] (Fig. 5A) and at the end of S phase.[19] Second, we showed that reducing the cellular abundance of Upf1 using RNA interference (RNAi) prevented efficient histone mRNA degradation following the inhibition of DNA

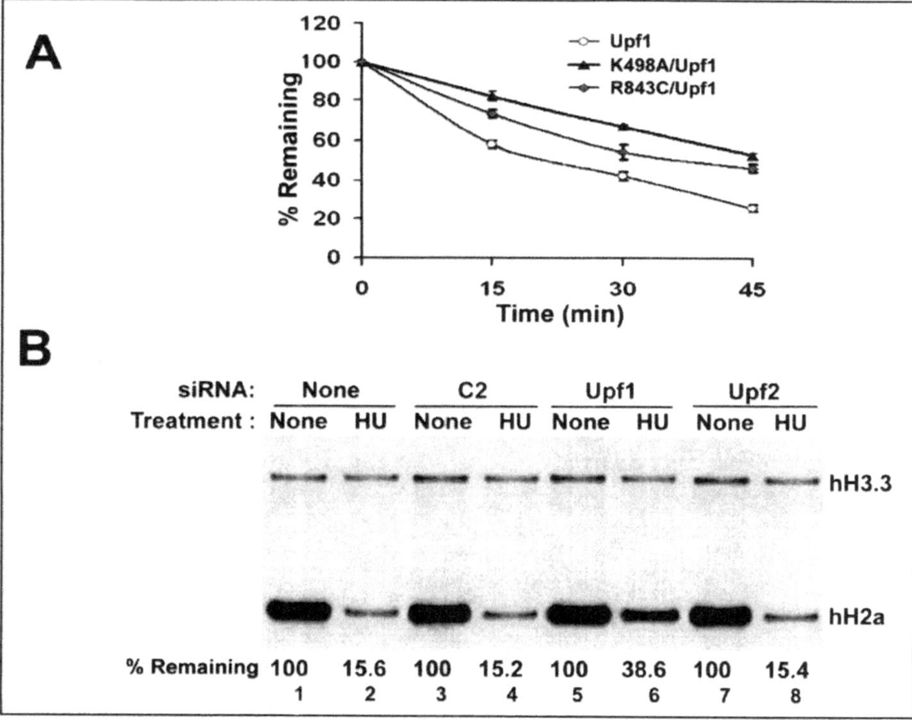

Figure 5. Upf1 is required for regulated histone mRNA degradation. A) HeLa cells stably transfected with wild-type Upf1, Upf1 harboring the R843C mutation, which renders the RNA helicase inactive, or Upf1 harboring the K498A mutation, which reduces ATPase activity, were treated with 5 mM HU. The level of endogenous hH2a mRNA was determined at the indicated times after treatment. The graph is representative of 5 independently performed experiments. Expression of wild-type Upf1 did not affect histone mRNA degradation following the inhibition of DNA synthesis compared with HeLa cells transfected with the empty expression vector (not shown). B) HeLa cells were transfected with nonspecific control siRNA (C2) (lanes 3 and 4), siRNA that down-regulates the level of Upf1 (lanes 5 and 6) or Upf2 (lanes 7 and 8), or nothing (lanes 1 and 2). Cells were then treated with 5 mM HU (lanes 2, 4, 6, and 8) for 45 min. Total-cell RNA was isolated, and the levels of hH2a and hH3.3 mRNAs were determined using an S1 nuclease protection assay. Lanes 1, 3, 5, and 7 measure hH2a and hH3.3 mRNA levels prior to treatment with HU. (Adapted from ref. 20.)

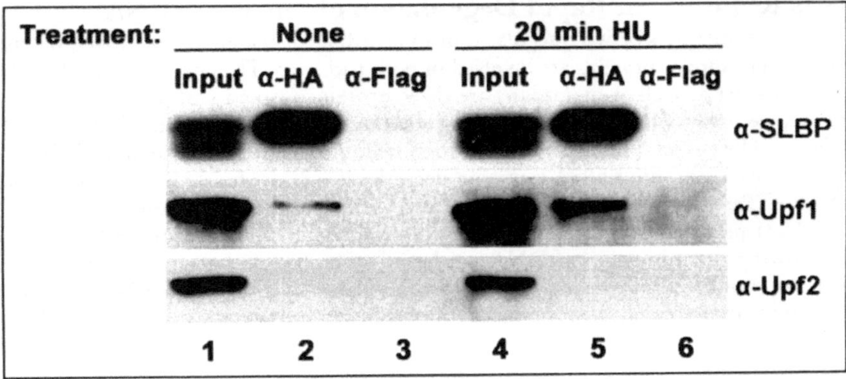

Figure 6. SLBP interacts with Upf1 in vivo. HeLa cells stably expressing HA-tagged wild-type SLBP were harvested before and after 20 min of HU treatment. Lysates were prepared and immunoprecipitated with anti-HA antibody or, as a control for nonspecific immunoprecipitation, anti-Flag antibody. Immunoprecipitates were resolved in 10% SDS-PAGE and analyzed by Western blotting using anti-SLBP antibody (top panel), anti-Upf1 antibody (middle panel), or anti Upf2-antibody (bottom panel). Input lysates (10%) (lanes 1 and 4), anti-HA precipitates (lanes 2 and 5), and anti-Flag precipitates (lanes 3 and 6) were analyzed. Note that the anti-HA antibody precipitates HA-tagged SLBP (upper panel, upper band in input lane) but not endogenous SLBP (upper panel, lower band in input lane) as expected. α specifies anti. (Adapted from ref. 20.)

synthesis, while reducing the cellular abundance of Upf2 using RNAi did not affect histone mRNA degradation following the inhibition of DNA synthesis[20] (Fig. 5B). Upf2 may not be required for histone mRNA degradation because histone pre-mRNAs do not contain any introns, and thereby they lack Upf2-containing EJCs. Consequently, histone mRNAs likely recruit Upf1 via a different mechanism that does not require either an EJC or Upf2.

SLBP Helps the Recruitment of Upf1 to the 3' End of Histone mRNAs

Since SLBP interacts with the stem-loop, it is an excellent candidate to participate in the recruitment of Upf1 to the 3' end of histone mRNAs. To determine whether SLBP and Upf1 interact in vivo, we performed coimmunoprecipitation experiments using lysates of HeLa cells stably expressing either HA-tagged wild-type SLBP (Fig. 6) or HA-tagged wild-type Upf1 (not shown). Results demonstrated that Upf1 and SLBP are present in the same protein complex.[19] More importantly, a brief treatment of cells with HU, which inhibits DNA synthesis and leads to the rapid degradation of histone mRNAs, stimulated the interaction between SLBP and Upf1 (Fig. 6).[20] In addition, treatment of the lysates with RNAse A before immunoprecipitation did not dissociate Upf1 from SLBP,[19] indicating that the interaction between SLBP and Upf1 is not bridged by RNA (not shown). However, Upf2 did not coimmunoprecipitate with SLBP[20] (Fig. 6), consistent with the fact that down-regulating Upf2 had no effect on histone mRNA degradation. Note that these experiments do not rule out the possibility that the interaction between SLBP and Upf1 is an indirect interaction and is mediated by other protein(s). We conclude that, following inhibition of DNA synthesis, Upf1 is recruited to the 3' end of histone mRNP by a protein complex that contains SLBP.[19]

Role of the DNA Damage Checkpoint Pathway in Regulated Histone mRNA Degradation

Replication of the eukaryotic chromosome requires not only faithful synthesis of DNA but also formation of proper chromatin structure, which requires the synthesis of a large

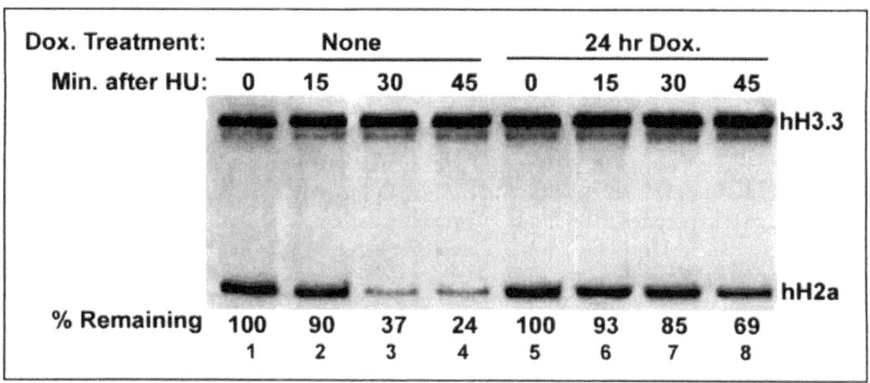

Figure 7. ATR is required for histone mRNA degradation following inhibition of DNA synthesis. U2OS cells stably expressing kinase-inactive ATR under control of the tetracycline regulatory system (Tet_{on})[50] were treated with HU for the indicated time before (lanes 1-4) or after the induction of kinase-inactive ATR with doxycycline for 24 hrs (lanes 5-8). Total-cell RNA was prepared, and the levels of both replication-dependent hH2a mRNA and replacement type hH3.3 mRNA were measured using an S1 nuclease protection assay. (Adapted from ref. 19.)

amount of histone proteins. In fact, inhibiting chromatin formation behind a replication fork activates a DNA damage checkpoint pathway that leads to the inhibition of DNA synthesis and cell-cycle arrest.[28] Candidates for mediating these signals are members of a family of kinases termed phosphoinositide 3-kinase (PI3-kinase)-related kinases (PIKKs). PIKKs detect chromosomal abnormalities, block cell-cycle progression, induce DNA-repair genes and, if the damage is irreparable, induce apoptosis.[29] Caffeine is a known inhibitor of PIKKs. The finding that caffeine pretreatment blocks regulated histone mRNA degradation following the inhibition of DNA synthesis suggests that a PIKK might be involved in histone mRNA degradation.[19] In addition, exposure of cells to genotoxic insults, including hydroxyurea, aphidicolin or UV light, all of which activate the ATR pathway,[29] results in histone mRNA degradation.[19] These results suggest that ATR might be required for regulated histone mRNA degradation. Consistent with this hypothesis, expression of a kinase-inactive ATR in U2OS cells prevents rapid histone mRNA degradation following the inhibition of DNA synthesis by hydroxyurea (Fig. 7).[19]

Discussion

The requirements for regulated histone mRNA degradation are reminiscent of the requirements for the NMD pathway. Both pathways require translation as well as the presence of a protein marker that functions downstream but not upstream of the translation termination codon. However, both the distance constraints and the protein markers differ in these two pathways. For NMD, the EJC, which is a large (>500-kDa) protein complex,[30] has to be located more than 20-24 nucleotides downstream of a termination codon. If an EJC is located less than 20-24 nucleotides downstream of a termination codon, then it is likely removed by translating ribosomes, and the mRNA is immune to NMD. However, there is apparently no maximum constraint for the distance between an EJC and a termination codon: Although the small size of the typical exon likely means an EJC would normally reside quite close to a termination codon, an EJC can be located over one-kb downstream of a termination codon and the mRNA is still rapidly degraded.[20] In contrast, for proper regulation of histone mRNA half-life, SLBP has to be located 22-77 nucleotides downstream of the termination codon. One reason for the differential distance constraints between the NMD pathway and histone mRNA degradation might be that SLBP plays a role in both translation termination of histone mRNAs and histone mRNA degradation.

Although Upf1, but not Upf2, is required for histone mRNA degradation, it is not known if other components of the NMD pathway are required. Proteins involved in both the phosphorylation (Smg1)[31] and dephosphorylation (Smg5, Smg6 and Smg7)[32] of Upf1 are required for NMD (see chapter by Yamashita et al). Upf1 has fourteen S/T-Q motifs, which is the consensus sequence phosphorylated by all PIKK family members except for mTOR, at the very C-terminus. hSmg1 phosphorylates hUpf1 on at least two of these sites in vivo.[31] The precise role of Upf1 phosphorylation by Smg1 in the NMD pathway is not known yet. However, it has been shown that hyperphosphorylated Upf1 preferentially associates with polyribosomes, whereas hypophosphorylated Upf1 does not.[33] Therefore, it is possible that phosphorylation of Upf1 by Smg1 is required for the association of Upf1 with the terminating ribosome, leading to the formation of the NMD surveillance complex and detection of abnormalities in mRNP structure or translation termination. On the other hand, dephosphorylation of Upf1 by the Smg5-7 complex might be important for recycling Upf1 so it can find new targets. Currently, we do not know whether Smg1 is also required for histone mRNA degradation. However, it is possible that the particular PIKK family member that phosphorylates Upf1 might be important in determining target specificity.

Is the Mechanism of Histone mRNA Degradation Like NMD in Yeast?

There are distinct similarities between NMD in yeast and regulation of the half-lives of metazoan replication-dependent histone mRNAs. Yeast mRNAs usually have very short 3'-UTRs so that the poly(A) tail is generally located less than 100 nucleotides downstream of the termination codon. The distance from the termination codon to the poly(A) tail is an important parameter for determining whether termination is premature or not (see chapters by Baker and Parker, and Amrani et al).

Recently, it has been shown using yeast extracts that ribosomes encountering a PTC cannot terminate translation efficiently and stall. Furthermore, recruitment of either poly(A) binding protein (PABP) or eukaryotic translation release factor (eRF3) to downstream of the PTC restores proper translation termination and stabilizes the mRNA.[34] These findings led to the faux UTR model, which states that proper translation termination and mRNA stability requires an interaction between the terminating ribosome and factor(s) associated with the 3'-UTR, like PABP,[34] (see chapters by Baker and Parker, and Amrani et al) and stresses the role of proper termination of translation in mRNA stability.

In all metazoan histone mRNAs, the distance between the termination codon and the stem-loop is highly conserved. The stem-loop starts 22 to 77 nucleotides downstream of the termination codon. More importantly, altering this distance affects both the constitutive half-life and the stability of histone mRNA following the inhibition of DNA synthesis. SLBP might be necessary for efficient translation termination and ribosome recycling, which is very similar to the role of PABP in yeast.[34] As a result, selective pressure has kept this distance short in all histone mRNAs because SLBP functions best if it is located at a proper distance from the termination codon.

Consequently, removal of SLBP by the terminating ribosome might result in inefficient translation termination and rapid histone mRNA degradation. Additionally, an explanation for the constitutive stability of histone mRNA containing a stem-loop situated far downstream of the termination codon might be that the proteins that interact with SLBP can no longer readily simultaneously interact with the terminating ribosome.

Signals That Activate Histone mRNA Degradation

It has been previously suggested that accumulation of excess histone proteins triggers the signal that feeds back on histone mRNA levels, resulting in rapid histone mRNA degradation.[8,35,36] However, inhibition of histone deposition, and hence chromatin formation, behind the replication fork[28,37] results in spontaneous DNA damage, activation of a DNA damage checkpoint, and inhibition of DNA synthesis.[28] These results are consistent with the possibility that the presence of excess histones per se does not create the signal for histone mRNA degradation. Rather, either the titration of histone chaperons, like Asf1 (anti-silencing factor 1)[38] or CAF-1 (chromatin assembly factor-1),[39] or the presence of abnormal chromatin structure created by excess histones, might lead to activation of a DNA damage checkpoint pathway and inhibition of DNA synthesis,[40] which in turn creates the signal for histone mRNA degradation. Recent evidence has emphasized that communication between the pathways of chromatin assembly and DNA replication[37] is essential for maintaining the balance between DNA and histone synthesis. Our present work is focused on determining how that signal is transferred to the 3' end of histone mRNP.

Upf1 Action in Histone mRNA Degradation

Two different models exist to explain the involvement of Upf1 in NMD: One posits that the recruitment of Upf1 is primarily due to inefficient translation termination, and the other proposes that Upf1 is recruited after translation termination by interacting with the downstream proteins, such as Upf2 of the EJC.[41] Similarly, Upf1 may be recruited to histone mRNA either by the terminating ribosome as a result of inefficient translation termination or by a SLBP-containing protein complex. The involvement of both ATR and Upf1 in histone mRNA decay can explain how rapid histone mRNA degradation and the inhibition of DNA synthesis are coordinated. Upf1 might simply be recruited to histone mRNP via SLBP. Activation of the ATR pathway may result in phosphorylation of Upf1 and/or SLBP, which has two S/T-Q motifs, resulting in more efficient recruitment of Upf1 to the 3' end of histone mRNAs (Fig. 8, model 1). Phosphorylation may then activate histone mRNA degradation.

Another way to recruit Upf1, as has been proposed for yeast, occurs when the ribosome stalls at a termination codon as a result of inefficient or slow translation termination (see chapters by Baker and Parker, and Amrani et al). In yeast, translation termination at the PTC is thought to be slow because the poly(A) binding protein is far away from the termination codon.[34] It is possible that after the inhibition of DNA synthesis, the 3' end of the histone mRNP might be altered so as to result in inefficient translation termination, possibly because of the modification and inactivation of SLBP and/or eRFs by the activated ATR pathway. Upf1 would then bind to the stalled ribosome, leading to the rapid degradation of histone mRNAs (Fig. 8, Model 2).

Recently, it has been shown that translation termination factors may play complex roles in the termination process, and that translation termination, like translation initiation or elongation, can also be regulated.[42] Regulating the efficiency of translation termination, like regulating unproductive splicing,[43] can provide eukaryotic cells with another way to modulate gene expression. The molecular details of Upf1 recruitment to mRNA are not understood for any organism yet. However, it is likely that binding of Upf1 to mRNP or the ribosome during translation termination is generally the critical step in triggering mRNA degradation. Upf1 interacts with the Dcp1 and Dcp2 decapping proteins and the Xrn1 5'-to-3' exonuclease, which are involved in degrading mRNA from its 5' end,[44,45] as well as poly(A) ribonuclease (PARN) and the Rrp4 and Rrp41 components of the exosome, which are involved in degrading mRNA from its 3' end.[45] Therefore, the recruitment of Upf1 to mRNA is required for mRNA degradation.

Similarly, recruitment of Upf1 to the 3' end of histone mRNAs is required for the recruitment of proteins involved in histone mRNA degradation. Recently, we identified a novel exonuclease, called 3'hExo, that interacts specifically with the 3' end of replication-dependent histone mRNAs in vitro.[14] It is possible that Upf1 might recruit 3'hExo to the 3' end of replication dependent histone mRNAs or activate 3'hExo by removing SLBP from the 3' end of histone mRNA after the inhibition of DNA synthesis. This could initiate histone mRNA degradation, which would subsequently be completed by the exosome (Fig. 8). Alternatively, Upf1 might bring either Xrn1, the exosome or both to histone mRNP and induce mRNA degradation as it does in NMD (Fig. 8).

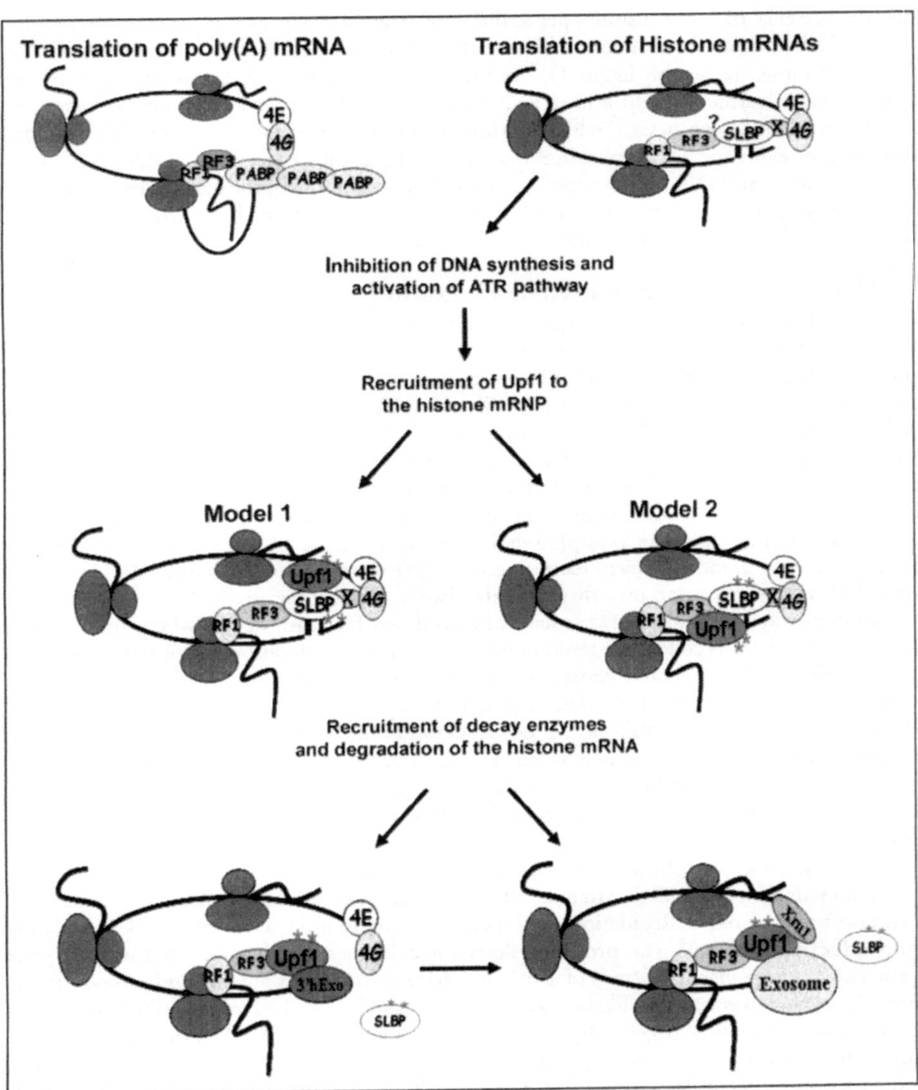

Figure 8. Model for histone mRNA degradation. Top) Mechanism of translation of polyadenylated mRNAs and histone mRNAs. Polyadenylated mRNAs are translated as circular templates, with eukaryotic translation initiation factor (eIF) 4E (4E) and poly(A) binding protein (PABP) both bound to the scaffold protein eIF4G (4G). This structure not only promotes translation initiation but is thought to help the recycling of ribosomes to the 5' end of the mRNA after translation termination. The interaction of eukaryotic translation release factor (RF3) with PABP may promote recycling. Histone mRNAs are also translated as circular mRNPs. eIF4E binds the cap, and SLBP bound to the 3' end of the mRNA interacts with eIF4G by binding to a novel adaptor protein (X; N. Gulseren-Cakmakci and W.F. Marzluff, unpublished data). It is possible that SLBP may also promote translation termination and ribosome recycling. Figure legend continued on the next page.

Figure 8, continued. Middle) We envisage two possible models for the degradation of histone mRNAs when DNA replication is inhibited and the ATR pathway is activated. Model 1: Activation of the ATR pathway results in phosphorylation of Upf1 and/or SLBP, which leads to the recruitment of Upf1 to the 3' end of histone mRNA. Upf1 could interact directly or indirectly with SLBP. Modification of one or both proteins (or the adaptor protein) results in recruitment of Upf1. Upf1 is then positioned to recruit and/or activate the factors required for mRNA decay. Model 2: The initial consequence of activating the ATR pathway may be phosphorylation of target proteins, resulting in the inhibition of efficient translation termination of histone mRNAs. In this case, Upf1 is recruited to stalled ribosomes, and binding may be stabilized by direct or indirect interaction with SLBP. Upf1 is then positioned to recruit and/or activate the factors required for mRNA decay. Bottom) Degradation of polyadenylated mRNAs during NMD can occur either by 5'-to-3' decay after Upf1 helps to recruit decapping enzymes[44,45] or by 3'-to-5' decay after Upf1 helps to recruit the exosome[45] to the mRNA 3' end. Similarly, we imagine that histone mRNA degradation could be initiated 3'-to-5' by recruitment of the 3'hExo,[14] which could then allow the exosome to complete degradation of the histone mRNA. Upf1 might also assist in removing SLBP from the 3' end of the mRNA, allowing 3'hExo to initiate degradation that could be followed by recruitment of the exosome. Alternatively, Upf1 may activate 5'-to-3' decay by recruiting decapping enzymes and/or the exosome after Upf1 helps remove SLBP from the histone 3' end.

It is clear that Upf1 has additional roles beyond the degradation of aberrant mRNAs or other NMD targets. As examples, mRNAs containing a Staufen1 binding site in the 3'-UTR[43] (see chapter by Kim and Maquat) and the replication-dependent histone mRNAs are normal cellular mRNAs that have half-lives regulated by Upf1. Down-regulating Upf1 in tissue-culture cells results in an increase in concentration of a number of mRNAs,[47] suggesting that there are a significant number of mRNAs that require Upf1 for degradation (see chapter by Sharifi and Dietz). Furthermore, it has been recently shown that exposure of cells to ionizing radiation (IR) results in phosphorylation of Upf1 and it has been proposed that both Upf1 and Smg1 are involved in the cellular response to genotoxic stress[48] (see chapter by Abraham et al). Following the exposure of cells to UV radiation, which activates ATR, but not to IR, which activates ATM, histone mRNA are degraded,[19] suggesting that Upf1 can be involved in the cellular response to genotoxic stress at least by regulating the levels of histone mRNAs. More importantly, Upf1 likely has additional targets, particularly following the exposure of cells to IR. It is possible that phosphorylation of Upf1 by different PIKK family members might determine which mRNAs are degraded in response to a particular type of genotoxic stress.

Acknowledgements

This work was supported by NIH grant GM29832 to W.F. Marzluff.

References

1. Marzluff WF. Metazoan replication dependent histone mRNAs: A unique class of RNA polymerase II transcripts. Current Opinion in Cell Biology 2005; 17:274-80.
2. Dominski Z, Zheng L-X, Sanchez R et al. The stem-loop binding protein facilitates 3' end formation by stabilizing U7 snRNP binding to the histone pre-mRNA. Mol Cell Biol 1999; 19:3561-70.
3. Mowry KL, Oh R, Steitz JA. Each of the conserved sequence elements flanking the cleavage site of mammalian histone pre-mRNAs has a distinct role in the 3'-end processing reaction. Mol Cell Biol 1989; 9:3105-8.
4. Whitfield ML, Kaygun H, Erkmann JA et al. SLBP is associated with histone mRNA on polyribosomes as a component of histone mRNP. Nucleic Acids Res 2004; 32:4833-42.
5. Erkmann JA, Wagner EJ, Dong J et al. Nuclear import of the stem-loop binding protein and localization during the cell cycle. Mol Biol Cell 2005; 16:2960-71.
6. Sanchez R, Marzluff WF. The stem-loop binding protein is required for efficient translation of histone mRNA in vivo and in vitro. Mol Cell Biol 2002; 22:7093-104.
7. Harris ME, Böhni R, Schneiderman MH et al. Regulation of histone mRNA in the unperturbed cell cycle: Evidence suggesting control at two posttranscriptional steps. Mol Cell Biol 1991; 11:2416-24.

8. Sariban E, Wu RS, Erickson LC et al. Interrelationships of protein and DNA syntheses during replication of mammalian cells. Mol Cell Biol 1985; 5:1279-86.
9. Zheng L-X, Dominski Z, Yang X et al. Phosphorylation of SLBP on two threonines triggers degradation of SLBP, the sole cell-cycle regulated factor required for regulation of histone mRNA processing, at the end of S-phase. Mol Cell Biol 2003; 23:1590-601.
10. Whitfield ML, Zheng L-X, Baldwin A et al. Stem-loop binding protein, the protein that binds the 3' end of histone mRNA, is cell cycle regulated by both translational and posttranslational mechanisms. Mol Cell Biol 2000; 20:4188-98.
11. Graves RA, Marzluff WF. Rapid, reversible alterations in histone gene transcription and histone mRNA levels in mouse myeloma cells. Mol Cell Biol 1984; 4:351-7.
12. Pandey NB, Williams AS, Sun J-H et al. Point mutations in the stem-loop at the 3' end of mouse histone mRNA reduce expression by reducing the efficiency of 3' end formation. Mol Cell Biol 1994; 14:1709-20.
13. Ross J, Peltz SW, Kobs G et al. Histone mRNA degradation in vivo: The first detectable step occurs at or near the 3' terminus. Mol Cell Biol 1986; 6:4362-71.
14. Dominski Z, Yang X, Kaygun H et al. A 3' exonuclease that specifically interacts with the 3' end of histone mRNA. Molecular Cell 2003; 12:295-305.
15. Ross J, Kobs G, Brewer G et al. Properties of the exonuclease activity that degrades H4 histone mRNA. J Biol Chem 1987; 262:9374-81.
16. Marzluff WF, Gongidi P, Woods KR et al. The human and mouse replication-dependent histone genes. Genomics 2002; 80:487-98.
17. Graves RA, Pandey NB, Chodchoy N et al. Translation is required for regulation of histone mRNA degradation. Cell 1987; 48:615-26.
18. Kaygun H, Marzluff WF. Translation termination is involved in histone mRNA degradation when DNA replication is inhibited. Mol Cell Biol 2005; 25:6879-88.
19. Kaygun H, Marzluff WF. Regulated degradation of replication-dependent histone mRNAs requires both ATR and Upf1. Nat Struct Mol Biol 2005; 12:794-800.
20. Maquat LE. Nonsense-mediated mRNA decay: Splicing, translation and mRNP dynamics. Nat Rev Mol Cell Biol 2004; 5:89-99.
21. Maquat LE, Li XJ. Mammalian heat shock p70 and histone H4 transcripts, which derive from naturally intronless genes, are immune to nonsense-mediated decay. RNA 2001; 7(3):445-56.
22. Stimac E, Groppi Jr VE, Coffino P. Inhibition of protein synthesis stabilizes histone mRNA. Mol Cell Biol 1984; 4:2082-7.
23. Goossen B, Hentze MW. Position is the critical determinant for function of iron- responsive elements as translational regulators. Mol Cell Biol 1992; 12:1959-66.
24. Gray NK, Hentze MW. Iron regulatory protein prevents binding of the 43S translation preinitiation complex to ferritin and eALAS mRNAs. EMBO J 1994; 13:3882-91.
25. Sun J-H, Pilch DR, Marzluff WF. The histone mRNA 3' end is required for localization of histone mRNA to polyribosomes. Nucleic Acids Res 1992; 20:6057-66.
26. Perlick HA, Medghalchi SM, Spencer FA et al. Mammalian orthologues of a yeast regulator of nonsense transcript stability. Proc Natl Acad Sci USA 1996; 93(20):10928-32.
27. Leeds P, Wood JM, Lee B-S et al. Gene products that promote mRNA turnover in Saccharomyces cerevisiae. Mol Cell Biol 1992; 12:2165-77.
28. Nabatiyan A, Krude T. Silencing of chromatin assembly factor 1 in human cells leads to cell death and loss of chromatin assembly during DNA synthesis. Mol Cell Biol 2004; 24:2853-62.
29. Sancar A, Lindsey-Boltz LA, Unsal-Kacmaz K et al. Molecular mechanisms of mammalian DNA repair and the DNA damage checkpoints. Annu Rev Biochem 2004; 73:39-85.
30. Le Hir H, Izaurralde E, Maquat LE et al. The spliceosome deposits multiple proteins 20-24 nucleotides upstream of mRNA exon-exon junctions. EMBO J 2000; 19:6860-9.
31. Yamashita A, Ohnishi T, Kashima I et al. Human SMG-1, a novel phosphatidylinositol 3-kinase-related protein kinase, associates with components of the mRNA surveillance complex and is involved in the regulation of nonsense-mediated mRNA decay. Genes Dev 2001; 15:2215-28.
32. Ohnishi T, Yamashita A, Kashima I et al. Phosphorylation of hUPF1 induces formation of mRNA surveillance complexes containing hSMG-5 and hSMG-7. Mol Cell 2003; 12:1187-200.
33. Pal M, Ishigaki Y, Nagy E et al. Evidence that phosphorylation of human Upfl protein varies with intracellular location and is mediated by a wortmannin-sensitive and rapamycin-sensitive PI 3-kinase-related kinase signaling pathway. RNA 2001; 7:5-15.
34. Amrani N, Ganesan R, Kervestin S et al. A faux 3'-UTR promotes aberrant termination and triggers nonsense-mediated mRNA decay. Nature 2004; 432:112-8.
35. Peltz SW, Ross J. Autogenous regulation of histone mRNA decay by histone proteins in a cell-free system. Mol Cell Biol 1987; 7:4345-56.

36. McLaren RS, Ross J. Individual purified core and linker histones induce histone H4 mRNA destabilization in vitro. J Biol Chem 1993; 268:14637-44.
37. Ye X, Franco AA, Santos H et al. Defective S phase chromatin assembly causes DNA damage, activation of the S phase checkpoint, and S phase arrest. Mol Cell 2003; 11:341-51.
38. Tyler JK, Adams CR, Chen SR et al. The RCAF complex mediates chromatin assembly during DNA replication and repair. Nature 1999; 402:555-560.
39. Kaufman PD, Kobayashi R, Kessler N et al. The p150 and p60 subunits of chromatin assembly factor I: A molecular link between newly synthesized histones and DNA replication. Cell 1995; 81:1105-1114.
40. Franco AA, Lam WM, Burgers PM et al. Histone deposition protein Asf1 maintains DNA replisome integrity and interacts with replication factor C. Genes Dev 2005; 19:1365-1375.
41. Baker KE, Parker R. Nonsense-mediated mRNA decay: Terminating erroneous gene expression. Curr Opin Cell Biol 2004; 16:293-9.
42. Le RF, Salehzada T, Bisbal C et al. A newly discovered function for RNase L in regulating translation termination. Nat Struct Mol Biol 2005; 12:505-12.
43. Wollerton MC, Gooding C, Wagner EJ et al. Autoregulation of polypyrimidine tract binding protein by alternative splicing leading to nonsense-mediated decay. Mol Cell 2004; 13:91-100.
44. Lykke-Andersen J. Identification of a human decapping complex associated with hUpf proteins in nonsense-mediated decay. Mol Cell Biol 2002; 22:8114-21.
45. Lejeune F, Li X, Maquat LE. Nonsense-mediated mRNA decay in mammalian cells involves decapping, deadenylating, and exonucleolytic activities. Mol Cell 2003; 12(3):675-87.
46. Kim YK, Furic L, LesGroseillers L et al. Mammalian Staufen1 recruits Upf1 to specific mRNA 3'UTRs so as to elicit mRNA decay. Cell 2005; 120:195-208.
47. Mendell JT, Sharifi NA, Meyers JL et al. Nonsense surveillance regulates expression of diverse classes of mammalian transcripts and mutes genomic noise. Nat Genet 2004; 36:1073-8.
48. Brumbaugh KM, Otterness DM, Geisen C et al. The mRNA surveillance protein hSMG-1 functions in genotoxic stress response pathways in mammalian cells. Mol Cell 2004; 14:585-98.
49. Battle DJ, Doudna JA. The stem-loop binding protein forms a highly stable and specific complex with the 3' stem-loop of histone mRNAs. RNA 2001; 7:123-32.
50. Nghiem P, Park PK, Kim Y et al. ATR inhibition selectively sensitizes G1 checkpoint-deficient cells to lethal premature chromatin condensation. Proc Natl Acad Sci USA 2001; 98:9092-7.

Chapter 19

S. cerevisiae Est1/*H. sapiens* SMG6 Protein Family Members Function in Telomere Metabolism

Claus M. Azzalin, Sophie Redon and Joachim Lingner*

Abstract

Telomeres define the physical ends of eukaryotic chromosomes. They protect chromosome termini from being recognized as damage-induced DNA breaks. Telomeric DNA sequences consist of short tandem G-rich repeats in the strand that forms the 3' end of the chromosome. Due to the end replication problem, telomeres shorten when replicated solely by semiconservative DNA replication. Telomerase is a cellular reverse transcriptase that can solve the end replication problem by iterative reverse transcription of its tightly associated telomerase RNA template onto chromosome ends. In the yeast *Saccharomyces cerevisiae*, the Est1 protein (p) enables telomere elongation, recruiting telomerase to chromosome ends. *S. cerevisiae* (Sc) Est1p bears sequence similarity to three human proteins, *Homo sapiens* (Hs) EST1A, Hs EST1B and Hs EST1C. As does Sc Est1p, Hs EST1A and Hs EST1B associate with telomerase in extracts. Overexpression of Hs EST1A alters telomere structure. These data support a role for Hs EST1A and Hs EST1B in telomere metabolism. Curiously, Hs EST1A, Hs EST1B and Hs EST1C are identical to Hs SMG6, Hs SMG5 and Hs SMG7, respectively, which have been shown to function in nonsense-mediated mRNA decay (NMD). In this review, we discuss the roles of Est1 proteins in telomere maintenance and speculate about possible links between telomere maintenance and RNA surveillance pathways.

Introduction

All eukaryotes maintain their genomes as linear DNA molecules. This requires special mechanisms to fully replicate DNA ends because of several reasons. First, DNA replication is semiconservative; DNA polymerases use a parental template strand to synthesize a complementary daughter molecule. Eukaryotic telomeres end with 3' protrusions at probably both chromosomal ends.[1-4] Thus, the 5' parental strand is resected and cannot provide a template for the synthesis of a 3' overhang.[5] Second, nucleolytic processing of telomere ends after semiconservative DNA replication contributes to telomere shortening. The presumed blunt end intermediate generated at the leading strand telomere is not detected and is probably quickly processed to regenerate a 3' overhang.[6-8]

The most common way out to the end replication problem occurs through the ribonucleoprotein enzyme telomerase, which was discovered in 1985 in the holotrichous ciliate *Tetrahymena thermophila.*[9] Telomerase adds short tandem telomeric repeats onto chromosome

*Corresponding Author: Joachim Lingner—Swiss Institute for Experimental Cancer Research, École Polytechnique Fédérale de Lausanne and National Center of Competence in Research "Frontiers in Genetics", CH-1066 Epalinges s/Lausanne, Switzerland. Email: joachim.lingner@isrec.ch

ends by reverse transcribing the template region of the telomerase RNAs. The telomerase reverse transcriptase polypeptide (TERT) is related in sequence and structure to reverse transcriptase of retroelements[10] and forms the catalytic core with the telomerase RNA moiety.[11-13] Genetic screens using yeast uncovered additional genes that are required for telomere maintenance in vivo but are not required for telomerase activity in vitro.[14] Sc Est1p was the first discovered polypeptide required for telomere maintenance.[15]

Est1 Protein in *S. cerevisiae*

The *EST1* (Ever Shorter Telomeres) gene was identified in budding yeast in a genetic screen for mutants defective in telomere elongation.[15] Null alleles of *EST1* and other *EST* genes gradually lose telomeric DNA, entering cellular senescence after approximately 60-80 generations. The *EST2* gene encodes the yeast TERT subunit, the *EST3* gene encodes a telomerase-associated protein of unknown function, and the *EST4* gene is identical to the *CDC13* gene, which encodes a protein that binds the single-strand telomeric DNA 3' overhang.[16] Epistasis analysis indicates that the four *EST* genes and the telomerase RNA gene *TLC1* function in the same genetic "pathway": telomere replication.[14] Interestingly, only Sc Est2p and TLC1 RNA are required for telomerase activity in vitro.[17,18]

Est1p associates in extracts with telomerase, binding probably directly to a bulged stem in TLC1 RNA.[19] This interaction is not disrupted in the absence of Est2p and Est3p,[20] while Est1p and Est2p do not coimmunoprecipitate in the absence of TLC1 RNA.[21] In vitro, recombinant Sc Est1p binds TLC1 RNA with good affinity (K_d = 50nM) but weak specificity.[22] Moreover, recombinant Sc Est1p has weak but specific affinity for single-stranded G-rich telomeric DNA (K_d = 250nM). The binding requires a free 3'-terminus as expected for a protein that binds to the 3' terminus of the chromosome. The domains of Sc Est1p that bind RNA and DNA have not been clearly defined. Deletion of an internal fragment of recombinant Sc Est1p between amino acids 435 and 565 abolished RNA and DNA binding.[22] On the other hand, point mutations near the N-terminus, some of which reside in the Sc Est1p-domain (Fig. 2; see below) of Sc Est1p, strongly perturbed association with telomerase in yeast extracts.[21]

Finally, Sc Est1p interacts physically with the single-strand telomere binding protein Sc Cdc13p as shown in two-hybrid and coimmunoprecipitation experiments.[23] Sc Cdc13p is required to protect the 5' end that contains the CA-rich telomeric strand from nucleolytic degradation,[24] and it mediates access of telomerase to the chromosomal terminus.[16] The recruitment function of Sc Cdc13p is reduced by a point mutation at amino acid 252 (Glu->Lys), which can be suppressed by a missense mutation at residue 444 of Sc Est1p (Lys->Glu). This reciprocal suppression depends on oppositely charged residues in Sc Cdc13p and Sc Est1p, providing excellent evidence for a direct physical interaction between these two proteins.[25] Consistent with a model in which the interaction between Sc Est1p and Sc Cdc13p mediates recruitment of telomerase (Fig. 1) are studies with fusion proteins. Strikingly, a fusion between Sc Cdc13p and Sc Est2p allows telomere maintenance even in the absence of Sc Est1p.[26]

Consistent with the recruitment model are also recent chromatin immunoprecipitation experiments that assessed association of different telomerase components throughout the cell cycle.[27] Results indicate that Sc Est1p binds to telomeres in late S phase, when telomeres are replicated by telomerase. A second recruitment mechanism is provided in G1 through the interaction of the telomere binding protein Ku with telomerase.[27] In the absence of Ku, telomeres are stable, albeit very short, demonstrating that the Ku-mediated recruitment of telomerase is not essential for telomerase activity.[28] Furthermore, Ku-mediated recruitment is not sufficient for telomerase recruitment as evidenced by the dependence of telomere-extension on the Sc Cdc13p-Sc Est1p interaction.

Discovery of Est1 Homologs

Since the discovery of Sc Est1p in 1989 (ref. 15), efforts failed for more than a decade to identify Sc Est1p-like proteins outside of budding yeast. However, a putative ortholog to Sc

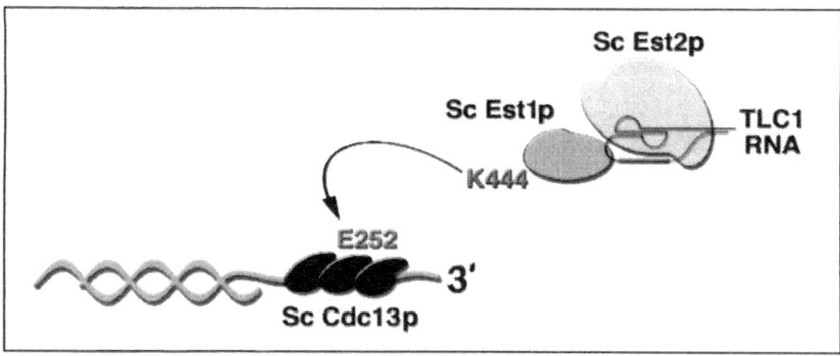

Figure 1. Proposed role for *S. cerevisiae* Est1p in telomerase recruitment. *S. cerevisiae* (Sc) Est1p mediates the interaction between telomere-bound Sc Cdc13p and telomerase, which has core components consisting of the catalytic subunit Sc Est2p and TLC1 RNA. Sc Est1p binds directly to TLC1 RNA. The binding of Sc Est1p and Sc Cdc13p involves an ionic interaction between Lys444 of Sc Est1p and Glu252 of Sc Cdc13p.

Est1p was identified in the closely related pathogenic yeast *Candida albicans*.[29] To identify Sc Est1p-like proteins in other phyla, we and others performed iterative profile searches.[30,31] First, a generalized profile using the conserved N-terminal regions of Sc Est1p and Sc Ebs1p was constructed. Sc Ebs1p is an Sc Est1p-related protein that has no well-defined function. Deletion of the *EBS1* gene results in slightly shorter but stable telomeres.[32] Database searches with the Sc Est1p/ Sc Ebs1p-profile led to the identification of Sc Est1p-homologs in *Schizosaccharomyces pombe*[33] and profile refinement. Subsequent database searches led to identification of additional Est1p-related sequences in various eukaryotes including *Caenorhabditis elegans*, *Drosophila melanogaster* and *Homo sapiens* (Fig. 2). Sequence comparison allowed definition of a conserved region of approximately 200 amino acids composed of a new domain named EST1 and a second domain comprising two tetratricopeptide repeats (TPRs). TPRs are alpha-helical rich protein-protein interaction modules found in proteins of diverse function in eukarya, bacteria and archea (reviewed in ref. 34).

In humans, three obvious Sc Est1p-related polypeptides exist named Hs EST1A, Hs EST1B and Hs EST1C (Fig. 2). The three proteins comprise 1419, 1016 and 1122 amino acids, respectively. Similarly to yeast Est1p, the EST1-TPR domains of Hs EST1B and Hs EST1C are located in the N-terminal region of the proteins. In contrast, the EST1-TRP domain of Hs EST1A is embedded in the middle of the protein. The 611 amino acids N-terminal extension of Hs EST1A is present in the putative Hs EST1A-orthologs of worm, mouse and zebrafish. Hs EST1A and Hs EST1B share a conserved C-terminal domain with low but significant homology to a PilT amino-terminal (PIN) domain region (PINc in Fig. 2). PIN domains contain conserved Asp and Glu residues, indicative of enzymes that bind divalent cations.[35]

Sc Est1 Homologs and Telomere Maintenance

C. albicans EST1

One putative Est1p-ortholog was found in *C. albicans*.[29] Homozygous *est1-Δ* strains experience mild loss of telomeric sequences upon the first passages in culture, as expected for a telomerase-null phenotype. However, upon further expansion, cells frequently undergo sudden dramatic telomere losses. This suggested a nonanticipated role for Ca Est1p in protecting telomere ends. Consistent with the observations using *S. cerevisiae*, extracts from *est1-Δ* strains of *C. albicans* contain telomerase activity, although defective extension of a subset of telomere primer substrates has been reported.[36]

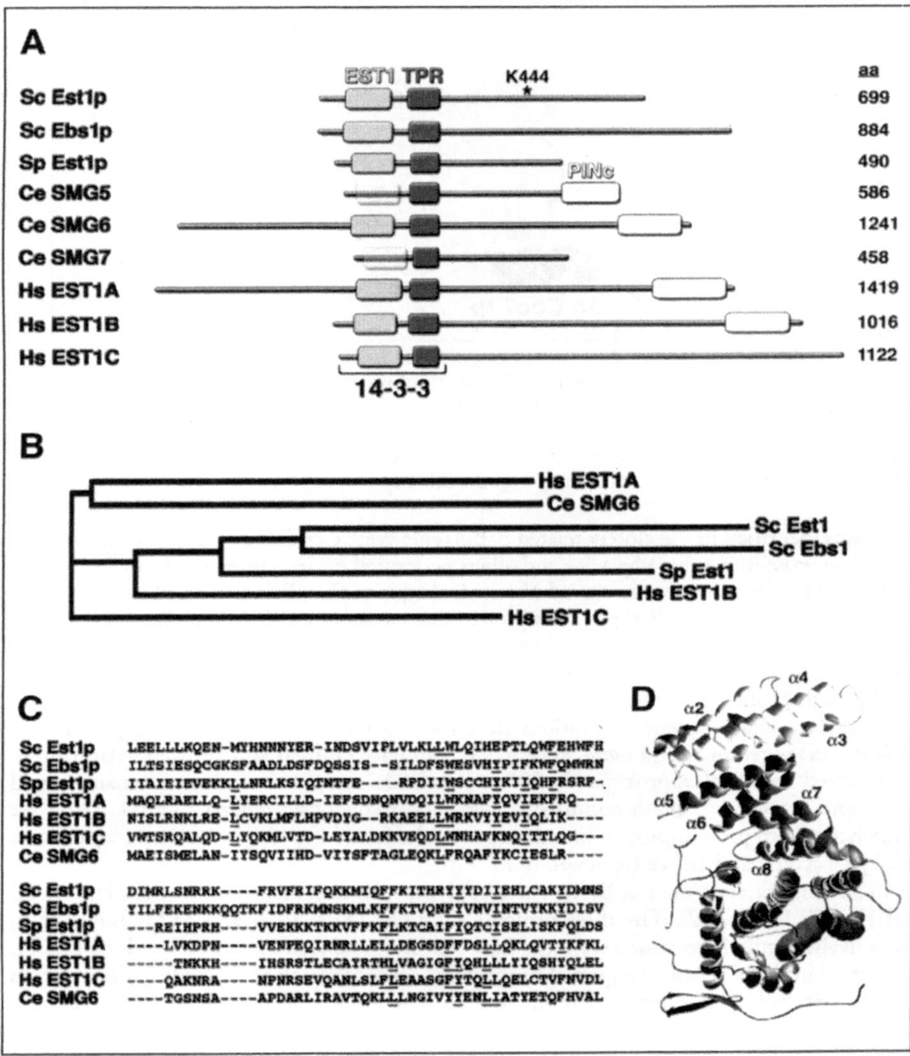

Figure 2. A) Architecture of proteins containing an EST1 domain from *S. cerevisiae* (Sc), *S. pombe* (Sp), *C. elegans* (Ce) and *H. sapiens* (Hs). EST1 domains are represented by the light-gray box; lightly shaded boxes for Ce SMG5 and Ce SMG7 indicate a lower level of homology. Tetratricopeptide (TPR) repeats are specified by dark boxes. PilT amino-terminal (PIN) domains are shown as unshaded boxes. The asterisk denotes the position of Lys444 (K444), which directly contacts Glu252 of Sc Cdc13p. The N-terminal region of Hs EST1C/SMG7, which has a 14-3-3 fold, is indicated.[44] Protein lengths are provided to the right (aa). B) Phylogenetic tree showing the distance relationship of the EST1 domains. The tree was generated using the ClustalW program, based on the alignment shown in (C). C) Alignment of the EST1 domains. Ce SMG5 and Ce SMG7 are omitted because their low homology did not allow a trustworthy alignment. Conserved aminoacids are underlined. D) Structure of the 55-kDa fragment comprising the first 497 amino acids of Hs EST1C/SMG7.[44] Helices α2-α4 comprise the EST1 domain, helices α5-α8 constitute the two tetratricopeptide repeats, and gaps in the structure correspond to disordered loops.

S. pombe EST1

S. pombe, like *S. cerevisiae*, contains two Est1p-paralogs. As in *S. cerevisiae*, only one of them is required for telomere maintenance, as evidenced by an *est*-minus phenotype and cellular senescence in the corresponding null-strain.[33] Also, Sp Est1p associates with telomerase activity, while it is not required for telomerase activity in vitro as is the case for other yeasts. Interestingly, in *S. pombe* cells that lack the major double-strand telomere binding protein Taz1p, Sp Est1p loss confers a lethal germination phenotype, while telomerase loss does not. This may be indicative of an additional role for Sp Est1p in telomere protection, as was observed using *C. albicans*.[33]

H. sapiens EST1

Immunoprecipitation experiments using antibodies raised against Hs EST1A and Hs EST1B revealed that these proteins associate with telomerase activity.[30,31] This is particularly evident for Hs EST1A, which appears to associate with about 70% of telomerase activity that is present in HeLa-cell extracts (Fig. 3 and ref. 30). Notably, we have found that the Hs EST1A-specific N-terminal extension interacts with telomerase (S. Redon, P. Reichenbach and J. Lingner, unpublished results).

To gain further insight into the role of Hs EST1A in telomere maintenance, we ectopically over-expressed Hs EST1A in the human fibrosarcoma-derived cell line HT1080. Strikingly, over-expressed Hs EST1A induced telomere-telomere associations that created chromosome

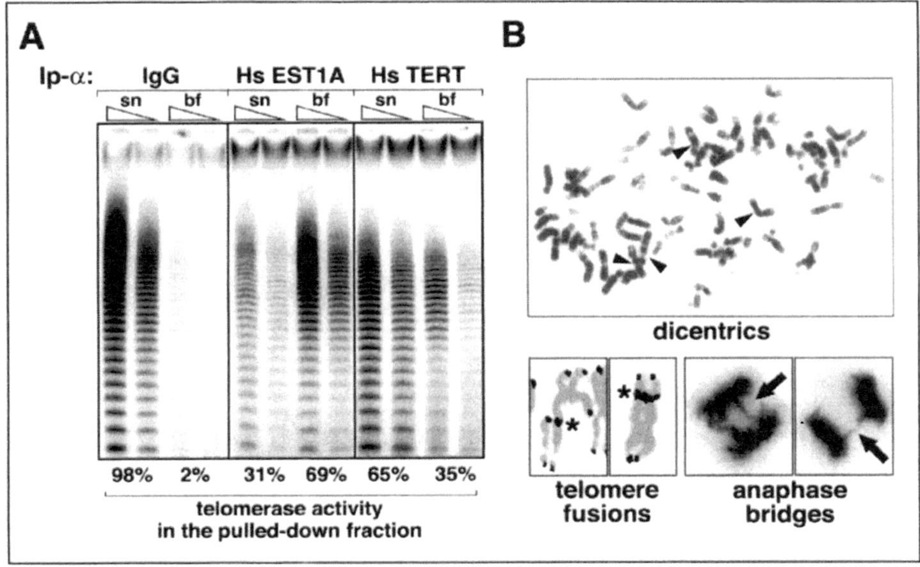

Figure 3. A) Coimmunoprecipitation of Hs EST1A and telomerase activity from HeLa cells. Telomerase-containing protein fractions were affinity purified from HeLa-cell nuclear extracts and incubated with: nonspecific IgG (left), antibodies directed against a Hs EST1A C-terminal peptide (center), or antibodies directed against Hs TERT (right). After immunoprecipitation, two dilutions of supernatants (sn) and bound fractions (bf) were assayed for telomerase activity. Numbers at the bottom indicate the relative percentage of telomerase activity that was immunopurified (pulled-down) in sn and bf. B) Metaphase and anaphase chromosomes were prepared from fibrosarcoma-derived HT1080 cells that over-express Hs EST1A. A representative metaphase spread is shown (upper), where arrowheads indicate dicentric chromosomes. Asterisks specify telomeric DNA (in black) that was detected at the site of fusions, and arrows point to anaphase chromosome bridges (lower).

bridges during anaphase (Fig. 3 and ref. 30) and, subsequently, a rapid apoptotic response. Telomeric fusions in Hs EST1A-overexpressing cells may stem from telomere uncapping followed by DNA break repair, reminiscent of the loss-of-function phenotype of the double-strand telomere binding protein Hs TRF2 (refs. 37,38). Thus, we hypothesize that endogenous Hs EST1A modulates telomere structure.

Over-expression of full-length Hs EST1A in kidney 293T cells leads to progressive telomere shortening.[31] This effect could be reversed by coexpression of Hs TERT, suggesting that shortening depends on telomerase action. Thus, Hs EST1A may also regulate telomere length.

Hs EST1A, B and C Are Identical to NMD Factors Hs SMG6, 5 and 7

The EST1-TPR domains of Hs EST1A, B and C are conserved in the three *C. elegans* SMG proteins (Ce SMG5, Ce SMG6 and Ce SMG7) that are involved in NMD (refs. 39,40; see chapter by Anderson). Indeed, Hs EST1A, B and C were independently also identified as the Hs SMG5, 6 and 7 proteins and shown to be involved in NMD in humans[41-43] (see chapter by Maquat). Based on the alignment of the EST1-TPR domains, Hs EST1A appears to be orthologous to Ce SMG6. Consistently, both Hs EST1A/SMG6 and Ce SMG6 possess the N-terminal extension that is not present in other EST1 domain-containing proteins. On the other hand, Ce SMG5 and Ce SMG7 do not have well-recognizable EST1 domains, making it difficult to unequivocally identify orthologous relationships. Nonetheless, the presence of the PIN domain in Hs EST1B and Ce SMG5 suggests that Hs EST1B corresponds to Hs SMG5 and Hs EST1C to Hs SMG7.

Ce SMG5, 6 and 7 are involved in the PP2A-mediated dephosphorylation of Ce UPF1 (also called Ce SMG2), which is a highly conserved 5'-3' helicase that plays a central role in NMD (see chapter by Andersen). Small interfering (si) RNA-mediated depletion of Hs SMG5-7/EST1A-C leads to impaired NMD,[42] over-expression of Hs EST1A/SMG6 promotes dephosphorylation of Hs UPF1[41] and Hs EST1B/SMG5, and Hs EST1C/SMG7 are found in complexes with phosphorylated Hs UPF1[43] (see chapter by Yamashita et al). Thus, the role of Hs SMG5-7 proteins in the phosphorylation cycle of Hs UPF1 appear to be conserved between worms and man.

It is worth pointing out that the sequence similarity of Est1p and Ebs1p in *S. cerevisiae* and *S. pombe* to SMG5, 6 and 7 proteins in *C. elegans and H. sapiens* challenges the notion that SMG5, 6 and 7 proteins are not conserved in yeast. However, functional assays measuring NMD *in S. cerevisiae* failed to reveal a function for Sc Est1p in NMD (C. Azzalin and J. Lingner, unpublished results).

The EST1-TPR Domain Has a 14-3-3 Fold

Recently, the crystal structure of the EST1-TPR domain of Hs EST1C/SMG7 was solved (ref. 44; Fig. 2D). Results revealed that the EST1 domain comprises three alpha-helices (helices 2-4) and confirmed the presence of two TPR domains forming typical coiled-coiled structures (helices 5-6 and 7-8). Interestingly, the EST1-TPR domain of Hs EST1C/SMG7 has a 14-3-3 shape, which is a structure that is known to bind phosphoserine/phosphothreonine-containing peptides.

We could not detect an interaction between Hs EST1C/SMG7 and telomerase in coimmunoprecipitation experiments. Nor did siRNA-mediated depletion of Hs EST1C have a strong immediate effect on telomere homeostasis (N. Hug, C. Azzalin and J. Lingner, unpublished results). Nevertheless, a possible role for the putative 14-3-3 folds of Hs EST1A and B in the regulation of telomerase and telomere maintenance can now be tested. Notably, expression of a dominant-negative 14-3-3 domain resulted in a redistribution of nuclear Hs TERT to the cytoplasm.[45] On the other hand, mutated Hs TERT that could not bind 14-3-3 was cytoplasmic. Whether Hs EST1A and/or Hs hEST1B use the 14-3-3 domain to interact with Hs TERT, thereby regulating its localization or activity at telomeres, can now be assessed.

Conclusion

The literature reviewed herein infers a sharing of factors between the mammalian telomere maintenance and NMD pathways. The involvement of Hs EST1A-C/SMG5-7 in NMD complicates any analysis of telomeric function. For example, it is difficult to prove whether telomeric dysfunctions that were observed in cells over-expressing Hs EST1A[30,31] are due to a direct disturbance of Hs EST1A function at telomeres. Our data demonstrating a physical interaction between Hs EST1A and telomerase support a direct role for this protein in telomere maintenance. However, an alternative explanation could be that Hs EST1A/SMG6 overexpression results in telomeric dysfunction due to effects on NMD that lead to deregulation of factors involved in telomere protection and lengthening. Indeed, NMD-deficient strains of *S. cerevisiae* have short telomeres and up-regulate, among many proteins, several telomeric factors that include Sc Est2p, Sc Est1p, Sc Est3p, Sc Stn1p, and Sc Ten1p.[46] However, expression profiles of Hs UPF1-depleted HeLa cells did not uncover deregulation of mRNAs that encode proteins involved in telomere metabolism.[47] One could still hypothesize that the stability of some mRNAs is modulated by NMD, specifically during DNA replication or in response to telomere shortening.

In conclusion, there are unexpected new roles for NMD factors in telomere integrity. This is in line with the recent observation that Hs SMG1, the human PI3-like protein kinase that is involved in NMD, plays a critical role in maintaining genome stability in response to DNA damage (ref. 48; see chapter by Abraham and Oliveira). Thus, components of the core NMD machinery may sit at the crossroads of DNA and RNA surveillance networks, and they may function in a complex to coordinate cellular genome stability. Thorough functional insight into telomere maintenance and NMD machineries will derive from future studies of different model organisms, which promise to provide clues to the evolution of mechanistic links between these and possibly other pathways. Loss-of-function studies of Hs EST1A-C, combined with mutational analysis, will reveal whether their multiple roles are genetically separable.

Acknowledgements

Work in the laboratory is supported by grants from the Swiss National Science Foundation, the Swiss Cancer League, the Human Frontier Science Program and the EU 6th Framework Programme.

References

1. Klobutcher LA, Swanton MT, Donini P et al. All gene-sized DNA molecules in four species of hypotrichs have the same terminal sequence and an unusual 3' terminus. Proc Natl Acad Sci 1981; 78(5):3015-3019.
2. Wellinger RJ, Wolf AJ, Zakian VA. Saccharomyces telomeres acquire single-strand TG1-3 tails late in S phase. Cell 1993; 72(1):51-60.
3. Makarov VL, Hirose Y, Langmore JP. Long G tails at both ends of human chromosomes suggest a C strand degradation mechanism for telomere shortening. Cell 1997; 88(5):657-666.
4. Huffman KE, Levene SD, Tesmer VM et al. Telomere shortening is proportional to the size of the G-rich telomeric 3'-overhang. J Biol Chem 2000; 275(26):19719-19722.
5. Lingner J, Promisel Cooper J, Cech TR. Telomerase and DNA end replication: No longer a lagging strand problem? Science 1995; 269:1533-1534.
6. Larrivee M, LeBel C, Wellinger RJ. The generation of proper constitutive G-tails on yeast telomeres is dependent on the MRX complex. Genes Dev 2004; 18(12):1391-1396.
7. Jacob NK, Kirk KE, Price CM. Generation of telomeric G strand overhangs involves both G and C strand cleavage. Mol Cell 2003; 11(4):1021-1032.
8. Sfeir AJ, Chai W, Shay JW et al. Telomereend processing the terminal nucleotides of human chromosomes. Mol Cell 2005; 18(1):131-138.
9. Greider CW, Blackburn EH. Identification of a specific telomere terminal transferase activity in Tetrahymena extracts. Cell 1985; 43(2 Pt 1):405-413.
10. Lingner J, Hughes TR, Shevchenko A et al. Reverse transcriptase motifs in the catalytic subunit of telomerase. Science 1997; 276:561-567.

11. Greider CW, Blackburn EH. A telomeric sequence in the RNA of Tetrahymena telomerase required for telomere repeat synthesis. Nature 1989; 337(6205):331-337.
12. Singer MS, Gottschling DE. TLC1: Template RNA component of Saccharomyces cerevisiae telomerase. Science 1994; 266(5184):404-409.
13. Feng J, Funk WD, Wang SS et al. The RNA component of human telomerase. Science 1995; 269(5228):1236-1241.
14. Lendvay TS, Morris DK, Sah J et al. Senescence mutants of Saccharomyces cerevisiae with a defect in telomere replication identify three additional EST genes. Genetics 1996; 144(4):1399-1412.
15. Lundblad V, Szostak JW. A mutant with a defect in telomere elongation leads to senescence in yeast. Cell 1989; 57(4):633-643.
16. Nugent CI, Hughes TR, Lue NF et al. Cdc13p: A single-strand telomeric DNA-binding protein with a dual role in yeast telomere maintenance. Science 1996; 274(5285):249-252.
17. Cohn M, Blackburn EH. Telomerase in yeast. Science 1995; 269(5222):396-400.
18. Lingner J, Cech TR, Hughes TR et al. Three Ever Shorter Telomere (EST) genes are dispensable for in vitro yeast telomerase activity. Proc Natl Acad Sci 1997; 94(21):11190-11195.
19. Seto AG, Livengood AJ, Tzfati Y et al. A bulged stem tethers Est1p to telomerase RNA in budding yeast. Genes Dev 2002; 16(21):2800-2812.
20. Hughes TR, Evans SK, Weilbaecher RG et al. The est3 protein is a subunit of yeast telomerase. Curr Biol 2000; 10(13):809-812.
21. Evans SK, Lundblad V. The Est1 subunit of Saccharomyces cerevisiae telomerase makes multiple contributions to telomere length maintenance. Genetics 2002; 162(3):1101-1115.
22. Virta-Pearlman V, Morris DK, Lundblad V. Est1 has the properties of a single-stranded telomere end-binding protein. Genes Dev 1996; 10(24):3094-3104.
23. Qi H, Zakian VA. The Saccharomyces telomerebinding protein Cdc13p interacts with both the catalytic subunit of DNA polymerase alpha and the telomerase-associated Est1 protein. Genes Dev 2000; 14(14):1777-1788.
24. Garvik B, Carson M, Hartwell L. Single-stranded DNA arising at telomeres in cdc13 mutants may constitute a specific signal for the RAD9 checkpoint. Mol Cell Biol 1995; 15(11):6128-6138.
25. Pennock E, Buckley K, Lundblad V. Cdc13 delivers separate complexes to the telomere for end protection and replication. Cell 2001; 104(3):387-396.
26. Evans SK, Lundblad V. Est1 and Cdc13 as comediators of telomerase access. Science 1999; 286(5437):117-120.
27. Fisher TS, Taggart AK, Zakian VA. Cell cycle-dependent regulation of yeast telomerase by Ku. Nat Struct Mol Biol 2004; 11(12):1198-1205.
28. Boulton SJ, Jackson SP. Components of the Ku-dependent nonhomologous end-joining pathway are involved in telomeric length maintenance and telomeric silencing. EMBO J 1998; 17:1819-1828.
29. Singh SM, Steinberg-Neifach O, Mian IS et al. Analysis of telomerase in Candida albicans: Potential role in telomere end protection. Eukaryot Cell 2002; 1(6):967-977.
30. Reichenbach P, Hoss M, Azzalin CM et al. A human homolog of yeast Est1 associates with telomerase and uncaps chromosome ends when overexpressed. Curr Biol 2003; 13(7):568-574.
31. Snow BE, Erdmann N, Cruickshank J et al. Functional conservation of the telomerase protein Est1p in humans. Curr Biol 2003; 13(8):698-704.
32. Zhou J, Hidaka K, Futcher B. The Est1 subunit of yeast telomerase binds the Tlc1 telomerase RNA. Mol Cell Biol 2000; 20(6):1947-1955.
33. Beernink HT, Miller K, Deshpande A et al. Telomere maintenance in fission yeast requires an Est1 ortholog. Curr Biol 2003; 13(7):575-580.
34. D'Andrea LD, Regan L. TPR proteins: The versatile helix. Trends Biochem Sci 2003; 28(12):655-662.
35. Clissold PM, Ponting CP. PIN domains in nonsense-mediated mRNA decay and RNAi. Curr Biol 2000; 10(24):R888-890.
36. Singh SM, Lue NF. Ever shorter telomere 1 (EST1)-dependent reverse transcription by Candida telomerase in vitro: Evidence in support of an activating function. Proc Natl Acad Sci 2003; 100(10):5718-5723.
37. van Steensel B, Smogorzewska A, de Lange T. TRF2 protects human telomeres from end-to-end fusions. Cell 1998; 92(3):401-413.
38. Celli GB, de Lange T. DNA processing is not required for ATM-mediated telomere damage response after TRF2 deletion. Nat Cell Biol 2005.
39. Cali BM, Kuchma SL, Latham J et al. smg-7 is required for mRNA surveillance in Caenorhabditis elegans. Genetics 1999; 151(2):605-616.

40. Pulak R, Anderson P. mRNA surveillance by the Caenorhabditis elegans smg genes. Genes Dev 1993; 7(10):1885-1897.
41. Chiu SY, Serin G, Ohara O et al. Characterization of human Smg5/7a: A protein with similarities to Caenorhabditis elegans SMG5 and SMG7 that functions in the dephosphorylation of Upf1. RNA 2003; 9(1):77-87.
42. Gatfield D, Unterholzner L, Ciccarelli FD et al. Nonsense-mediated mRNA decay in Drosophila: At the intersection of the yeast and mammalian pathways. EMBO J 2003; 22(15):3960-3970.
43. Ohnishi T, Yamashita A, Kashima I et al. Phosphorylation of hUPF1 induces formation of mRNA surveillance complexes containing hSMG-5 and hSMG-7. Mol Cell 2003; 12(5):1187-1200.
44. Fukuhara N, Ebert J, Unterholzner L et al. SMG7 is a 14-3-3-like adaptor in the nonsense-mediated mRNA decay pathway. Mol Cell 2005; 17(4):537-547.
45. Seimiya H, Sawada H, Muramatsu Y et al. Involvement of 14-3-3 proteins in nuclear localization of telomerase. EMBO J 2000; 19(11):2652-2661.
46. Dahlseid JN, Lew-Smith J, Lelivelt MJ et al. mRNAs encoding telomerase components and regulators are controlled by UPF genes in Saccharomyces cerevisiae. Eukaryot Cell 2003; 2(1):134-142.
47. Mendell JT, Sharifi NA, Meyers JL et al. Nonsense surveillance regulates expression of diverse classes of mammalian transcripts and mutes genomic noise. Nat Genet 2004; 36(10):1073-1078.
48. Brumbaugh KM, Otterness DM, Geisen C et al. The mRNA surveillance protein hSMG-1 functions in genotoxic stress response pathways in mammalian cells. Mol Cell 2004; 14(5):585-598.

Chapter 20

Nonsense-Associated Altered Splicing

Zuo Zhang and Adrian R. Krainer*

Abstract

Frameshift and nonsense mutations can influence several aspects of gene expression, including mRNA stability and splicing fidelity. The mechanisms through which premature termination codons (PTCs) can apparently affect splice-site selection remain elusive and controversial, although many examples are attributable to mutation of exonic sequences involved in exon definition. Here we review the evidence for and against translational reading frame and exonic splicing regulatory sequences influencing splice-site choice.

Introduction

The high fidelity of eukaryotic gene expression requires multiple quality-control mechanisms, which involve physical interactions among components of the transcriptional, pre-mRNA processing, mRNA transport, and mRNA translational machineries.[1] Many of the steps in the pathway of gene expression are mechanistically connected, even when they occur in different cellular compartments. mRNA surveillance pathways can prevent synthesis of aberrant mRNAs or destroy them to avoid potentially deleterious effects to the cell.[2-5]

How eukaryotic cells respond to premature termination codons (PTCs) has been well documented in the past several years.[2,5-7] It has been noted that about 30% of mutations that contribute to human disease introduce a PTC, either by directly creating it (nonsense mutation) or by causing a frameshift that results in a PTC (see chapters by Holbrook et al and Keeling et al).[8,9] In many cases, PTCs influence gene expression by affecting mRNA stability in a translation-dependent manner. This process, called nonsense-mediated mRNA decay (NMD), is an important cellular RNA surveillance mechanism, through which eukaryotic cells can specifically degrade mRNAs with PTCs.[5] Many molecular details about the mechanisms of NMD in mammalian cells have been obtained recently (see chapters by Maquat, Singh and Lykke-Andersen, and Yamashita et al). Although NMD and many of its constituent factors are evolutionarily conserved, there are significant species-specific mechanistic differences. In mammals, both pre-mRNA splicing and mRNA translation have been shown to be required for NMD.[5,10]

As a crucial step in eukaryotic gene expression, pre-mRNA splicing by the spliceosomal machinery precisely removes introns through recognition of splicing signals present in the pre-mRNA sequence.[11,12] Alternative splicing, which involves flexible and/or regulated recognition of these splicing signals by various RNA-binding proteins, plays a key role in proteome diversity.[13] Many human diseases have been attributed to defects in pre-mRNA splicing caused by various mutations.[14] In the last decade, several studies have suggested that PTCs also affect pre-mRNA splicing in a translational open reading frame (ORF)-dependent manner, giving rise to nonsense-associated altered splicing (NAS).[15-20] NAS results in exon skipping, activation of cryptic or latent splice sites, or intron retention. In contrast to NMD, how an in-frame

*Corresponding Author: Adrian R. Krainer—Cold Spring Harbor Laboratory, P.O. Box 100, Cold Spring Harbor, New York 11724, U.S.A. Email: krainer@cshl.edu

Nonsense-Mediated mRNA Decay, edited by Lynne E. Maquat. ©2006 Eurekah.com.

PTC influences pre-mRNA splicing has been controversial, because some form of translation would have to be involved in recognition of the ORF before mRNA splicing is completed. Not only does translation occur after pre-mRNA splicing in the gene expression pathway, but because gene expression in eukaryotes is highly compartmentalized, it is difficult to imagine how pre-mRNA splicing, which is a nuclear event, can be regulated by what is presumably cytoplasmic translation. Although there is published evidence in support of nuclear translation or nuclear scanning,[21,22] the link between these processes and splicing is poorly understood and their existence remains questionable.[23,24]

Examples That Support NAS

In the published instances of NAS, the exons with PTCs are excluded from the mature mRNA, either by exon skipping or by use of alternative splice sites to maintain the ORF.

The studies of fibrillin (FBN1) RNA by Dietz and colleagues provided the first instance of NAS.[16] Mutations in the FBN1 gene cause Marfan syndrome (MFS), an inherited disease of connective tissue. An allele with a T to G mutation at position 26 in exon 51 of FBN1 introduces an in-frame amber stop codon (TAG), and results in skipping of this exon during pre-mRNA splicing and restoration of the downstream ORF (Fig. 1A). Further analysis showed that other nonsense codons, such as TAA (ochre), at the same position also induce exon skipping, whereas a silent mutation that generates a TAC codon does not. An upstream frameshift mutation, which places the original amber mutation out of frame, significantly reduces exon skipping. Thus, a mechanism involving nuclear scanning of the ORF within pre-mRNA, as well as ORF-dependent regulation of alternative splicing, was proposed to exist in order to maintain the integrity of the ORF within spliced mRNA.

Another line of evidence supporting an influence of the ORF during splicing came from studies by Pintel and colleagues of the minute virus of mice (MVM), an autonomous parvovirus with a 5-kbp DNA genome.[15,25] A nonsense mutation within the NS1/2 common exon causes retention of the downstream intron, and this effect depends on the upstream initiation codon (Fig. 1B). In contrast, a missense mutation at the same position does not show the same effect. Similarly, a nonsense but not a missense mutation in the NS2-specific exon also causes retention of this intron, an effect that can be suppressed by frameshift mutations that take the nonsense mutation out of frame. The authors concluded that there is recognition of the ORF or communication between the exons within a pre-mRNA in the nucleus before, or concomitant with, splicing. A similar effect of intron retention caused by nonsense mutations has been observed for immunoglobulin μ (Ig-μ) and T-cell receptor β (TCR-β) transcripts.[6]

Recent results indicate that in-frame nonsense mutations can affect splice-site selection during the splicing of carbamoylphosphate synthetase, aspartate transcarbamylase, and dihydroorotase (CAD) pre-mRNA (Fig. 1C), a phenomenon that was termed suppression of splicing (SOS).[19] A database survey found that in >90% of human genes sampled, there is at least one in-frame stop codon between the normal 5' splice site (ss) and an intronic "latent" 5'ss that is rarely used.[26] In several cases, eliminating the stop codon(s) results in activation of the latent intronic 5'ss, although the mechanism is not yet understood.[19] Li and coworkers also found that the intronic 5'ss can be used if the cells are subject to heat shock, which demonstrates that these sites are intrinsically functional. Further work from the same group showed that the mechanism of SOS appears to be novel, in that it differs from NMD, which otherwise could have accounted for these observations. They found that several conditions that abrogate NMD do not affect SOS.[27] Among these conditions was the use of protein-synthesis inhibitors, raising the question of how the nonsense codons are recognized before intron removal and without the involvement of ribosomes. With respect to the potential prevalence of SOS, a different bioinformatic analysis detected no statistically significant enrichment of stop codons between normal 5'ss and latent 5'ss.[28,29] The discrepancy between these two statistical analyses is due to different assumptions.

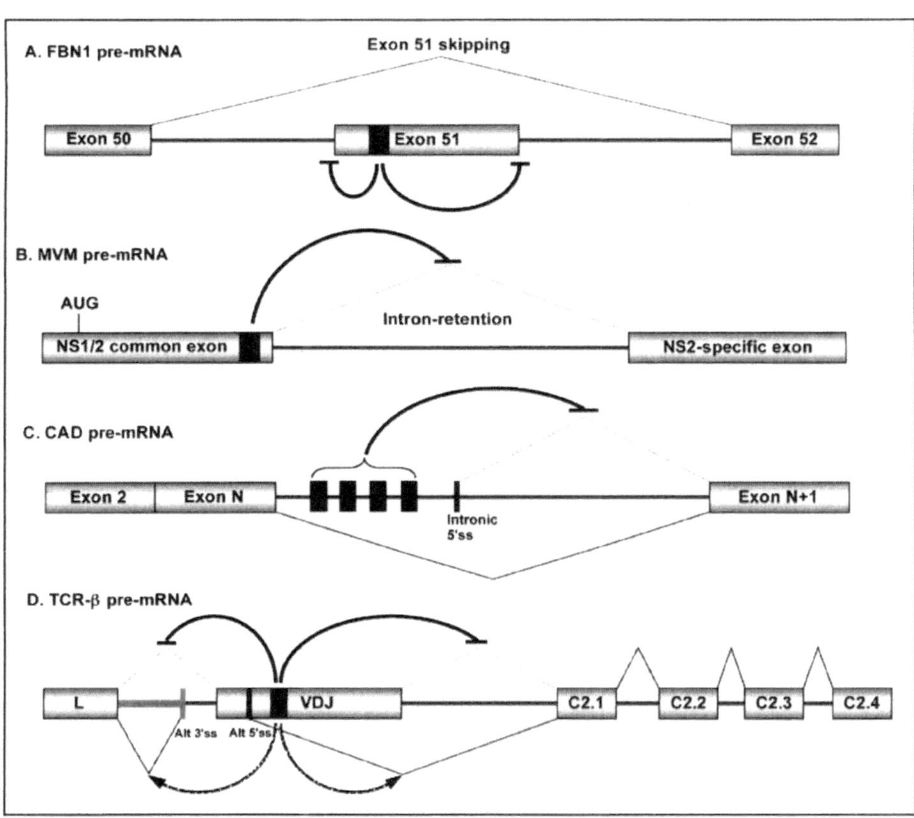

Figure 1. Examples of nonsense-associated altered splicing (NAS). A) Certain nonsense mutations in exon 51 of the *FBN1* gene cause exon 51 skipping; B) A nonsense mutation within the NS1/2 common exon of the minute virus of mice causes intron retention; C) Intronic in-frame stop codons inhibit the use of a downstream latent intronic 5' splice site (ss) in a CAD minigene; D) Nonsense mutations within the VDJ exon of the TCR-β gene increase the level of an alternatively spliced transcript lacking the nonsense mutations by promoting use of alternative (Alt) splice sites, thus restoring the ORF. Black boxes represent nonsense mutations.

A particularly interesting report of NAS came from studies of the mouse TCR-β gene.[18,20] Several different nonsense mutations within the VDJ exon were found to significantly increase the level of an alternatively spliced transcript, which is normally expressed at a very low level (Fig. 1D). However, several missense or silent mutations at the same position had no effect. The alternatively spliced mRNA is generated by use of a minor intronic 3'ss and an exonic 5'ss. This mRNA lacks all of the nonsense mutations and terminates translation normally. The possibility that the mutations disrupt the function of exonic splicing enhancers (ESEs) or exonic splicing silencers (ESSs) appears to be ruled out because a 10-nucleotide (nt) insertion but not a 9-nt insertion or a 10-nt insertion with a compensating 1-nt deletion was also found to increase the level of NAS products. NAS appears to be dependent on translation because cycloheximide, a stem-loop structure in the 5'-untranslated region, or expression of a suppressor tRNA inhibited up-regulation of the alternatively spliced transcript in the presence of nonsense mutations. The recognition of nonsense codons appears to take place after splicing, as an intron-split nonsense codon can also elicit NAS in this system.[20] Interestingly, Dietz and coworkers recently showed that NAS in the case of the TCR-β gene requires Upf1 but not Upf2, although both are key factors in NMD.[30] The mechanism of TCR-β NAS is still unknown. An interesting cytoplasm-to-nuclear feedback model has been proposed, though evidence against this model was provided in a recent study from Mühlemann and co-workers[31] (see below).

Examples That Do Not Support NAS

Although reports of ORF-dependent NAS generated significant interest in the RNA processing field, several studies have provided evidence that alternative mechanisms not involving ORF recognition can account for specific instances of NAS.

In addition to the conventional splicing signals spanning exon-intron boundaries, other cis-acting elements in pre-mRNA that regulate splicing have been extensively studied.[7] These elements are prevalent in both exons and introns, and can stimulate or inhibit the use of particular splice sites. Many studies have focused on ESE and ESS elements, which are involved in both constitutive and alternative splicing (Fig. 2). The wide distribution of ESEs in the transcribed genome has been demonstrated by both experimental and bioinformatic approaches.[32,33] Disruption of an ESE can lead to abnormal splicing, which is the underlying cause of various human diseases.[14,34] ESEs are specifically recognized by various RNA-binding proteins, the most prominent of which are members of the SR protein family of pre-mRNA splicing factors. SR proteins have one or two RNA-recognition motifs (RRMs) at the N-terminus, and a C-terminal arginine/serine-rich (RS) domain with many consecutive Arg-Ser dipeptides.[35] The RRMs mediate RNA binding, and the RS domain, which is extensively phosphorylated, is believed to be involved in protein-protein interactions and, in some cases, binding the intronic branchpoint sequence.[36] Most SR proteins accumulate in the cell nucleus, but a subset of them shuttles between the nucleus and cytoplasm.[37] The functional role of shuttling by certain SR proteins is not well understood, but some of the shuttling SR proteins have been found to play a role in mRNA export[38,39] and translation.[40] The recent discovery that SR proteins not only

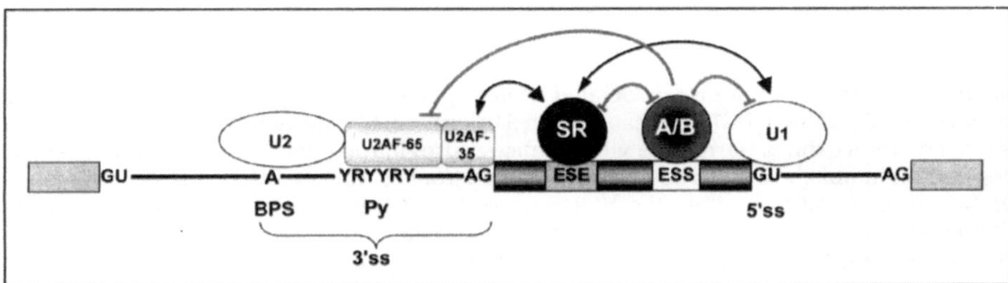

Figure 2. Model for the action of ESE and ESS elements in exon definition and splicing regulation. Typically, an exon is flanked by loosely conserved 3' splice site (ss) and 5'ss sequences. An SR protein binds to an exonic splicing enhancer (ESE) through its RNA-recognition motif(s) and contacts the splicing factor U2AF and/or U1 small ribonucleoprotein particle (snRNP) at the adjacent splice sites through its RS domain. The U2AF splicing factor consists of two subunits, U2AF65 and U2AF35. The large subunit, U2AF65, binds to the polypyrimidine (Py) tract, which consists mostly of pyrimidines (Y) but is often interrupted by purines (R). U2AF65 also facilitates binding of U2 snRNP to the branchpoint sequence (BPS) (A). U2AF35 recognizes the 3'ss AG dinucleotide. U1 snRNP binds to the 5'ss through base pairing of U1 snRNA. An exon is defined by several splicing factor/pre-mRNA interactions, which are strengthened by protein-protein interactions mediated by the RS domain of SR proteins. Many exonic splicing silencers (ESSs) are recognized by the hnRNP A/B family of proteins. ESS-bound hnRNP A/B protein antagonizes exon definition by inhibiting the binding of SR proteins and other splicing factors, such as U2AF and U1 snRNP.

regulate splice-site choice but also play a role in the surveillance of the resulting alternatively spliced mRNA has added another dimension to the functions of this conserved protein family.[41] Recently, systematic studies of ESS elements have been carried out by several groups, and pointed to a significant contribution of ESS elements and the factors that recognize them in the regulation of splicing. In contrast to ESEs, which are often bound by SR proteins, the most studied type of ESS is recognized by the hnRNP A/B family of proteins.[42]

ESEs are discrete 6- to 8-nt sequences that are usually highly degenerate. Single point mutations—missense, nonsense, or silent—can sometimes sufficiently disrupt recognition of a critical ESE by its cognate SR protein, which in turn leads to an abnormal splicing pattern. A study carried out by Liu et al. provided detailed evidence in support of this scenario.[34] By analyzing an example of NAS in the BRCA1 gene, it was shown that inappropriate exon skipping actually results from disruption of an SF2/ASF-dependent ESE in the coding sequence. ESEs can be disrupted by single nonsense, missense or silent mutations regardless of the ORF. Analysis of a dataset of 50 single-base mutations in human genes that cause exon-skipping in vivo showed that more than half of the mutations reduce or eliminate at least one predicted ESE.[34,43] The discovery that exon skipping correlates with the disruption of a functional ESE, but not with the nature of the mutations, argues against a regulatory role in splicing by nonsense mutations in a ORF-dependent manner. Another systematic study of the splicing of the SMN1 and SMN2 genes by Cartegni et al. also provided evidence that the disruption of an ESE by a translationally silent point mutation is sufficient to cause exon skipping.[44]

Regarding the NAS of the FBN1 transcript, a study by Francke and colleagues found that a silent mutation at position 41 of exon 51 also induces efficient exon skipping,[45] which together with the original study from the Dietz lab[16] suggests that the sequences at positions 26 and 41 may serve as cis-acting elements (ESE or ESS) controlling splicing of exon 51. Some of the reported missense mutations at position 26 that did not induce exon 51 skipping may not have disrupted the presumptive cis-element. Based on this information, Caputi et al[46] demonstrated that a nonsense mutation at position 26 of the FBN1 gene exon 51 induces NMD of the target mRNA. Most importantly, this mutation disrupts an SC35-specific ESE, correlating with the skipping of this exon. Taken together, these data show that NAS in the case of this allele of the FBN1 gene does not depend on the translational reading frame and is due to the disruption of an SC35-dependent ESE.

A nuclear translation or ribosome scanning model has been proposed to explain some of the observed instances of NAS, but this model has recently been questioned.[23,24] An alternative mechanism that was proposed to explain NAS, such as for the TCR-β gene, does not involve nuclear translation.[23] Rather, a cytoplasm-to-nucleus feedback model postulates that a shuttling splicing factor could signal the nucleus, perhaps by abnormally accumulating in the cytoplasm due to the increased level of cytoplasmic PTC-containing (+) transcripts. In an attempt to test this model, Mohn et al[31] coexpressed PTC-lacking (-) and PTC+ TCR-β minigenes in Hela cells. According to the model, splicing of PTC- pre-mRNA should be affected by coexpression of its PTC+ counterpart, which would lead to increased levels of alternatively spliced PTC- mRNA. Their results were incompatible with the feedback model, in that the alternatively spliced mRNA levels from the PTC- allele do not increase when the PTC+ minigene is coexpressed, and therefore, NAS is allele-specific. In addition, after systematic investigation of the ORF-dependence of TCR-β NAS, Mohn and coworkers reported that the requirement for a PTC[20] could not be reproduced in their hands: they observed upregulation of the alternatively spliced transcript both for nonsense and silent mutations. These data provide evidence contradicting the previously reported NAS, and showing that alternative splicing induced by nonsense mutations in the TCR-β gene is not PTC-specific. Furthermore, a computational analysis suggested the potential involvement of ESEs in this regulation. On the other hand, a mechanism based solely on interference with splicing signals cannot readily account for the observation that NAS of the TCR-β gene requires Upf1.[30]

NAS of the Ig gene has also been challenged by recent results from two different groups.[47,48] Bühler and Mühlemann demonstrated that a previously unreported alternative splicing event induced by nonsense mutations in the Ig-μ VDJ exon is independent of premature termination of the ORF. The resulting alternatively spliced isoform is translated and is subject to NMD when it has a PTC. However, the alternative splicing event itself is independent of the Upf1 NMD factor. The nonsense mutations tested appear to interfere with potential ESEs, suggesting that this alternative splicing event is not ORF-dependent. Lytle and Steitz used quantitative RT-PCR and RNase protection to measure the in vivo steady-state levels of every exon-intron junction in wild-type, nonsense, and missense mutant transcripts of the Ig-μ gene. They found that the rate of intron removal is not significantly influenced by the presence of a PTC in a neighboring exon in Ig-μ pre-mRNA. These studies indicate that in many cases nonsense mutations or the presence of a PTC do not affect pre-mRNA splicing in the nucleus in an ORF-dependent manner.

The NAS model has been tested in various ways for more than a decade, but recent evidence increasingly indicates that in many cases NAS is a consequence of disruption of ESE or ESS elements by nonsense or frameshift mutations. This mechanism clearly explains at least some instances of NAS.[17,34,44,46,49] However, other examples of NAS events appear to be dependent on translational frame, and cannot easily be explained by abrogation of ESEs, such as the SOS model based on splicing of CAD transcripts.[19] In order to test the generality of SOS, we designed experiments using model substrates to address the underlying mechanisms in vivo and in vitro.[41] Our basic strategy was to generate β-globin gene derivatives with a duplicated first-intron 5'ss, and with or without in-frame stop codons between the duplicated 5'ss. These model substrates allow a choice between two alternative 5'ss in essentially identical sequence contexts. Increasing the levels of various SR proteins can promote use of the proximal (downstream) 5'ss.[50,51] We found that the in-frame nonsense mutations did not play a direct role in selection of the distal (upstream) or proximal 5'ss. A reduction in the use of the proximal 5'ss in the presence of an upstream in-frame stop codon was indeed observed, but we determined that this reduction was an indirect effect of NMD, rather than being due to NAS. These results argued against a general function of nonsense mutations in splice-site selection in a reading frame-dependent manner. However, it is possible that certain specific features of a pre-mRNA substrate—in addition to an in-frame stop codon upstream of a target 5'ss—are required to trigger SOS, and that such features are absent from the model β-globin pre-mRNA. These studies also uncovered an unexpected function of SR proteins in efficiently targeting mRNA bearing a PTC to the NMD pathway.[41]

The mechanism of SOS, which appears to be independent of translation and distinct from NMD,[27] has not been elucidated, and SOS might be a gene-specific process. The study by Wachtel et al provided evidence that SOS may be a novel mechanism distinct from the known RNA surveillance mechanisms. SOS differs from NMD because it is not dependent on translation and is not affected by RNA interference (i)-mediated down-regulation of the Upf1 and Upf2 NMD factors. Moreover, a dominant-negative variant of Upf1, which was shown to abrogate NMD, does not activate latent splicing. In contrast, it was previously shown by Mendell et al. using RNAi-mediated knockdown, that Upf1, but not Upf2, is required for NAS of the TCR-β gene.[30] Thus, there appear to be mechanistic differences between NMD and at least these instances of NAS and SOS.

Recently, a systematic in vivo selection and identification of ESS elements has been reported by Wang and coworkers.[52] This genome-wide analysis of ESSs has not only suggested their role in pseudo exon suppression and alternative splicing, but also provided some insights that may be relevant to NAS. Among 133 unique ESSs, 59 (~44%) harbored one or more PTCs, which is considerably higher than expected by chance (~16%). To address the possibility that these sequences function as silencers through the process of NAS, they constructed three different vectors for each of three PTC-containing ESS decamers by inserting one to three bases before the decamer insertion site. All nine constructs were observed to cause exon skipping in transient transfection assays, regardless of whether they generated PTCs, consistent with direct ESS activity for these decamers. The result is inconsistent with models involving ORF-dependent NAS. Of the 62 PTCs in the ESS decamers, 55 were TAG, compared to only four occurrences of TGA and three of TAA. The counts of out-of-frame triplets were also far higher for TAG than for TGA or TAA in both alternate reading frames. TAG triplets appear to be more abundant in ESS elements. Thus, the experimental and statistical analyses found no evidence for effects of an ORF on splicing in the reporter system. These observations suggest that, in addition to frequent disruption of ESEs, some nonsense mutations (especially amber mutations) could create ESSs, providing a plausible alternative explanation for some apparent cases of NAS. One alternative explanation for SMN1/2 exon 7 skipping is the creation of an hnRNP-A1-dependent ESS,[53] although results from our laboratory argue against this model in the specific case of SMN2 (L. Cartegni et al., submitted). In general, the reason some nonsense mutations cause efficient exon skipping may be because they simultaneously disrupt an ESE and create an ESS.

Conclusions

In addition to NMD, a well-established cellular mRNA surveillance system, NAS has been suggested to be another different but intriguing quality-control mechanism based on pre-mRNA splicing level. However, recent detailed studies of several reported examples of NAS have questioned this model by obtaining different results or providing alternative explanations that are independent of an ORF, such as disruption of ESEs or indirect effects of NMD. With increasing evidence for these more easily envisioned explanations, it appears that NAS, if it exists at all, does not play a general role in pre-mRNA splicing in a ORF-dependent manner. Although NAS may provide a selective advantage by skipping the PTC-containing exon to allow the synthesis of coded protein with residual function, instead of making a completely inactive one, this situation of producing a partially functional protein by skipping a constitutive exon is not a common event. In many cases, exon-skipping causes a frameshift and generates one or more PTCs in the downstream ORF, which triggers NMD. Even if the ORF can be maintained, producing inactive or dominant-negative forms of proteins will still be a likely outcome. While NAS has been proposed to generate proteins with diverse functions, in the case of TCR-β NAS, the upregulated alternatively spliced transcript resulting from nonsense mutations may not be functional at all.[17] Thus, it is hard to envisage a selective advantage for NAS if skipping a PTC-containing exon only rarely results in the generation of functional protein. The exons affected by nonsense mutations appear to be mostly flanked by weak splice sites. These weakly defined exons are more likely to require other cis-elements for splicing, and hence are apt to be more sensitive to point mutations.[7] So far all the cases of reported NAS have been associated with particular transcripts. Therefore, NAS may not be part of a general mechanism for regulating pre-mRNA splicing.

Recent studies have revealed that the genetic basis of many human diseases reflects the influence of certain mutations on various aspects of mRNA metabolism, such as pre-mRNA splicing, rather than their expected effects on the structure and function of the encoded proteins.[7,9,14] Growing evidence argues that the effect of coding sequence single-nucleotide polymorphisms (cSNPs) on pre-mRNA splicing can account for a significant fraction of disease variation and the evolution of a new gene function.[7] With more studies of cSNPs at a whole-genome level, we can expect more observations consistent with NAS to be reported. However, each case needs to be carefully examined by a combination of experimental and bioinformatics approaches. Based on currently available information, we believe that most nonsense mutations do not directly influence pre-mRNA splicing in a translation-dependent manner.

References

1. Maniatis T, Reed R. An extensive network of coupling among gene expression machines. Nature 2002; 416(6880):499-506.
2. Hentze MW, Kulozik AE. A perfect message: RNA surveillance and nonsense-mediated decay. Cell 1999; 96(3):307-310.
3. Maquat LE, Carmichael GG. Quality control of mRNA function. Cell 2001; 104(2):173-176.
4. Lykke-Andersen J. mRNA quality control: Marking the message for life or death. Curr Biol 2001; 11(3):R88-91.
5. Maquat LE. Nonsense-mediated mRNA decay: Splicing, translation and mRNP dynamics. Nat Rev Mol Cell Biol 2004; 5(2):89-99.
6. Mühlemann O, Mock-Casagrande CS, Wang J et al. Precursor RNAs harboring nonsense codons accumulate near the site of transcription. Mol Cell 2001; 8(1):33-43.
7. Cartegni L, Chew SL, Krainer AR. Listening to silence and understanding nonsense: Exonic mutations that affect splicing. Nat Rev Genet 2002; 3(4):285-298.
8. Philips AV, Cooper TA. RNA processing and human disease. Cell Mol Life Sci 2000; 57(2):235-249.
9. Frischmeyer PA, Dietz HC. Nonsense-mediated mRNA decay in health and disease. Hum Mol Genet 1999; 8(10):1893-1900.
10. Thermann R, Neu-Yilik G, Deters A et al. Binary specification of nonsense codons by splicing and cytoplasmic translation. EMBO J 1998; 17(12):3484-3494.
11. Black DL. Mechanisms of alternative premessenger RNA splicing. Annu Rev Biochem 2003; 72:291-336.
12. Hastings ML, Krainer AR. Pre-mRNA splicing in the new millennium. Curr Opin Cell Biol 2001; 13(3):302-309.
13. Graveley BR. Alternative splicing: Increasing diversity in the proteomic world. Trends Genet 2001; 17(2):100-107.
14. Faustino NA, Cooper TA. Pre-mRNA splicing and human disease. Genes Dev 2003; 17(4):419-437.
15. Gersappe A, Burger L, Pintel DJ. A premature termination codon in either exon of minute virus of mice P4 promoter-generated pre-mRNA can inhibit nuclear splicing of the intervening intron in an open reading frame-dependent manner. J Biol Chem 1999; 274(32):22452-22458.
16. Dietz HC, Valle D, Francomano CA et al. The skipping of constitutive exons in vivo induced by nonsense mutations. Science 1993; 259(5095):680-683.
17. Maquat LE. NASty effects on fibrillin pre-mRNA splicing: Another case of ESE does it, but proposals for translation-dependent splice site choice live on. Genes Dev 2002; 16(14):1743-1753.
18. Wang J, Chang YF, Hamilton JI et al. Nonsense-associated altered splicing: A frame-dependent response distinct from nonsense-mediated decay. Mol Cell 2002; 10(4):951-957.
19. Li B, Wachtel C, Miriami E et al. Stop codons affect 5' splice site selection by surveillance of splicing. Proc Natl Acad Sci USA 2002; 99(8):5277-5282.
20. Wang J, Hamilton JI, Carter MS et al. Alternatively spliced TCR mRNA induced by disruption of reading frame. Science 2002; 297(5578):108-110.
21. Brogna S, Sato TA, Rosbash M. Ribosome components are associated with sites of transcription. Mol Cell 2002; 10(4):93-104.
22. Iborra FJ, Jackson DA, Cook PR. Coupled transcription and translation within nuclei of mammalian cells. Science 2001; 293(5532):1139-1142.
23. Dahlberg JE, Lund E, Goodwin EB. Nuclear translation: What is the evidence? RNA 2003; 9(1):1-8.
24. Nathanson L, Xia T, Deutscher MP. Nuclear protein synthesis: A reevaluation. RNA 2003; 9(1):9-13.
25. Gersappe A, Pintel DJ. A premature termination codon interferes with the nuclear function of an exon splicing enhancer in an open reading frame-dependent manner. Mol Cell Biol 1999; 19(3):1640-1650.
26. Miriami E, Sperling R, Sperling J et al. Regulation of splicing: The importance of being translatable. RNA 2004; 10(1):1-4.
27. Wachtel C, Li B, Sperling J et al. Stop codon-mediated suppression of splicing is a novel nuclear scanning mechanism not affected by elements of protein synthesis and NMD. RNA 2004; 10(11):1740-1750.
28. Zhang X, Lee J, Chasin LA. The effect of nonsense codons on splicing: A genomic analysis. RNA 2003; 9(6):637-639.
29. Zhang XH, Chasin LA. Latent splice sites and stop codons revisited. RNA 2004; 10(1):5-6.
30. Mendell JT, ap Rhys CM, Dietz HC. Separable roles for rent1/hUpf1 in altered splicing and decay of nonsense transcripts. Science 2002; 298(5592):419-422.

31. Mohn F, Bühler M, Mühlemann O. Nonsense-associated alternative splicing of T-cell receptor beta genes: No evidence for frame dependence. RNA 2005; 11(2):147-156.
32. Liu HX, Zhang M, Krainer AR. Identification of functional exonic splicing enhancer motifs recognized by individual SR proteins. Genes Dev 1998; 12(13):1998-2012.
33. Fairbrother WG, Yeh RF, Sharp PA et al. Predictive identification of exonic splicing enhancers in human genes. Science 2002; 297(5583):1007-1013.
34. Liu HX, Cartegni L, Zhang MQ et al. A mechanism for exon skipping caused by nonsense or missense mutations in BRCA1 and other genes. Nat Genet 2001; 27(1):55-58.
35. Graveley BR. Sorting out the complexity of SR protein functions. RNA 2000; 6(9):1197-1211.
36. Shen H, Kan JL, Green MR. Arginine-serine-rich domains bound at splicing enhancers contact the branchpoint to promote prespliceosome assembly. Mol Cell 2004; 13(3):367-376.
37. Cáceres JF, Screaton GR, Krainer AR. A specific subset of SR proteins shuttles continuously between the nucleus and the cytoplasm. Genes Dev 1998; 12(1):55-66.
38. Huang Y, Gattoni R, Stévenin J et al. SR splicing factors serve as adapter proteins for TAP-dependent mRNA export. Mol Cell 2003; 11(3):837-843.
39. Huang Y, Steitz JA. Splicing factors SRp20 and 9G8 promote the nucleocytoplasmic export of mRNA. Mol Cell 2001; 7(4):899-905.
40. Sanford JR, Gray NK, Beckmann K et al. A novel role for shuttling SR proteins in mRNA translation. Genes Dev 2004; 18(7):755-768.
41. Zhang Z, Krainer AR. Involvement of SR proteins in mRNA surveillance. Mol Cell 2004; 16(4):597-607.
42. Smith CW, Valcárcel J. Alternative pre-mRNA splicing: The logic of combinatorial control. Trends Biochem Sci 2000; 25(8):381-388.
43. Valentine CR. The association of nonsense codons with exon skipping. Mutat Res 1998; 411(2):87-117.
44. Cartegni L, Krainer AR. Disruption of an SF2/ASF-dependent exonic splicing enhancer in SMN2 causes spinal muscular atrophy in the absence of SMN1. Nat Genet 2002; 30(4):377-384.
45. Liu W, Qian C, Francke U. Silent mutation induces exon skipping of fibrillin-1 gene in Marfan syndrome. Nat Genet 1997; 16(4):328-329.
46. Caputi M, Kendzior Jr RJ, Beemon KL. A nonsense mutation in the fibrillin-1 gene of a Marfan syndrome patient induces NMD and disrupts an exonic splicing enhancer. Genes Dev 2002; 16(14):1754-1759.
47. Lytle JR, Steitz JA. Premature termination codons do not affect the rate of splicing of neighboring introns. RNA 2004; 10(4):657-668.
48. Bühler M, Mühlemann O. Alternative splicing induced by nonsense mutations in the immunoglobulin mu VDJ exon is independent of truncation of the open reading frame. RNA 2005; 11(2):139-146.
49. Maquat LE. The power of point mutations. Nat Genet 2001; 27(1):5-6.
50. Ge H, Manley JL. A protein factor, ASF, controls cell-specific alternative splicing of SV40 early pre-mRNA in vitro. Cell 1990; 62(1):25-34.
51. Krainer AR, Conway GC, Kozak D. The essential pre-mRNA splicing factor SF2 influences 5' splice site selection by activating proximal sites. Cell 1990; 62(1):35-42.
52. Wang Z, Rolish ME, Yeo G et al. Systematic identification and analysis of exonic splicing silencers. Cell 2004; 119(6):831-845.
53. Kashima T, Manley JL. A negative element in SMN2 exon 7 inhibits splicing in spinal muscular atrophy. Nat Genet 2003; 34(4):460-463.

Index

Note: Distinctions are not made between proteins and genes in some of the terms.

N

P

Q

R

S

T

U